Veterinary Entomology

Veterinary Entomology

Edited by:

Patricia Marques

Delve PUBLISHING

www.delvepublishing.com

Veterinary Entomology

Patricia Marques

Delve Publishing
224 Shoreacres Road
Burlington, ON L7L 2H2
Canada
www.delvepublishing.com
Email: orders@arclereducation.com

ISBN: 978-1-77407-783-2 (Hardcover)

DECLARATION

ABOUT THE EDITOR

Patricia obtained her PhD form University College Dublin in 2010. Her interests are on Microbiology and Parasitology. She is currently working as a Postdoc at University of Maryland Baltimore, USA on Chlamydia infections.

TABLE OF CONTENTS

List of Contributors .... *xvii*
List of Abbreviations .... *xxxi*
Preface .... *xxxiii*

Section 1: Ectoparasites

Chapter 1 Nestedness of Ectoparasite-Vertebrate Host Networks 3
Abstract 3
Introduction 4
Results 5
Discussion 10
Methods 13
Acknowledgments 16
References 17

Chapter 2 Molecular Detection of Rickettsia Typhi in Cats and Fleas 25
Abstract 25
Introduction 26
Methods 27
Results 31
Discussion 37
Conclusions 40
References 41

Chapter 3 Molecular Detection of Acinetobacter Species in Lice and Keds of Domestic Animals in Oromia Regional State, Ethiopia 45
Abstract 45
Introduction 46
Materials And Methods 49

Results 54
Discussion 61
Acknowledgments 64
References 65

Chapter 4 **Ticks Infesting Domestic Dogs in the UK: A Large-scale Surveillance Programme 71**
Abstract 71
Background 72
Methods 74
Results 75
Discussion 80
Conclusions 84
References 85

Chapter 5 **Use of Big Data in the Surveillance of Veterinary Diseases: Early Detection of Tick Paralysis in Companion Animals 89**
Abstract 89
Background 90
Methods 92
Results 94
Discussion 101
Conclusion 103
References 105

Chapter 6 **The Microbiota of Hematophagous Ectoparasites Collected from Migratory Birds 107**
Abstract 108
Introduction 108
Materials and Methods 109
Results 113
Discussion 120
Acknowledgments 124
References 125

Section 2: Vectors

Chapter 7 **Disentangling Vector-Borne Transmission Networks: A Universal DNA Barcoding Method to Identify Vertebrate Hosts from Arthropod Bloodmeals... 135**

Abstract.. 135

Introduction.. 136

Results... 137

Discussion.. 141

Materials And Methods.. 143

Acknowledgments... 146

References.. 148

Chapter 8 **Correction: Disentangling Vector-Borne Transmission Networks: A Universal DNA Barcoding Method to Identify Vertebrate Hosts from Arthropod Bloodmeals... 151**

Chapter 9 **One Health: the Importance of Companion Animal Vector-Borne Diseases.. 153**

Abstract.. 153

One Health: History And Scope.. 154

Companion Animals In One Health.. 156

Companion Animal Vector-Borne Diseases In One Health.............. 157

Companion Animal One Health: The Wsava One Health Committee..... 161

References.. 163

Chapter 10 **Immune Responses of a Native and an Invasive Bird to Buggy Creek Virus (Togaviridae: Alphavirus) and Its Arthropod Vector, the Swallow Bug (Oeciacus vicarius) 167**

Abstract.. 167

Introduction.. 168

Methods.. 170

Elisas... 174

Results... 176

Discussion.. 178

Acknowledgments... 180

Author Contributions.. 180

References.. 181

Chapter 11 **Bluetongue Disease Risk Assessment Based on Observed and Projected Culicoides obsoletus spp. Vector Densities 187**
Abstract 187
Introduction 188
Methods 190
Results 193
Discussion 199
References 202

Chapter 12 **Worldwide Niche and Future Potential Distribution of Culicoides imicola, a Major Vector of Bluetongue and African Horse Sickness Viruses 207**
Abstract 207
Introduction 208
Methods 211
Climex 212
Results 214
Discussion 216
Acknowledgments 219
Author Contributions 219
References 220

Chapter 13 **Correction: Worldwide Niche and Future Potential Distribution of Culicoides imicola, a Major Vector of Bluetongue and African Horse Sickness Viruses 227**
Reference 229

Section 3: Mosquitoes

Chapter 14 **Environmental Drivers of Culicoides Phenology: How Important Is Species-Specific Variation When Determining Disease Policy? 233**
Abstract 234
Introduction 234
Materials And Methods 237
Results 243
Discussion 251
Acknowledgments 256
Author Contributions 256

References...257

Chapter 15 Culicoides Midge Bites Modulate the Host Response and Impact on Bluetongue Virus Infection in Sheep 263

Abstract...264

Introduction...264

Materials And Methods...266

Results...270

Discussion...277

Acknowledgments...280

Author Contributions...281

References...282

Chapter 16 A Bayesian Hierarchical Model for Estimation of Abundance and Spatial Density of Aedes aegypti 287

Abstract...287

Introduction...288

Methods...290

Results...297

Discussion...304

Acknowledgments...307

Author Contributions...307

References...308

Chapter 17 Latitudinal Diversity of Culex pipiens Biotypes and Hybrids in Farm, Peri-Urban, and Wetland Habitats in Europe 315

Abstract...315

Introduction...316

Materials & Methods...317

Results...320

Discussion...323

Conclusions...325

Acknowledgments...325

References...327

Section 4: Ticks

Chapter 18 Bowman A, Nuttall P: Ticks: Biology, Disease and Control 333

Review 333

Chapter 19 Multi-source Analysis Reveals Latitudinal and Altitudinal shifts in range of Ixodes Ricinus at its northern Distribution Limit 337

Abstract 337

Background 339

Materials And Methods 340

Results 345

Discussion 350

Conclusion 354

Acknowledgements 354

Authors' Contributions 355

References 356

Chapter 20 Abiotic and Biotic Factors Associated with Tick Population Dynamics on a Mammalian Host: Ixodes hexagonus Infesting Otters, Lutra lutra 361

Abstract 361

Introduction 362

Results 363

Discussion 367

Materials And Methods 371

Acknowledgments 375

Author Contributions 375

References 376

Chapter 21 DNA of Piroplasms of Ruminants and Dogs in Ixodid Bat Ticks 381

Abstract 382

Introduction 382

Materials And Methods 384

Results 387

Discussion 393

Conclusions 398

Acknowledgments 398

References 399

Chapter 22 **Molecular Ecological Insights into Neotropical Bird–Tick Interactions 405**

Abstract 406

Introduction 406

Results 409

Discussion 416

Materials And Methods 419

Acknowledgments 424

Author Contributions 425

References 426

Section 5: Flies

Chapter 23 **A Bayesian Geostatistical Moran Curve Model for Estimating Net Changes of Tsetse Populations in Zambia 435**

Abstract 435

Introduction 436

Methods 438

Results 444

Discussion 447

Conclusions 449

Acknowledgments 450

Author Contributions 450

References 451

Index 455

LIST OF CONTRIBUTORS

Sean P. Graham
Department of Biological Sciences, Auburn University, Auburn, Alabama, United States of America,

Hassan K. Hassan
Global Health Infectious Disease Research Program, Department of Global Health, University of South Florida, Tampa, Florida, United States of America,

Nathan D. Burkett-Cadena
Department of Entomology and Plant Pathology, Auburn University, Auburn, Alabama, United States of America

Craig Guyer
Department of Biological Sciences, Auburn University, Auburn, Alabama, United States of America,

Thomas R. Unnasch
Global Health Infectious Disease Research Program, Department of Global Health, University of South Florida, Tampa, Florida, United States of America,

Maria Mercedes Nogueras
Department of Infectious Diseases. Corporació Sanitària i Universitària Parc Taulí - Institut Universitari Parc Taulí – Autonomous University of Barcelona, Sabadell, Barcelona, Spain,

Immaculada Pons
Department of Infectious Diseases. Corporació Sanitària i Universitària Parc Taulí - Institut Universitari Parc Taulí – Autonomous University of Barcelona, Sabadell, Barcelona, Spain,

Ana Ortuño
Department of Animal Health, Autonomous University of Barcelona, Bellaterra, Barcelona, Spain,

Jaime Miret
Animal Shelter Company (Vallès Oriental), Granollers, Barcelona, Spain,

Julia Pla
Clinical Veterinary, Sabadell, Barcelona, Spain,

Joaquim Castellà
Department of Animal Health, Autonomous University of Barcelona, Bellaterra, Barcelona, Spain,

Ferran Segura
Department of Infectious Diseases. Corporació Sanitària i Universitària Parc Taulí - Institut Universitari Parc Taulí– Autonomous University of Barcelona, Sabadell, Barcelona, Spain,
Department of Medicine, Universitat Autònoma de Barcelona, Bellaterra, Barcelona, Spain,
University of Minnesota, United States of America,

Bersissa Kumsa
Department of Parasitology, College of Veterinary Medicine and Agriculture, Addis Ababa University, Bishoftu, Ethiopia,
Aix Marseille Universite´, URMITE, UM63, CNRS 7278, IRD 198, Inserm 1095, Marseille, France

Cristina Socolovschi
Aix Marseille Universite´, URMITE, UM63, CNRS 7278, IRD 198, Inserm 1095, Marseille, France

Philippe Parola
Aix Marseille Universite´, URMITE, UM63, CNRS 7278, IRD 198, Inserm 1095, Marseille, France

Jean-Marc Rolain
Aix Marseille Universite´, URMITE, UM63, CNRS 7278, IRD 198, Inserm 1095, Marseille, France

Didier Raoult
Aix Marseille Universite´, URMITE, UM63, CNRS 7278, IRD 198, Inserm 1095, Marseille, France

Swaid Abdullah
Veterinary Parasitology and Ecology Group, School of Biological Sciences, Life Sciences Building, University of Bristol, Bristol, UK.

Chris Helps
Molecular Diagnostic Unit, Langford Veterinary Services and School of Veterinary Sciences, University of Bristol, Bristol, UK.

Severine Tasker
Molecular Diagnostic Unit, Langford Veterinary Services and School of Veterinary Sciences, University of Bristol, Bristol, UK.

Hannah Newbury
MSD Animal Health, Walton Manor, Walton, Milton Keynes, UK.

Richard Wall
Veterinary Parasitology and Ecology Group, School of Biological Sciences, Life Sciences Building, University of Bristol, Bristol, UK.

Vanina Guernier
School of Veterinary Science, The University of Queensland, Gatton Campus, Gatton 4343, QLD, Australia.

Gabriel J. Milinovich
School of Veterinary Science, The University of Queensland, Gatton Campus, Gatton 4343, QLD, Australia.

Marcos Antonio Bezerra Santos
School of Veterinary Science, The University of Queensland, Gatton Campus, Gatton 4343, QLD, Australia.
Federal Rural University of Pernambuco, Academic Unit of Garanhuns, Garanhuns 55292-270, PE, Brazil.

Mark Haworth
School of Veterinary Science, The University of Queensland, Gatton Campus, Gatton 4343, QLD, Australia.

Glen Coleman
School of Veterinary Science, The University of Queensland, Gatton Campus, Gatton 4343, QLD, Australia.

Ricardo J. Soares Magalhaes
School of Veterinary Science, The University of Queensland, Gatton Campus, Gatton 4343, QLD, Australia.
Child Health Research Centre, The University of Queensland, Herston 4006, QLD, Australia.

Francesco Cerutti
S.S. Genetica e Immunobiochimica, Istituto Zooprofilattico Sperimentale del Piemonte, Liguria e Valle d'Aosta, Torino, Italy,

Paola Modesto
S.S. Genetica e Immunobiochimica, Istituto Zooprofilattico Sperimentale del Piemonte, Liguria e Valle d'Aosta, Torino, Italy,

Francesca Rizzo
S.S. Laboratorio Specialistico Diagnostica Molecolare Virologica e Ovocoltura, Istituto Zooprofilattico Sperimentale del Piemonte, Liguria e Valle d'Aosta, Torino, Italy,

Alessandra Cravero
S.S. Microbiologia Molecolare e Analisi Genomiche, Istituto Zooprofilattico Sperimentale del Piemonte, Liguria e Valle d'Aosta, Torino, Italy,

Irena Jurman
IGA Technology Services, Udine, Italy,

Stefano Costa
Laboratorio Chimico Camera Commercio Torino, Torino, Italy,

Mauro Giammarino
Department of Prevention, ASL CN 1, Racconigi (CN), Italy

Maria Lucia Mandola
S.S. Laboratorio Specialistico Diagnostica Molecolare Virologica e Ovocoltura, Istituto Zooprofilattico Sperimentale del Piemonte, Liguria e Valle d'Aosta, Torino, Italy,

Mariella Goria
S.S. Microbiologia Molecolare e Analisi Genomiche, Istituto Zooprofilattico Sperimentale del Piemonte, Liguria e Valle d'Aosta, Torino, Italy,

Slobodanka Radovic
IGA Technology Services, Udine, Italy,

Federica Cattonaro
IGA Technology Services, Udine, Italy,

Pier Luigi Acuti
S.S. Genetica e Immunobiochimica, Istituto Zooprofilattico Sperimentale del Piemonte, Liguria e Valle d'Aosta, Torino, Italy,

Simone Peletto
S.S. Genetica e Immunobiochimica, Istituto Zooprofilattico Sperimentale del Piemonte, Liguria e Valle d'Aosta, Torino, Italy,

Miguel Alcaide
Estacio´n Biolo´gica de Don˜ ana (CSIC), Isla de la Cartuja, Sevilla, Spain,

Ciro Rico
Estacio´n Biolo´gica de Don˜ ana (CSIC), Isla de la Cartuja, Sevilla, Spain,

Santiago Ruiz
Diputacio´n de Huelva, A´rea de Medio Ambiente, Huelva, Spain

Ramo´ n Soriguer
Estacio´n Biolo´gica de Don˜ ana (CSIC), Isla de la Cartuja, Sevilla, Spain,

Joaquı´n Mun˜oz
Estacio´n Biolo´gica de Don˜ ana (CSIC), Isla de la Cartuja, Sevilla, Spain,

Jordi Figuerola
Estacio´n Biolo´gica de Don˜ ana (CSIC), Isla de la Cartuja, Sevilla, Spain,

Michael J Day
School of Veterinary Sciences, University of Bristol, Langford BS40 5DU, UK

Carol A. Fassbinder-Orth
Biology Department, Creighton University, Omaha, Nebraska, United States of America,

Virginia A. Barak
Biology Department, Creighton University, Omaha, Nebraska, United States of America,

Charles R. Brown
Department of Biological Sciences, University of Tulsa, Tulsa, Oklahoma, United States of America,
Blood Systems Research Institute, United States of America,

Katharina Brugger
Institute for Veterinary Public Health, University of Veterinary Medicine Vienna, Vienna, Austria

Franz Rubel
Institute for Veterinary Public Health, University of Veterinary Medicine Vienna, Vienna, Austria

Sylvain Guichard
Cirad, UR AGIRs, F-34398, Montpellier, France,

InSTePP, University of Minnesota, St. Paul, MN, United States of America,

He'le`ne Guis
Cirad, UMR CMAEE, F-34398, Montpellier, France,
INRA, UMR1309 CMAEE, F-34398, Montpellier, France,

Annelise Tran
Cirad, UR AGIRs, F-34398, Montpellier, France,
Cirad, UMR15 TETIS, F-34398, Montpellier, France,

Claire Garros
Cirad, UMR CMAEE, F-34398, Montpellier, France,
INRA, UMR1309 CMAEE, F-34398, Montpellier, France,

Thomas Balenghien
Cirad, UMR CMAEE, F-34398, Montpellier, France,
INRA, UMR1309 CMAEE, F-34398, Montpellier, France,

Darren J. Kriticos
InSTePP, University of Minnesota, St. Paul, MN, United States of America,
CSIRO Agriculture Flagship and Biosecurity Flagship, Canberra, ACT 2601, Australia

Kate R. Searle
Centre for Ecology and Hydrology, Bush Estate, Penicuik, Midlothian, United Kingdom,

James Barber
The Pirbright Institute, Ash Road, Pirbright, Woking, United Kingdom,

Francesca Stubbins
The Pirbright Institute, Ash Road, Pirbright, Woking, United Kingdom,

Karien Labuschagne
PVVD, ARC-Onderstepoort Veterinary Institute, Private Bag X05, Onderstepoort, South Africa,

Simon Carpenter
The Pirbright Institute, Ash Road, Pirbright, Woking, United Kingdom,

Adam Butler
Biomathematics and Statistics Scotland, James Clerk Maxwell Building, The King's Buildings, Edinburgh, United Kingdom,

Eric Denison
The Pirbright Institute, Ash Road, Pirbright, Woking, United Kingdom,

Christopher Sanders
The Pirbright Institute, Ash Road, Pirbright, Woking, United Kingdom,

Philip S. Mellor
The Pirbright Institute, Ash Road, Pirbright, Woking, United Kingdom,

Anthony Wilson
The Pirbright Institute, Ash Road, Pirbright, Woking, United Kingdom,

Noel Nelson
The Met Office, Based at the Pirbright Institute, Pirbright, Woking, Surrey, United Kingdom

Simon Gubbins
The Pirbright Institute, Ash Road, Pirbright, Woking, United Kingdom,

Bethan V. Purse
Centre for Ecology and Hydrology, Bush Estate, Penicuik, Midlothian, United Kingdom,

Nonito Pages
Centre de Recerca en Sanitat Animal CReSA, Universitat Auto`noma de Barcelona – Institut de Recerca i Tecnologia Agroalimentaries, Bellaterra, Spain,

Emmanuel Bre
Virologie, Unite´ Mixte de Recherche UMR1161, Agence Nationale de Se´curite´ Sanitaire de l'Alimentation, de l'Environnement et du Travail – Institut National de la Recherche Agronomique – Ecole Nationale Ve´te´rinaire d'Alfort, Maisons-Alfort, France,

Ce´line Urien
Virologie et Immunologie Mole´culaires, Unite´ de Recherche UR892, Institut National de la Recherche Agronomique, Jouy-en-Josas, France

Sandra Talavera
Centre de Recerca en Sanitat Animal CReSA, Universitat Auto`noma de Barcelona – Institut de Recerca i Tecnologia Agroalimentaries, Bellaterra, Spain,

Cyril Viarouge
Virologie, Unite´ Mixte de Recherche UMR1161, Agence Nationale de Se´curite´

Sanitaire de l'Alimentation, de l'Environnement et du Travail – Institut National de la Recherche Agronomique – Ecole Nationale Ve´te´rinaire d'Alfort, Maisons-Alfort, France,

Cristina LorcaOro
Centre de Recerca en Sanitat Animal CReSA, Universitat Auto`noma de Barcelona – Institut de Recerca i Tecnologia Agroalimentaries, Bellaterra, Spain,

Luc Jouneau
Virologie et Immunologie Mole´culaires, Unite´ de Recherche UR892, Institut National de la Recherche Agronomique, Jouy-en-Josas, France

Bernard Charley
Virologie et Immunologie Mole´culaires, Unite´ de Recherche UR892, Institut National de la Recherche Agronomique, Jouy-en-Josas, France

Ste´phan Zientara
Virologie, Unite´ Mixte de Recherche UMR1161, Agence Nationale de Se´curite´ Sanitaire de l'Alimentation, de l'Environnement et du Travail – Institut National de la Recherche Agronomique – Ecole Nationale Ve´te´rinaire d'Alfort, Maisons-Alfort, France,

Albert Bensaid
Centre de Recerca en Sanitat Animal CReSA, Universitat Auto`noma de Barcelona – Institut de Recerca i Tecnologia Agroalimentaries, Bellaterra, Spain,

David Solanes
Centre de Recerca en Sanitat Animal CReSA, Universitat Auto`noma de Barcelona – Institut de Recerca i Tecnologia Agroalimentaries, Bellaterra, Spain,

Joan Pujols
Centre de Recerca en Sanitat Animal CReSA, Universitat Auto`noma de Barcelona – Institut de Recerca i Tecnologia Agroalimentaries, Bellaterra, Spain,

Isabelle Schwartz-Cornil
[1]Centre de Recerca en Sanitat Animal CReSA, Universitat Auto`noma de Barcelona – Institut de Recerca i Tecnologia Agroalimentaries, Bellaterra, Spain,

Daniel A. M. Villela
Fundação Oswaldo Cruz, Programa de Computação Científica—Rio de Janeiro, Brazil,

Claudia T. Codeço
Fundação Oswaldo Cruz, Programa de Computação Científica—Rio de Janeiro, Brazil,

Felipe Figueiredo
Fundação Oswaldo Cruz, Programa de Computação Científica—Rio de Janeiro, Brazil,

Gabriela A. Garcia
Fundação Oswaldo Cruz, Departamento de Entomologia, Laboratório de Transmissores de Hematozoários—Rio de Janeiro, Brazil

Rafael Maciel-de-Freitas
Fundação Oswaldo Cruz, Departamento de Entomologia, Laboratório de Transmissores de Hematozoários—Rio de Janeiro, Brazil

Claudio J. Struchiner
Fundação Oswaldo Cruz, Programa de Computação Científica—Rio de Janeiro, Brazil,
Fundação Oswaldo Cruz, Departamento de Entomologia, Laboratório de Transmissores de Hematozoários—Rio de Janeiro, Brazil

Chantal B. F. Vogels
Laboratory of Entomology, Wageningen University and Research centre, Wageningen, The Netherlands,

Tim W. R. Mo¨hlmann
Laboratory of Entomology, Wageningen University and Research centre, Wageningen, The Netherlands,
IFM Theory and Modelling, Linko¨ping University, Linko¨ping, Sweden,

Diede Melsen
Laboratory of Entomology, Wageningen University and Research centre, Wageningen, The Netherlands,

Guido Favia
School of Biosciences and Veterinary Medicine, University of Camerino, Camerino, Italy

Uno Wennergren
IFM Theory and Modelling, Linko¨ping University, Linko¨ping, Sweden,

Constantianus J. M. Koenraadt
Laboratory of Entomology, Wageningen University and Research centre, Wageningen, The Netherlands,

Alan R Walker
Royal (Dick) School of Veterinary Studies, Summerhall Place, Edinburgh, EH9 1QH, UK

Solveig Jore
Norwegian Veterinary Institute, Oslo, Norway

Hildegunn Viljugrein
Norwegian Veterinary Institute, Oslo, Norway
Centre for Ecological and Evolutionary Synthesis (CEES), University of Oslo, Oslo, Norway

Merete Hofshagen
Norwegian Veterinary Institute, Oslo, Norway

Hege Brun-Hansen
Norwegian School of Veterinary Science, Oslo, Norway

Anja B Kristoffersen
Norwegian Veterinary Institute, Oslo, Norway
University of Oslo, Oslo, Norway

Karin Nygård
Norwegian Institute of Public Health, Oslo, Norway

Edgar Brun
Norwegian Veterinary Institute, Oslo, Norway

Preben Ottesen
Norwegian Institute of Public Health, Oslo, Norway

Bente K Sævik
Norwegian School of Veterinary Science, Oslo, Norway

Bjørnar Ytrehus
Norwegian Veterinary Institute, Oslo, Norway

Ellie Sherrard-Smith
Organisms and the Environment, Cardiff University, Cardiff, United Kingdom

Elizabeth Chadwick
Organisms and the Environment, Cardiff University, Cardiff, United Kingdom

Joanne Cable
Organisms and the Environment, Cardiff University, Cardiff, United Kingdom

Sándor Hornok
Department of Parasitology and Zoology, University of Veterinary Medicine, Budapest, Hungary

Krisztina Szőke
Department of Parasitology and Zoology, University of Veterinary Medicine, Budapest, Hungary

Dávid Kováts
Department of Evolutionary Zoology and Human Biology, Debrecen University, Debrecen, Hungary

Péter Estók
Department of Zoology, Eszterházy Károly University, Eger, Hungary

Tamás Görföl
Department of Zoology, Hungarian Natural History Museum, Budapest, Hungary

Sándor A. Boldogh
Department of Nature Conservation, Aggtelek National Park Directorate, Jósvafő, Hungary

Nóra Takács
Department of Parasitology and Zoology, University of Veterinary Medicine, Budapest, Hungary

Jenő Kontschán
Plant Protection Institute, Centre for Agricultural Research, Hungarian Academy of Sciences, Budapest, Hungary

Gábor Földvári
Department of Parasitology and Zoology, University of Veterinary Medicine, Budapest, Hungary

Levente Barti
Romanian Bat Protection Association, Satu Mare, Romania

Alexandra Corduneanu
Department of Parasitology and Parasitic Diseases, University of Agricultural Sciences and Veterinary Medicine, Cluj-Napoca, Romania

Attila D. Sándor
Department of Parasitology and Parasitic Diseases, University of Agricultural Sciences and Veterinary Medicine, Cluj-Napoca, Romania

Matthew J. Miller
Smithsonian Tropical Research Institute, Panama City, Republic of Panama
Department of Biology, Villanova University, Villanova, Pennsylvania, United States of America

Helen J. Esser
Smithsonian Tropical Research Institute, Panama City, Republic of Panama
Department of Environmental Sciences, Wageningen University, Wageningen, the Netherlands

Jose R. Loaiza
Smithsonian Tropical Research Institute, Panama City, Republic of Panama
Centro de Biodiversidad y Descubrimiento de Drogas, Instituto de Investigaciones Científicas y Servicios de Alta Tecnología (INDICASAT-AIP), Panama City, Republic of Panama
Centralamerican Master's Program in Entomology, University of Panama, Panama City, Republic of Panama

Edward Allen Herre
Smithsonian Tropical Research Institute, Panama City, Republic of Panama

Celestino Aguilar
Smithsonian Tropical Research Institute, Panama City, Republic of Panama
Centro de Biodiversidad y Descubrimiento de Drogas, Instituto de Investigaciones Científicas y Servicios de Alta Tecnología (INDICASAT-AIP), Panama City, Republic of Panama

Diomedes Quintero
G. B. Fairchild Invertebrate Museum, University of Panama, Panama City, Republic of Panama

Eric Alvarez
Smithsonian Tropical Research Institute, Panama City, Republic of Panama
Centralamerican Master's Program in Entomology, University of Panama, Panama City, Republic of Panama

Eldredge Bermingham
Smithsonian Tropical Research Institute, Panama City, Republic of Panama
Patricia and Phillip Frost Museum of Science, Miami, Florida, United States of America

Luigi Sedda
Department of Zoology, University of Oxford, Oxford, United Kingdom,

Cornelius Mweempwa
Department of Veterinary and Livestock Development, Ministry of Agriculture and Cooperatives, Lusaka, Zambia,

Els Ducheyne
Avia-GIS, Zoersel, Belgium,

Claudia De Pus
Animal Health Department, Institute of Tropical Medicine, Antwerp, Belgium

Guy Hendrickx
Avia-GIS, Zoersel, Belgium,

David J. Rogers
Department of Zoology, University of Oxford, Oxford, United Kingdom,

LIST OF ABBREVIATIONS

AIC	Akaike Information Criterion
BOLD	Barcode of Life Data Systems
BGS	Biogents Sentinel
BTV	Bluetongue virus
BHI	Brain-heart infusion
CIs	Confidence intervals
CCM	Cross correlation map
DLST	Day-time land surface temperature
EEEV	Eastern equine encephalitis virus
FAO	Food and Agriculture Organization
GAM	Generalized additive model
GLM	Generalized Linear Model
HIV	Human immunodeficiency virus
IFA	Indirect immunofluorescence assay
ISPRA	Institute for Environmental Protection and Research
IACUC	Institutional Animal Care and Use Committee
IPCC	Intergovernmental Panel on Climate Change
LIV	Louping ill virus
LB	Lyme borreliosis
MRR	Mark-release-recapture
MASL	Metres above sea level
MCMC	Monte Carlo Markov Chain
MMLP	Mosquito Magnet Liberty Plus
MT	Murine typhus
NSW	New South Wales
NLST	Night-time land surface temperature

NDVI	Normalized Difference Vegetation Index
NAO	North Atlantic Oscillation
NCHRS	Norwegian Cattle Health recording system
NIPH	Norwegian Institute of Public Health
NVI	Norwegian Veterinary Institute
OVI	Onderstepoort Veterinary Institute
OD	Optical density
OIE	Organization for Animal Health
PBS	Phosphate buffered saline
PCR	Polymerase Chain Reaction
PCA	Principal Component analysis
RH	Relative humidity
RT-PCR	Reverse transcription polymerase chain reaction
RMSF	Rocky Mountain spotted fever
SBV	Schmallenberg virus
SVFP	Seasonal vector free period
SACs	Species-accumulation curves
SFG	Spotted fever group
TMB	Tetramethylbenzidine
TBE	Tick-borne encephalitis
TNF	Tuskegee National Forest
WNV	West Nile virus
WHO	World Health Organization
WSAVA	World Small Animal Veterinary Association
WN	Wrapped Normal

PREFACE

Entomology (from Ancient Greek ἔντομον (entomon), meaning 'insect', and -λογία (-logia), meaning 'study of') is the scientific study of insects, a branch of zoology. In the past the term "insect" was vaguer, and historically the definition of entomology included the study of terrestrial animals in other arthropod groups or other phyla, such as arachnids, myriapods, earthworms, land snails, and slugs. This wider meaning may still be encountered in informal use.

In this book starts by the study of Ectoparasites (Section 1), which include fleas, lice, ticks and other insects and arthropods. Ectoparasites are a taxonomically diverse group of organisms that infest the skin of human beings and other animals. Ectoparasitic arthropods and nematodes are similar in that an individual organism can produce skin lesions that are large enough to see with the unaided eye. This ectoparasites can infect several animal hosts, including dogs, cats, birds and bats.

Section 2 looks into insects and arthropods as vectors of disease. Transmitting bacteria, parasites and viruses. A disease vector is any agent which carries and transmits an infectious pathogen into another living organism; most agents regarded as vectors are organisms, such as intermediate parasites or microbes.

The next chapters are dedicated to the study of specific arthropods, mosquitoes (Section 3), ticks (Section 4) and flies (Section 5).

Mosquitoes comprise a group of about 3,500 species of small insects that are flies (order Diptera). Within Diptera they constitute the family Culicidae (from the Latin culex meaning "gnat"). The word "mosquito" (formed by mosca and diminutive -ito) is Spanish for "little fly". Mosquitoes have a slender segmented body, one pair of wings, one pair of halteres, three pairs of long hair-like legs, and elongated mouthparts. The mosquito life cycle consists of egg, larva, pupa, and adult stages. Eggs are laid on the water surface; they hatch into motile larvae that feed on aquatic algae and organic material. The adult females of most species have tube-like mouthparts (called a proboscis) that can pierce the skin of a host and feed on blood, which contains protein and iron needed to produce eggs. Thousands of mosquito species feed on the blood of various hosts—vertebrates, including mammals, birds, reptiles, amphibians, and some fish; along with some invertebrates, primarily other arthropods. This loss of blood is seldom of any importance to the host. The mosquito's saliva is transferred to the host during the bite, and can cause an itchy rash. In addition, many species can ingest pathogens while biting, and transmit them to future hosts. In this way, mosquitoes are important vectors of diseases such as malaria, yellow fever, Chikungunya, West Nile, dengue fever, filariasis, Zika and other arboviruses.

Ticks (Ixodida) are arachnids, typically 3 to 5 mm long, part of the superorder Parasitiformes. Along with mites, they constitute the subclass Acari. Ticks are external parasites, living by feeding on the blood of mammals, birds, and sometimes reptiles and amphibians. Ticks evolved by the Cretaceous period, the most common form of fossilisation being amber immersion. Ticks are widely distributed around the world, especially in warm, humid climates. Almost all ticks belong to one of two major families, the Ixodidae or hard ticks, and the Argasidae or soft ticks. Adults have ovoid or pear-shaped bodies, which become engorged with blood when they feed, and eight legs. In addition to having a hard shield on their dorsal surfaces, hard ticks have a beak-like structure at the front containing the mouthparts, whereas soft ticks have their mouthparts on the underside of their bodies. Both families locate a potential host by odour or from changes in the environment. Ticks have four stages to their lifecycle, namely egg, larva, nymph, and adult. Ixodid ticks have three hosts, taking at least a year to complete their lifecycle. Argasid ticks have up to seven nymphal stages (instars), each one requiring a blood meal. Because of their habit of ingesting blood, ticks are vectors of many diseases that affect humans and other animals.

True flies are insects of the order Diptera, the name being derived from the Greek δι- di- "two", and πτερόν pteron "wings". Insects of this order use only a single pair of wings to fly, the hindwings having evolved into advanced mechanosensory organs known as halteres, which act as high-speed sensors of rotational movement and allow dipterans to perform advanced aerobatics. Diptera is a large order containing an estimated 1,000,000 species including horse-flies, crane flies, hoverflies and others, although only about 125,000 species have been described. Flies have a mobile head, with a pair of large compound eyes, and mouthparts designed for piercing and sucking (mosquitoes, black flies and robber flies), or for lapping and sucking in the other groups. Their wing arrangement gives them great maneuverability in flight, and claws and pads on their feet enable them to cling to smooth surfaces. Flies undergo complete metamorphosis; the eggs are laid on the larval food-source and the larvae, which lack true limbs, develop in a protected environment, often inside their food source. The pupa is a tough capsule from which the adult emerges when ready to do so; flies mostly have short lives as adults. Diptera is one of the major insect orders and of considerable ecological and human importance. Flies are important pollinators, second only to the bees and their Hymenopteran relatives. Flies may have been among the evolutionarily earliest pollinators responsible for early plant pollination. Fruit flies are used as model organisms in research, but less benignly, mosquitoes are vectors for malaria, dengue, West Nile fever, yellow fever, encephalitis, and other infectious diseases; and houseflies, commensal with humans all over the world, spread food-borne illnesses. Flies can be annoyances especially in some parts of the world where they can occur in large numbers, buzzing and settling on the skin or eyes to bite or seek fluids. Larger flies such as tsetse flies and screwworms cause significant economic harm to cattle. Blowfly larvae, known as gentles, and other dipteran larvae, known more generally as maggots, are used as fishing bait and as food for carnivorous animals. They are also used in medicine in debridement to clean wounds.

SECTION 1: ECTOPARASITES

1

Nestedness of Ectoparasite-Vertebrate Host Networks

Sean P. Graham[1], Hassan K. Hassan[2], Nathan D. Burkett-Cadena[3], Craig Guyer[1], Thomas R. Unnasch[2]

[1] Department of Biological Sciences, Auburn University, Auburn, Alabama, United States of America,
[2] Global Health Infectious Disease Research Program, Department of Global Health, University of South Florida, Tampa, Florida, United States of America,
[3] Department of Entomology and Plant Pathology, Auburn University, Auburn, Alabama, United States of America

ABSTRACT

Determining the structure of ectoparasite-host networks will enable disease ecologists to better understand and predict the spread of vector-borne diseases. If these networks have consistent properties, then studying the structure of well-understood networks could lead to extrapolation of these properties to others, including those that support emerging pathogens. Borrowing a quantitative measure of network structure from studies of mutualistic relationships between plants and their pollinators,

Citation: Graham SP, Hassan HK, Burkett-Cadena ND, Guyer C, Unnasch TR (2009) Nestedness of Ectoparasite-Vertebrate Host Networks. PLoS ONE 4(11): e7873. https://doi.org/10.1371/journal.pone.0007873

we analyzed 29 ectoparasite-vertebrate host networks—including three derived from molecular bloodmeal analysis of mosquito feeding patterns—using measures of nestedness to identify non-random interactions among species. We found significant nestedness in ectoparasite-vertebrate host lists for habitats ranging from tropical rainforests to polar environments. These networks showed non-random patterns of nesting, and did not differ significantly from published estimates of nestedness from mutualistic networks. Mutualistic and antagonistic networks appear to be organized similarly, with generalized ectoparasites interacting with hosts that attract many ectoparasites and more specialized ectoparasites usually interacting with these same "generalized" hosts. This finding has implications for understanding the network dynamics of vector-born pathogens. We suggest that nestedness (rather than random ectoparasite-host associations) can allow rapid transfer of pathogens throughout a network, and expand upon such concepts as the dilution effect, bridge vectors, and host switching in the context of nested ectoparasite-vertebrate host networks.

INTRODUCTION

Increased focus on pathogens that emerge from their usual host populations and affect novel hosts (e.g., humans) is warranted. Several strategies have been used to characterize these pathogens, many of which stem from epidemiological paradigms [1], [2]. Many studies focus on a single species acknowledged to play a key role in a pathogen's transmission. Rarely do such studies consider vector-borne pathogen transmission cycles as properties of whole parasite-host networks [3], [4]. The growing field of disease ecology has begun to resolve this discrepancy by merging the full repertoire of field and experimental research with epidemiological models [5]. Numerous studies have demonstrated the utility of this approach to understand disease dynamics [6]–[8], and these studies have become increasingly important due to the alarming increase in emerging pathogens in recent decades [9]–[12].

Ecologists have developed several measures to characterize ecological network dynamics, including the concepts of modularity [13], connectance [14], degree-Strength asymmetry [15], quantitative link density [16], and nestedness [17]–[19]. Nestedness is a factor used to measure the extent to which specialized species (i.e., those with few interactions with other species) interact with increasingly large subsets of generalists (i.e., those with many interactions). Analyses of nestedness are practical tools that require presence/absence data for species interactions at a particular place.

Nestedness has been used to examine communities among habitat fragments [20], as well as to characterize network structure of interacting organisms within communities of mutualists [17]–[19], commensals [21], [22], and parasites [23], [24].

At the scale of an infracommunity (within individual hosts) or component community (among populations of a single host species) [25], host-parasite relationships have often been described as being dominated by random associations [26]–[28]. Recent studies have documented nested patterns of interactions [29]–[32]. However, community-level analyses of entire ectoparasite-vertebrate host networks are rare. Although a recent study analyzed the distribution of specialization in host-parasite networks for selected taxa (fleas and rodents, fish and monogeneans) across a single geographic region (north temperate zone) [33], no study has examined entire ectoparasite-vertebrate host networks across a variety of sites.

It has been suggested that interactions based on mutual benefits trend toward nestedness, while those resulting from antagonistic relationships will lead to specialization and compartmentalization [34]. While most large ecological interaction networks have compartments [13], [23], [34], there is disagreement whether parasite-host networks are organized and develop in such a way that might lead to a nested pattern, and a comparative study between these networks at the scale of whole communities (e.g., as in mutualistic networks) [17] is needed [35]. Strong patterns of nestedness in other network interaction types occur due to the presence of generalized species which interact with specialists as well as other generalists. Extreme generalist ectoparasites are known, as well as very specialized species [36], [37]. Thus, we test the hypothesis that generalized and specialized ectoparasites and their vertebrate hosts interact in a way consistent with the pattern of nestedness observed in other network types. Finally, we suggest that nestedness vs. randomness in ectoparasite-vertebrate host networks has the capacity to facilitate the spread of pathogens across taxa.

RESULTS

Most of our networks (17 of 27) were more nested than the null model simulations using NODF (Table 1). There was no interaction between geographic area or species richness of a network and NODF nestedness (area: $R^2=0.11$; $p>0.05$; richness: $R^2=0.14$; $p>0.05$). A positive relationship was documented between nestedness (N) and species richness of ectoparasite-vertebrate host networks ($R^2=0.417$; $F=18.777$; $p<0.0005$). Collectively, these data suggest that the completeness of the matrix (in terms of number of species included in the analysis), rather than the size of the area studied, was an important factor influencing nestedness.

Table 1. Ectoparasite-vertebrate host networks analyzed for this study.

Network Location	*Nestedness* (N)	*Nestedness* (NODF)	*p=*	*Total # species*	*Latitude*	*Host type(s)*	*Ectoparasite type(s)*	*Ref.*
Auburn, AL, USA	0.8494	21.08	0.01	49	Temperate	M†, B, A, R	Mosquitoes	This study
Tuskegee N.F., AL, USA	0.9353	12.58	<0.000	93	Temperate	M†, B, A, R	Mosquitoes	This study
Mississippi, USA	0.8909	41.83	ns	32	Temperate	M,B, A, R	Mosquitoes	This study
St. Catherine's Island, GA, USA	0.9314	17.8	<0.000	92	Temperate	M†, B, R	Ticks, lice, mites, fleas	[68]
Sapelo Island, GA, USA	0.7612	15.15	ns	30	Temperate	M, B, R	Ticks, lice, mites, fleas	[68]
Jekyll Island, GA, USA	0.7902	13.24	ns	36	Temperate	M, B, R	Ticks, lice, mites, fleas	[68]
Cumberland Island, GA, USA	0.8987	20.75	0.01	100	Temperate	M, B, R	Ticks, lice, mites, fleas	[68]
Indiana, USA	0.8913	41.83	<0.000	259	Temperate	M†	Ticks, lice, chiggers, mites, fleas	[60]
Alaska, USA	0.9332	10.06	<0.000	110	Polar	M, B	Ticks, lice, mites, fleas	[69], [70], [71], [72]
New Zealand	0.9636	6.07	<0.000	238	Temperate	M†, B, R	Ticks, lice, mites, fleas mosquitoes, flies	[73], [74]
New Mexico, USA	0.9163	9.35	<0.000	217	Temperate (arid)	M†	Fleas, lice	[75]
Panama	0.9812	2.13	<0.000	645	Tropical	M†, B, R	Ticks, lice, chiggers, mites, fleas, bat bugs, bat flies	[59]
Atlantic Forest, Brazil	0.9763	6.2	<0.000	177	Tropical	M, B	Ticks, lice, mites, fleas, bat bugs, bat flies	[76], [77], [78], [79], [80], [81]
Mountain Zebra N.P., South Africa	0.7864	7.19	<0.000	39	Mediterranean	M	Ticks	[82]
Sweden	0.8314	9.68	ns	43	Polar	M†, B	Black flies	[83]

Slovakia	0.883	37.29	<0.000	51	Temperate	M	Fleas	[84]
Paraguay	0.769	10.7	ns	51	Tropical	M	Bat flies	[85]
Britain	0.9049	5.63	ns	137	Temperate	M	Ticks, fleas, chiggers, lice, mites	[86], [87]
Uganda	0.9552	7.97	<0.000	256	Tropical	M†, B, R	Ticks, fleas, lice, tsetse flies	[88], [89], [90], [91]
Gannet Islands, Labarador	0.4864	57.49	ns	21	Polar	B	Ticks, lice	[92]
New South Wales	0.6281	38.93	ns	59	Temperate (marine)	F	Monogeneans, isopods,copep-ods	[93]
Pacific Canada	0.6314	11.79	ns	58	Temperate (marine)	F	Monogeneans, isopods,copep-ods	[93]
South Caro-lina, USA	0.8555	6.07	ns	81	Temperate	M, B, R	Ticks, fleas, mites, chig-gers	[94]
Utah, USA	0.9519	7.88	<0.000	244	Temperate (arid)	M, B, R	Ticks, chig-gers, fleas, mites	[95], [96], [97], [98], [99]
Queensland, Australia	0.977	4.81	<0.000	389	Trropical	M†, B, R	Ticks, fleas, chiggers, batflies	[100], [101], [102], [103]
Tasmania, Australia	0.9197	9.84	<0.000	164	Temperate	M†, B, R	Ticks, fleas, chiggers, batflies	[100], [101], [102], [103]
Madagascar	0.9023	6.52	0.02	88	Tropical	M†	Ticks	[104]

M=mammals, B=birds, R=reptiles, A=amphibians, and F=fish.

†Indicates that network specifically included humans.

https://doi.org/10.1371/journal.pone.0007873.t001

In our comparison to previous work using the nestedness estimate *N*, We found a significant main effect of network type on degree of nestedness ($F_{2,13}$=6.20; p=0.003) such that nestedness of ectoparasite-vertebrate host networks did not differ significantly from mutualistic plant-pollinator or seed dispersal networks, but did differ from food webs (Fig. 1). However, this analysis was confounded by a significant interaction between species richness and network type ($F_{2,13}$=9.26; p<0.0001). There were significant positive correlations between species richness and degree of nestedness for the networks based on antagonistic relationships (ectoparasite-vertebrate host: R^2=0.61; F=39.3; p<0.0001; food web: R^2=0.305; F=5.273; p=0.04), but no correlation for networks based on positive relationships (mutualism: R^2=0.043; F=2.216; p>0.05; Fig. 2). Comparison to plant-pollinator networks available through the Interaction Web Database also yielded no significant differences using the nestedness indicator NODF (F=0.768; p=0.469).

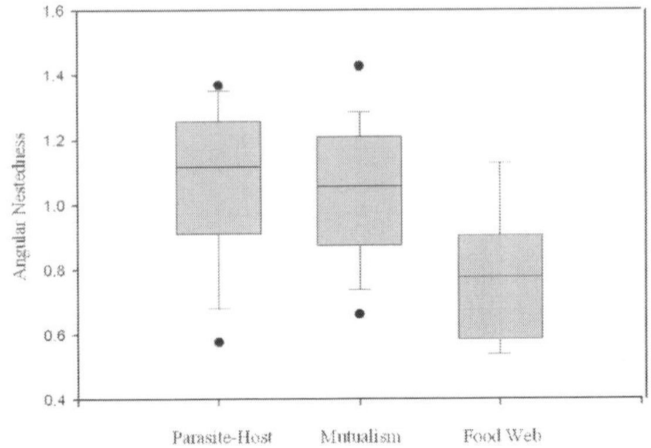

Figure 1. Box-plots of mean nestedness (*N*). Comparison between previously published calculations of plant-pollinator/seed dispersal (mutualism) networks, food webs [17], and ectoparasite-vertebrate host networks (this study).

https://doi.org/10.1371/journal.pone.0007873.g001

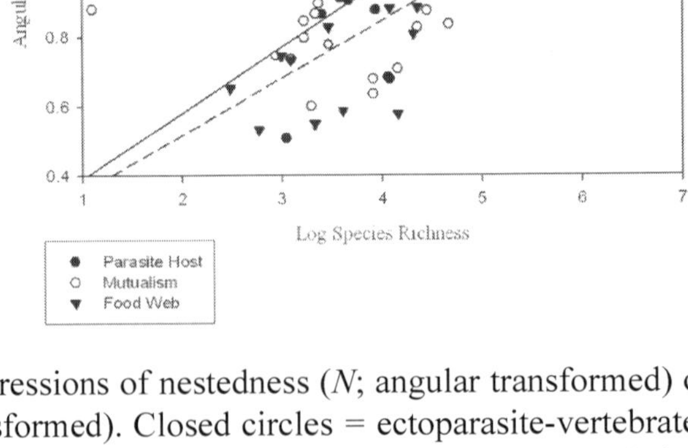

Figure 2. Regressions of nestedness (*N*; angular transformed) on species richness (log transformed). Closed circles = ectoparasite-vertebrate host networks (this study); open circles = mutualisms (plant-pollinator and plant-seed disperser networks) [17]; closed triangles = food web networks [17].

https://doi.org/10.1371/journal.pone.0007873.g002

Within the hypothetical, perfectly nested TNF network, only a single mosquito was needed to retain connections to all hosts (Fig. 3). Randomized associations of this same matrix reduced dominance of any one host or parasite and, in the example presented here (Fig. 4), required 15 parasites to retain connections to all hosts. The observed diagram for this network (Fig. 5) was more similar to the extreme of complete nestedness than it was to the extreme of random expectation. This resulted from a dominant mosquito (*Culex erraticus*) that visited most hosts, and a dominant host (White-tailed deer) that was visited by more mosquitoes than any other host. Because of significant nesting, only six of the 17 mosquitoes in this network were required to maintain connections to all known hosts.

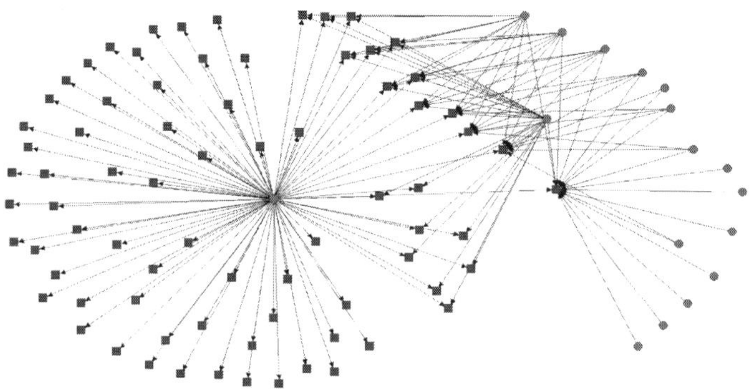

Figure 3. Hypothetical network of perfect nestedness for 17 ectoparasites and 76 hosts. Drawn using UCINET.

https://doi.org/10.1371/journal.pone.0007873.g003

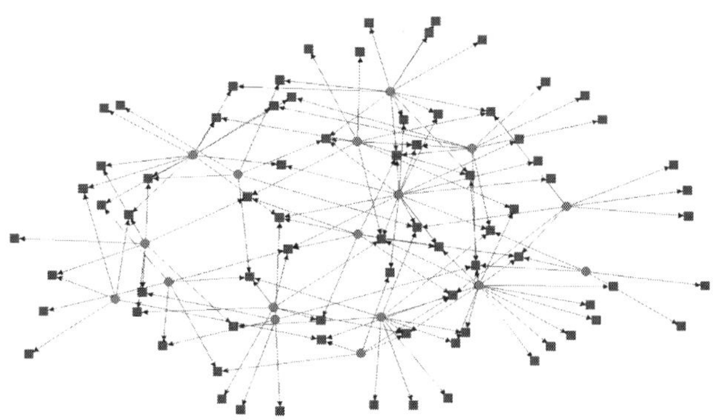

Figure 4. Hypothetical network of random interactions for 17 ectoparasites and 76 hosts. Drawn using UCINET.

https://doi.org/10.1371/journal.pone.0007873.g004

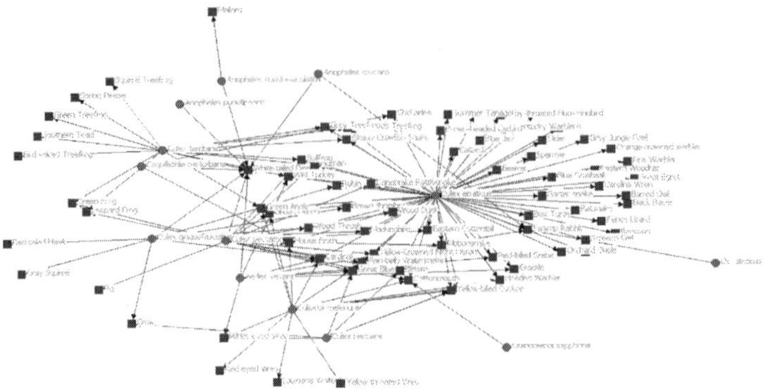

Figure 5. Observed network of mosquitoes and their vertebrate hosts. Network from Tuskegee National Forest, Alabama, USA, representing 17 actual mosquito species and their 76 vertebrate hosts. Red dots represent mosquitoes and blue squares represent vertebrate hosts. Drawn using UCINET.

https://doi.org/10.1371/journal.pone.0007873.g005

DISCUSSION

We detected significant nestedness in most ectoparasite-vertebrate host networks. This indicates that specialized ectoparasites usually interact with hosts that attract many ectoparasites ('generalist' hosts), and generalist ectoparasites interact with these hosts as well as those that attract fewer ectoparasites ('specialist' hosts). Networks analyzed were as species rich (21–645 species), and were as geographically diverse (including networks from five continents) as previous attempts to characterize nestedness in other types of species interactions [17]–[19]. Despite the diversity of latitudinal, ecosystem, and taxonomic affiliations of these networks, considerable nestedness appears to be a consistent rule in ectoparasite-vertebrate host networks.

Moreover, our results were robust with respect to passive sampling and other problems associated with nestedness estimates, which can inflate the results of nestedness calculations [38], [39]. Sampling artifacts could still be responsible for a portion of the trends observed because parasite species richness and host abundance covary in a way that is difficult to tease apart [40]. However, biologically meaningful patterns occur between hosts and parasites despite this dilemma [33], [41], and nestedness analyses are robust to variation in sampling effort [42]. Our mosquito-host networks were also statistically nested despite the fact that mosquitoes were the collectors of host samples. This offers an interesting qualitative comparison between host versus ectoparasite sampling and their potential biases; in both situations, nestedness detected was significantly higher than null model simulations.

These results underscore the remarkably similar properties that have been attributed to networks with interacting components, from human social systems to mutualistic networks of plants and pollinators [43]. These networks are often characterized by asymmetric connections between network members, with generalist species interacting with each other, and specialized species (e.g., those with few interactions) usually interacting with generalists [34], [35]. A study of flea-rodent networks from north temperate zones revealed these patterns hold for ectoparasites [33], and our data corroborate and expand upon this. Mutualistic and parasitic networks are on opposite ends of the interaction spectrum in terms of the relationship of the participants (from +/+ to +/−), and nonetheless they appear to share a similar network structure. Although counterintuitive, the overall similarity and low interaction intimacy of plant/host visitation by pollinators/ectoparasites—and their subsequent co-evolutionary interactions—probably lead to this structure [44].

Various factors suggested to play a role in the development of nested mutualistic networks [45], [46] probably also contribute to the nested structure documented here. Much of the nestedness detected in our networks is probably due to abundance and sampling interactions [33]; the most abundant species typically are responsible for large cores of interactions, and rarer, more specialized species are more likely to "sample" the most abundant hosts. Phenotypic trait matching between ectoparasites and host defenses could contribute to nested patterns. The mouthparts, attachment organs, behavior, and morphology of ectoparasites exhibit generalized and more specialized patterns for obtaining blood from a variety of hosts or specific hosts [47]. Hosts have also developed variable defenses which could lead to nested preferences among ectoparasites [48], [49]. Spatiotemporal distribution of ectoparasites and their hosts (e.g., some ectoparasites and potential hosts are not coincident in space or time) and the existence of 'forbidden interactions' (mosquitoes do not feed upon fish or salamanders at our study site) [50] may possibly play a role as well. Finally, there is evidence of phylogenetic concordance in ectoparasite-host interactions [51], and future studies which incorporate phylogenetic information will likely determine that this is an essential factor in the nesetedness of these networks

[52].

Our comparison using a network analysis which takes into account varying network properties (NODF) showed no differences between ectoparasite-vertebrate host networks and plant-pollinator networks. This analysis and others have determined nestedness in parasite-host networks [23], [24], suggesting that a dichotomy between network structure may occur between different interaction intimacies [44], rather than differences in the interaction types (i.e., +/+ vs. +/−). It is therefore likely that nonsymbiotic mutualistic and antagonistic interaction networks are simultaneously compartmentalized and nested [13], [17], [23], [44]. Our results confirm the nestedness of antagonistic nonsymbiotic networks, although the networks we analyzed represent a range of interaction intimacies ranging from low (mosquitoes) to relatively high (ticks, lice). Qualitatively, we suggest that these networks are compartmentalized, since most ectoparasite groups appear to form their own nested subcompartments. Future research should seek to determine if ectoparasite-vertebrate host networks are made up of large nested networks composed of nested subcompartments linked by generalists, and focus should also be directed toward endoparasite-host networks to confirm that symbiotic vs. nonsymbiotic networks exhibit different properties.

Our results have implications for understanding disease transmission cycles. In the specific case of mosquitoes at Tuskegee National Forest (TNF), nesting assures that novel diseases, such as West Nile virus, have the potential to spread widely among vertebrate hosts. This prediction is associated with the fact that nesting results from a core ectoparasite (*Culex erraticus*) that feeds on the majority of vertebrate hosts and a core host (White-tailed deer) that is fed upon by the majority of mosquitoes. These core taxa greatly reduce the number of linkages needed to unite all taxa within the network. Thus, only six mosquitoes are required to retain all interactions of the TNF network whereas 15 mosquitoes are required for a randomized version of the same network. The concept of compartmentalized nestedness predicts that pathogens likely coevolve with hosts and ectoparasites within specific compartments, but the possibility for escape into additional or all other compartments is possible due to the high connectance of the network. Zoonoses (diseases caused by pathogens which normally cycle through wildlife) can thus be the result of compartment-spanning or compartment-shifting pathogens [53], [54], and network analysis may lead to discovery of likely conditions that exacerbate such phenomena (e.g., during peaks of host or ectoparasite abundance).

Increased nestedness in larger ectoparasite-vertebrate host networks implies that species richness actually leads to increased connectance between individual species in the network, greatly facilitating the spread of disease unless member hosts, especially core hosts, have reduced capacity to amplify and allow transmission of a pathogen (e.g., decreased reservoir competence). Thus, the transmission pattern of vector-born pathogens is likely determined by high connectivity between hosts and ectoparasites due to nestedness, counterbalanced by the existence of nested compartments of resistant hosts that are dead ends for pathogens (expanding the "dilution effect" hypothesis) [4].

Finally, disease ecologists often attempt to understand and mitigate vector-born pathogens from the host point of view, and network analyses are a new way of looking at this network interaction instead as a double-edged sword. For example, vector-born diseases are often thought of as the result of ectoparasites that have broad host preferences (so-called bridge vectors), which occasionally spread diseases to humans [55]–[57]. Network analyses reveal the whole story: bridge vectors are usually abundant and catholic in their host preference, however, attention from these ectoparasites comes as the inevitable consequence of being a large and/or abundant potential host. Current theory regarding host switching by ectoparasites as a driver for disease emergence [53], [57] should consider this; host switching may have less to do with ectoparasite host switching and more to do with host population dynamics (e.g., surges in host abundance and/or availability may lead to recruitment of new ectoparasites). Fortunately, since ectoparasite-vertebrate host networks have consistent network properties, findings from well-studied model systems (e.g., lyme disease, ticks, and their hosts) [58] should have reliable crossover to less well-studied networks which harbor vector-born diseases. Thus, nestedness analyses should be seen as a new step toward understanding the complexity of these networks in such a way that new threats—such as emerging infectious diseases—can more quickly be characterized, understood, and effectively managed.

METHODS

Ectoparasite-Vertebrate Host Networks

Twenty-six ectoparasite-vertebrate host matrices were assembled from lists of interacting species available from the literature (Table 1). Each list was generated for a specified contiguous region. Methodology for these studies typically included capturing vertebrates and identifying ectoparasites extracted from them [59], [60]. These studies described regions that ranged in size from entire countries (e.g. Panama, South Africa) to smaller regions within countries (e.g. Gannet Islands, Labrador). Therefore, these matrices were composed of multiple component communities [25]. This approach is analogous to the methodology of previous nestedness analyses of plants and their pollinators or seed dispersers [17].

Ectoparasites and their vertebrate hosts were included within interaction matrices only if species-level identifications were available. This approach probably underestimates total species richness within each network, since small ectoparasites frequently are identified only to genus. This procedure might underestimate nestedness since many specialist ectoparasites probably remain undescribed.

To these literature records we added three mosquito-host networks derived from our studies of vectors of West Nile virus (WNV) and eastern equine encephalitis virus (EEEV) in Mississippi and Alabama, USA. In these studies we developed extensive host lists for diverse mosquitoes using molecular analysis of mosquito bloodmeals. In brief, we collected blood-engorged female mosquitoes from field sites and identified species of vertebrate hosts of individual mosquitoes using Polymerase Chain Reaction (PCR) amplification of host DNA in blood meals. Resulting sequences were matched to sequences in GENBANK via BLAST to determine the likely host species (98% sequence match). Additional details of the procedures used to identify host bloodmeals can be found elsewhere [50], [61], [62].

Measure of Nestedness

Nestedness was determined from each network using ANINHADO 3.0 software [63]. This program maximally packs each matrix representing a network, and then calculates an isocline of perfect nestedness for the given size and shape of a matrix. Within this matrix the distance of each interaction (fill) from the isocline is calculated, averaged, and used to create an index (T) that varies from 0 (perfectly nested) and 100. Following 17, we then calculated $N=(100-T)/100$ so that increasing values of the index represented increasing nestedness and to facilitate comparison to analyses of other published networks.

There has been a recent surge of interest in refining nestedness analyses such that they better reflect the original concept of nestedness [63]. Therefore, we also determined a new measure of nestedness (e.g., NODF) using

ANINHADO 3.0 software [63]. This measure accounts for certain flaws in other nestedness estimates that have inappropriate interactions with matrix properties such as size and shape. NODF considers matrix row and column fill differences, as well as the overlap of row and column presences [63]. Although this metric is clearly superior to absolute nestedness temperature estimates, we provide analyses using *N* to allow comparison to previous work. However, due to statistical problems associated with *N* [63], we considered analyses involving NODF the test of our hypothesis, and only these analyses are considered as such in our results and discussion. We compared *N* of our networks to published estimates of nestedness for mutualisms and food webs from 17. NODF values were compared to 32 plant-pollinator networks from around the world available from the Interaction Web Database [64], [65].

Data Analysis

Calculations of nestedness were tested against a null model generated by Monte-Carlo simulations (1000 permutations) within program ANINHADO [63]. Nestedness might result from passive sampling rather than assemblage structuring [39], especially for ectoparasite-vertebrate host networks in which common ectoparasites and hosts might be sampled more frequently than rare ones. To account for this, we chose a null model which considers passive sampling [39]. This null model simulation randomly fills cells in proportion to the row and column totals of each interacting species.

Two extreme outliers (nestedness values were > two standard deviations above the mean value) [66] were removed from the dataset leaving a sample of 27 networks. These were ectoparasite-vertebrate host networks with highest and lowest species richness. Nestedness values were angular transformed and the corresponding value of species richness of each network was log transformed to achieve normality [18]. Potential confounding effects of geographic area and species richness on nestedness were analyzed using regression. Ectoparasite-vertebrate host networks were compared to other ecological networks [17] using an ANCOVA with network type (host-parasite, food web, or plant-pollinator/disperser) as the main effect, nestedness as the dependent variable, and species richness as the covariate. NODF values for ectoparasite-vertebrate host networks were similarly compared to plant-pollinator networks available from the Interaction Web Database.

To visualize potential patterns of interactions at the Tuskegee National Forest site, we used UCINET software to create ball-and-stick models for a fully nested, a random, and the observed set of interacting species. We then calculated the minimum number of mosquitoes required to sample blood from the vertebrate hosts in fully nested, random, and the observed mosquito-host networks [67]. In all analyses, α was set at 0.05.

ACKNOWLEDGMENTS

We thank Robert Unnasch, Mark Liu, Kyle Barrett, Christina Romagosa, David Steen, Valerie Johnson, Geoff Sorrell, Shannon Hoss, Sharon Hermann, and John Peterson for comments and suggestions on earlier versions of this manuscript. We are also grateful to P. Guimarães for discussions and troubleshooting ANINHADO software with us. We thank the editorial staff of PLoS ONE, M. Almeida-Neto, and A. Nielsen for reviewing this manuscript and providing extremely helpful criticisms and suggestions which vastly improved this document.

REFERENCES

1. Busenberg S, Driessch PVD (1990) Analysis of a disease transmission model in a population with varying size. J Math Biol 28: 1432–1416.
2. Antonovics J, Iwasa Y, Hassell MP (1995) A generalized model of parasitoid, venereal, and vector-based transmission processes. Am Nat 5: 661–675.
3. Gilot B, Laforge ML, Pinchot J, Raoult D (1990) Relationships between the *Rhipicephalus sanguineus* complex ecology and Mediterranean spotted fever epidemiology in France. Europ J Epidemiol 6: 357–362.
4. Ostfeld RS, Keesing F (2000) The function of biodiversity in the ecology of vector-borne zoonotic diseases. Can J Zool 78: 2061–2078.
5. Collinge SK, Ray C (2006) Disease Ecology: Community Structure and Pathogen Dynamics. Oxford: Oxford University Press.
6. Yates TL, Mills JN, Parmenter CA, Ksiazek TG, Parmenter RR, et al. (2002) The ecology and evolutionary history of an emergent disease: hantavirus pulmonary syndrome. Bioscience 52: 989–998.
7. Allan BF, Keesing F, Ostfield RS (2002) Effect of forest fragmentation on lyme disease risk. Cons Biol 17: 267–272.
8. Unnasch RS, Sprenger T, Katholi CR, Cupp EW, Hill GE, et al. (2006) A dynamic transmission model of eastern equine encephalitis virus. Ecological Modeling 192: 425–440.
9. Daszak P, Cunningham AA, Hyatt AD (2000) Emerging infectious diseases of wildlife—threats to biodiversity and human health. Science 287: 443–449.
10. Taylor LH, Latham SM, Woolhouse MEJ (2001) Risk factors for human disease emergence. Phil Trans Roy Soc Lon B 356: 983–989.
11. Anderson PK, Cunningham AA, Patel NG, Morales FJ (2004) Emerging infectious diseases of plants: pathogen pollution, climate change, and agrotechnology. Trends Ecol Evol 19: 535–544.
12. Fauci AS (2004) Emerging infectious diseases: a clear and present danger to humanity. JAMA 292: 1887–1888.
13. Olesen JM, Bascompte J, Dupont YL, Jordano P (2007) The modularity of pollination networks. Proc Natl Acad Sci USA 104: 19891–19896.
14. Olesen , JM , Bascompte J, Dupont YL, Jordano P (2006) The smallest of all worlds: Pollination networks. J Theor Biol 240: 270–276.

15. Bascompte J, Jordano P, Olesen JM (2006) Asymmetric coevolutionary neworks facilitate biodiversity maintenance. Science 312: 431–433.
16. Bersier LF, Banasek-Richter C, Cattin MF (2002) Quantitative descriptors of food-web matrices. Ecology 83: 2394–2407.
17. Bascompte JP, Jordano CJ, Melián JM, Olesen JM (2003) The nested assembly of plant-animal mutualistic networks. Proc Natl Acad Sci USA100: 9383–9387.
18. Guimarães PR, Sazima C, Reis SF, Sazima I (2007) The nested structure of marine cleaning symbiosis: is it like flowers and bees? Biol Let 3: 51–54.
19. Ollerton J, McCollin D, Fautin DG, Allen GR (2007) Finding NEMO: nestedness engendered by mutualistic organization in anemonefish and their hosts. Proc R Soc B 217: 591–598.
20. Atmar W, Patterson BD (1993) The measure of order and disorder in the distribution of species in fragmented habitat. Oecologia 96: 373–382.
21. Verdú M, Valiente-Banuet A (2008) The nested assembly of plant facilitation networks prevents species extinctions. Am Nat 173: 751–760.
22. Blick R, Burns KC (2009) Network properties of arboreal plants: are epiphytes, mistletoes, and lianas structured similarly? Persp Plant Ecol 11: 41–52.
23. Vacher C, Piou D, Desprez-Loustau M (2008) Architecture of an antagonistic tree/fungus network: the asymmetric influence of past evolutionary history. PLoS ONE 3: 1–10.
24. Löwenberg-Neto P (2008) The structure of the parasite-host interactions between *Philornis* (Diptera: Muscidae) and neotropical birds. J Trop Ecol 24: 575–580.
25. Poulin R (1997) Species richness of parasite communities: evolution and pattern. Annu Rev Ecol Evol Syst 28: 341–358.
26. Worthen WB, Rhode K (1996) Nested subset analyses of colonization-dominated communities: metazoan ectoparasites of marine fishes. Oikos 75: 471–478.
27. Guégan JF, Huegany B (1994) A nested parasite species subset pattern in tropical fish: host as a major determinant of parasite infracommunity structure. Oecologia 100: 184–189.

28. Rhode KC, Heap M (1998) Latitudinal differences in species and community richness and in community structure of metazoan endo- and ectoparasites of marine teleost fish. Int J Parasitol 28: 461–474.
29. Krasnov BR, Stanko M, Morand S (2006) Are ectoparasite communities structured? Species co-occurrence, temporal variation and null models. J Anim Ecol 75: 1330–1339.
30. Poulin R, Guégan JF (2000) Nestedness, anti-nestedness, and the relationship between prevalence and intensity in ectoparasite assemblages of marine fish: a spatial model of species coexistence. Int J Parasitol 30: 1147–1152.
31. Poulin R, Valtonen ET (2001) Nested assemblages resulting from host size variation: the case of endoparasite communities in fish hosts. Int J Parasitol 31: 1194–1204.
32. Morand SK, Rhode K, Hayward C (2002) Order in ectoparasite communities of marine fish is explained by epidemiological processes. Parasitol 124: S57–S63.
33. Vásquez DP, Poulin R, Krasnov BS, Shenbrot GI (2005) Species abundanceand distribution of specialization in host-parasite interaction networks. J An Ecol 74: 946–955.
34. Thompson JN (2006) Mutualistic webs of species. Science 312: 372–373.
35. Thébault E, Fontaine C (2008) Does asymmetric specialization differ between mutualistic and trophic networks? Oikos 117: 555–563.
36. McCoy KD, Boulinier T, Tirard C, Michalakis C (2001) Host specificity of a generalist parasite: genetic evidence of sympatric host races in the seabird tick *Ixodes uriae*. J Evol Biol 14: 395–405.
37. Dick CW, Patterson BD (2007) Against all odds: explaining high host specificity in dispersal-prone parasites. Int J Parasitol 37: 871–876.
38. Almeida-Neto PR, Guimarães PR Jr, Guimarães PR, Loyola RD, Ulrich W (2008) A consistent metric for nestedness analysis in ecological systems: reconciling concept and measurement. Oikos 117: 1227–1239.
39. Fischer J, Lindemayer DB (2002) Treating the nestedness temperature calculator as a "black box" can lead to false conclusions. Oikos 99: 193–199.

40. Guégan JF, Kennedy CR (1996) Parasite richness/sampling effort/ host range: the fancy three-piece jigsaw puzzle. Parasitol Today 12: 367–369.
41. Poulin R (1995) Phylogeny, ecology, and the richness of parasite communities in vertebrates. Ecol Monogr 65: 283–302.
42. Nielsen A, Bascompte J (2007) Ecological networks, nestedness and sampling effort. J Ecol 95: 1134–1141.
43. Jordano PJ, Bascompte J, Olesen JM (2003) Invariant properties in co-evolutionary networks of plant-animal interactions. Ecol Letters 6: 69–81.
44. Guimarães PR, et al. (2007) Interaction intimacy affects structure and coevolutionary dynamics in mutualistic networks. Current Biology 17: 1–7.
45. Bascompte J, Jordano P (2007) Plant-animal mutalistc networks: the architechture of biodiversity. Ann Rev Ecol Evol Syst 38: 567–596.
46. Vázquez DP, Blüthgen N, Cagnolo L, Chacoff NP (2009) Uniting pattern and process in plant-animal mutualistic networks: a review. Ann Bot 103: 1445–1457.
47. Šimková A, Ondračková M, Gelnar M, Morand S (2002) Morphology and coexistence of congeneric ectoparasite species: reinforcement of reproductive isolation? Biol J Linnean Soc 76: 125–135.
48. Wikel SK (1999) Modulation of the host immune system by ectoparasitic arthropods. BioScience 49: 311–320.
49. Wikel SK, Alarcon-Chaidez FJ (2001) Progress toward molecular characterization of ectoparasite modulation of host immunity. Vet Parasitol 101: 275–287.
50. Burkett-Cadena ND, Graham SP, Hassan HK, Guyer C, Eubanks MD, et al. (2008) Blood feeding patterns of potential arbovirus vectors of the genus *Culex* targeting ectothermic hosts. Am J Trop Med Hyg 79: 809–815.
51. Hafner MS, Page RD (1995) Molecular phylogenies and host-parasite cospeciation: gophers and lice as a model system. Phil Trans Roy Soc Lon B 349: 77–83.
52. Rezende EL, Lavabre JE, Guimarães PR, Jordano P, Bascompte J (2007) Non-random coextinctions in phlyogenetically structured mutualistic networks. Nature 448: 925–928.

53. Ricklefs RE, Fallon SM (2002) Diversification and host switching in avian malaria parasites. Proc R Soc Lond B 269: 885–892.
54. Kilpatrick AM, Kramer LD, Campbell SR, Alleyne EO, Dobson AP, et al. (2005) West Nile virus risk assessment and the bridge vector paradigm. Emerging Infectious Diseases 11: 425–429.
55. Gratz NG (2004) Critical review of the vector status of *Aedes albopictus*. Med Vet Entomol 18: 215–227.
56. Hamer GL, Kitron UD, Brawn JD, Loss SR, Ruiz MO, et al. (2008) *Culex pipiens* (Diptera: Culicidea): a bridge vector of West Nile virus to humans. J Med Entomol 45: 125–128.
57. Kilpatrick AM, Kramer LD, Jones JM, Marra , PP , Daszak P (2006) West Nile virus epidemics in North America are driven by shifts in mosquito behavior. PLoS Biol 4: 606–610.
58. LoGuiduce K, Ostfield RS, Schmidt KA, Keesing F (2003) The ecology of infectious disease: effects of host diversity and community composition on lyme disease risk. Proc Natl Acad Sci USA 100: 567–571.
59. Wenzel RL, Tipton VJ, editors. (1966) Ectoparasites of Panama (Field Museum of Natural History, Chicago).
60. Whitaker JO (1982) Ectoparasites of Mammals of Indiana (The Indiana Academy of Science, Indianapolis).
61. Cupp EW, Zhang D, Yue X, Cupp MS, Guyer C, et al. (2004) Identification of reptilian and amphibian blood meals from mosquitoes in an eastern equine encephalomyelitis virus focus in Central Alabama. Am J Trop Med Hyg 71: 272–276.
62. Hassan HK, et al. (2003) Avian host preference by vectors of eastern equine encephalitis virus. Am J Trop Med Hyg 69: 641–647.
63. Guimarães PR Jr, Guimarães PR (2006) Improving the analyses of nestedness for large sets of matrices. Environ Modl Software 21: 1521–1513.
64. Vazquez DP, Melian CJ (2008) Interaction Web Database. Available: www. hceas. ucsb. edu/interactionweb/index. html.
65. Bezerra ELS, Machado IC, Mello M (2009) Pollination networks of oil-flowers: the smallest of all worlds. J Anim Ecol 78: 1096–1101.
66. Zar JH (1996) Biostatistical Analysis, third ed. (Prentice Hall, New Jersey).

67. Vane-Wright RI, Humphries CJ, Williams PH (1991) What to protect? Systematics and the agony of choice. Biol Cons 55: 235–254.

68. Wilson N, Durden L (2003) Ectoparasites of terrestrial vertebrates inhabiting the Georgia Barrier Islands, USA: an inventory and preliminary biogeographical analysis. J Biogeogr 30: 1207–1220.

69. Philip CB (1938) A parasitological reconnaissance in Alaska with particular reference to varying hares. II. Parasitological data. J Parasitol 24: 483–488.

70. Choe JA, Kim KC (1987) Community structure of arthropod parasites on Alaskan seabirds. Can J Zool 65: 2998–3005.

71. Murrel BP, Durden LA, Cook JA (2003) Host associations of the tick, *Ixodes angustus* (Acari: Ixodidae), on Alaskan mammals. J Med Entomol 40: 682–685.

72. Haas GE, Kucera JR, Runck AM, MacDonald SO, Cook JA (2005) Mammal fleas (Siphonaptera: Ceratophyllidae) new for Alaska and the southeastern mainland collected during seven years of a field survey of small mammals. J Entomol Soc Brit Columbia 102: 65–75.

73. Heath ACG (1977) Zoogeography of the New Zealand tick fauna. Tuatara 23: 26–39.

74. Tenquist JD, Charleston WAG (2001) A revision of the annotated checklist of ectoparasites of terrestrial mammals in New Zealand. J Roy Soc New Zealand 31: 481–542.

75. Ford PL, Fagerlund RA, Duszynski DW, Polechla PJ (2004) Fleas and Lice of Mammals in New Mexico. Rocky Mountain Research Station (US Department of Agriculture), General Technical Report.

76. Barros-Battesti DM, Arzua M, Linardi PM, Bothelho JR, Sbalqueiro IJ (1998) Interrelationships between ectoparasites and wild rodents from Tijucas do Sul, State of Parana, Brazil. Mem Inst Oswaldo Cruz 96: 719–725.

77. Barros-Battesti DM, Yoshinari NH, Bonoldi VLN, Gomes AD (2000) Parasitism by *Ixodes didelphis* and *I. loricatus* (Acari: Ixodidae) on small wild animals from an Atlantic Forest in the state of Sao Paulo, Brazil. J Med Entomol 37: 820–827.

78. Martins-Hatano FD, Gettinger D, Bergallo HG (2002) Ecology and host specificity of Laelapine mites (Acari: Laelapidae) of small mammals in an Atlantic Forest area of Brazil. J Parasitol 88: 36–40.

79. Bittencourt EB, Rocha CFD (2003) Host-ectoparasite specificity in a small mammal community in an area of Atlantic Rain Forest (Ilha Grande, State of Rio de Janeiro), southeastern Brazil. Mem Inst Oswaldo Cruz 98: 793–798.
80. De Lyra-Neves RM, Isidro de Farias AM, Telino-Junior WR (2003) Ecological relationships between feather mites (acari) and wild birds Emberizidae (Aves) in a fragment of Atlantic Forest in northeastern Brazil. Revista Brasileira de Zoologia 20: 481–485.
81. Labruna MB, et al. (2007) Ticks collected on birds in the state of Sao Paulo, Brazil. Exp Appl Acarol 43: 147–160.
82. Horak IG, Fourie LJ, Novellie PA, Williams EJ (1991) Parasites of domestic and wild animals in South Africa XXVI. The mosaic of Ixodid tick infestations on birds and mammals in the Mountain Zebra National Park. Onderstepoort J Vet Res 58: 125–136.
83. Malmqvist B, Strasevicius D, Hellgren O, Adler PH, Bensch S (2003) Vertebrate host specificity of wild-caught blackflies revealed by mitochondrial DNA in blood. Proc R Soc Lond B 271: Suppl152–155.
84. Stanko MD, Mikilisová J, De Bellocq G, Morand S (2002) Mammal diversity and patterns of ectoparasite species richness and abundance. Oecologia 131: 289–295.
85. Dick CW, Gettinger D (2005) A faunal survey of streblid flies (diptera: streblidae) associated with bats in Paraguay. J Parasitol 91: 1015–1024.
86. Thompson GB (1939) A check-list and host-list of the ectoparasites recorded from British birds and mammals. Part I. Mammals (excluding bats). Trans Soc Brit Entomol 6: 1–22.
87. Baker AS, Craven JC (2003) Checkist of the mites (Arachnida: Acari) associated with bats (Mammalia: Chiroptera) in the British Isles. Syst Appl Aracol Sp Publ 14: 1–20.
88. Thompson GB (1935) An ectoparasite census of ducks and geese (Anatidae) in Uganda. J Anim Ecol 4: 192–194.
89. Matthysse JG, Colbo MH (1990) The Ixodid ticks of Uganda. Exp Ap Aracol 8: 221–222.
90. Clausen PH, Adeyemi I, Bauer B, Breloeer M, Salchow F, et al. (1998) Host preferences of tsetse (Diptera: Glossinidae) based on blood meal identifications Med Vet Entomol 12: 169–180.

91. Price MA, Graham OH (1997) Chewing and Sucking Lice as Parasites of Mammals and Birds. Agricultural Research Service (US Department of Agriculture), Technical Bulletin 1849.
92. Muzaffar SB (2000) Ectoparasites of auks (alcidae) at the gannet islands, Labrador: diversity, ecology, and host-parasite interactions. Unpubl MS thesis (Memorial University of Newfoundland, St. John's).
93. Rhode KC, Hayward C, Heap M (1995) Aspects of the ecology of metazoan ectoparasites of marine fishes. Int J Parastitol 25: 945–970.
94. Nelder MP, Reeves WK (2005) Ectoparasites of road-killed vertebrates in northwestern South Carolina, USA. Vet Parasitol 129: 313–322.
95. Edmunds LR (1951) A checklist of the ticks of Utah. Pan-Pacific Entomol 27: 23–26.
96. Brennan JM, Beck DE (1955) The chiggers of Utah (Acarina: Trombiculidae). Great Basin Nat 15: 1–26.
97. Stark HE (1958) The Siphonaptera of Utah. (US Dept of Health, Education and Welfare, Communicable Disease Center, Atlanta).
98. Parker DD, Howell JF (1959) Host-flea relationships in the Great Salt Lake desert. J Parasitol 45: 597–604.
99. Allred DM, Bece E (1966) Mites of Utah Mammals (Brigham Young University Science Bulletin, Biological Service, Provo).
100. Roberts FHS (1970) Australian Ticks (Commonwealth Scientific and Industrial Research Organization, Australia).
101. Maa TC (1971) Studies in batflies (Diptera: Streblidae, Ncyteribiidae) Part I. Pac Insects Monogr 28: 1–247.
102. Dunnet GM, Mardon DK (1974) A monograph of Australian Fleas (Siphonaptera). Aust J Zool Suppl Ser 30: 1–174.
103. Dumrow R, Lester LN (1985) Chiggers of Australia (Acari: Trombiculidae): an Annotated Checklist, Keys and Bibliography. Aust J Zool Suppl 114: 1–111.
104. Uilengberg G, Hoogstraal H, Klein JM (1979) The ticks of Madagascar (Ixodoidea) and their role as vectors. Archives de l'Institut Pasteur de Madagascar (numéro spécial).

2

Molecular Detection of Rickettsia Typhi in Cats and Fleas

Maria Mercedes Nogueras[1,] Immaculada Pons[1], Ana Ortuño[2,] Jaime Miret[3,] Julia Pla[4], Joaquim Castellà[2] and Ferran Segura[1, 5]

[1] Department of Infectious Diseases. Corporació Sanitària i Universitària Parc Taulí - Institut Universitari Parc Taulí – Autonomous University of Barcelona, Sabadell, Barcelona, Spain,
[2] Department of Animal Health, Autonomous University of Barcelona, Bellaterra, Barcelona, Spain,
[3] Animal Shelter Company (Vallès Oriental), Granollers, Barcelona, Spain,
[4] Clinical Veterinary, Sabadell, Barcelona, Spain,
[5] Department of Medicine, Universitat Autònoma de Barcelona, Bellaterra, Barcelona, Spain,
University of Minnesota, United States of America,

ABSTRACT

Background

Rickettsia typhi is the etiological agent of murine typhus (MT), a disease transmitted by two cycles: rat-flea-rat, and peridomestic cycle. Murine

---

Citation: Nogueras MM, Pons I, Ortuño A, Miret J, Pla J, Castellà J, et al. (2013) Molecular Detection of Rickettsia typhi in Cats and Fleas. PLoS ONE 8(8): e71386. https://doi.org/10.1371/journal.pone.0071386

typhus is often misdiagnosed and underreported. A correct diagnosis is important because MT can cause severe illness and death. Our previous seroprevalence results pointed to presence of human *R. typhi* infection in our region; however, no clinical case has been reported. Although cats have been related to MT, no naturally infected cat has been described. The aim of the study is to confirm the existence of *R. typhi* in our location analyzing its presence in cats and fleas.

Methodology/Principal Findings

221 cats and 80 fleas were collected from Veterinary clinics, shelters, and the street (2001-2009). Variables surveyed were: date of collection, age, sex, municipality, living place, outdoor activities, demographic area, healthy status, contact with animals, and ectoparasite infestation. IgG against *R. typhi* were evaluated by indirect immunofluorescence assay. Molecular detection in cats and fleas was performed by real-time PCR. Cultures were performed in those cats with positive molecular detection. Statistical analysis was carried out using SPSS. A $p < 0.05$ was considered significant.

Thirty-five (15.8%) cats were seropositive. There were no significant associations among seropositivity and any variables. *R. typhi* was detected in 5 blood and 2 cultures. High titres and molecular detection were observed in stray cats and pets, as well as in spring and winter. All fleas were *Ctenocephalides felis*. *R. typhi* was detected in 44 fleas (55%), from shelters and pets. Co-infection with *R. felis* was observed.

Conclusions

Although no clinical case has been described in this area, the presence of *R. typhi* in cats and fleas is demonstrated. Moreover, a considerable percentage of those animals lived in households. To our knowledge, this is the first time *R. typhi* is detected in naturally infected cats.

INTRODUCTION

Murine typhus is one of the most prevalent Rickettsioses worldwide distributed, and it is endemic in coastal areas and ports [1]. Its aetiological agent is *Rickettsia typhi*, which belongs to typhus group rickettsiae. Two cycles are involved in *R. typhi* transmission: a classical cycle rat-flea-rat, and a peridomestic cycle involving cats, dogs, opossums, sheep and their fleas [1].

Murine typhus is often acute and mild [2]. However, it can cause severe illness and death [1,3–6]. Although the lack of a correct diagnosis increases the risk of severity, murine typhus is usually misdiagnosed and underreported and its prevalence is unknown in many countries. Murine typhus may be mistaken for other diseases because of its non-specific symptoms [2–4]. In addition, epidemiological criteria are not always present [3–5].

In Spain, murine typhus has been known for more than 25 years [7,8]. Most clinical cases occur in two regions: *Anda luciae* [5,7,8] and the Canary Islands [9–11]. However, sero-epidemiological studies have demonstrated the presence of *R. typhi* infection throughout the country [12–15]. For instance, even though no clinical cases have been described in our region, the seroprevalence of *R. typhi* infection in human population was 8.8%. Consequently, murine typhus could be misdiagnosis in Spain. For this reason, it is necessary to demonstrate the presence of *R. typhi* as well as to detect reservoirs, vectors, and risk factors.

An association between infected cats and murine typhus human cases has been described [4,16]. Our previous results showed that subjects who reported contact with pets tended to have higher *R. typhi* seroprevalence [12]. Thus, a peridomestic cycle may be present in our region and cats and fleas could be involved. There are few studies on fleas related to *R. typhi*. Moreover, although cats could be susceptible to *R. typhi* subclinical infection [16,17], no naturally infected cats have been described up to know. The objectives of the study were to detect *R. typhi* in fleas and naturally infected cats, and to demonstrate the presence of *R. typhi* in a region where no clinical cases has been reported and, thus, murine typhus may be misdiagnosed.

METHODS

Geographical Area

The study was undertaken on the central coast of Catalonia, a predominantly urban region in Northeastern Spain. A total of twenty municipalities, which belonged to six areas, participated in the study (Figure 1).

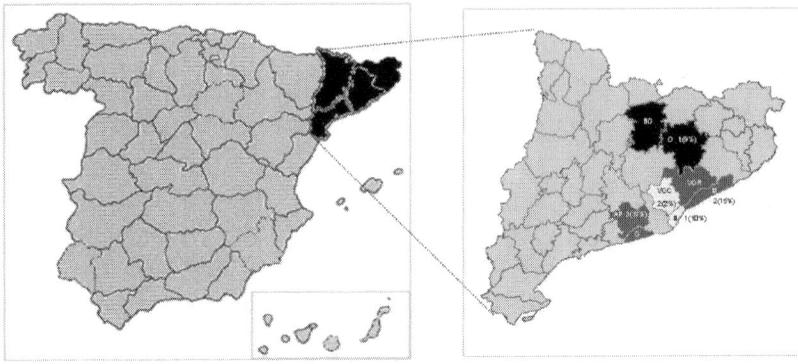

Figure 1. Geographical area of the study. Predominantly urban areas: Barcelonés [B], Baix Llobregat [BL], Vallés Occidental [VOC]. Predominantly suburban areas: Maresme [M], Selva [S], Vallés Oriental [VOR].

https://doi.org/10.1371/journal.pone.0071386.g001

Samples

Two hundred and twenty-one cats were analysed (January 2001 to March 2009). Forty-four were stray cats subjected to health and reproductive control by municipalities. Their samples were provided by the municipal veterinarians. One hundred and seventy-seven cats were attended at different veterinary clinics. One of these clinics worked together with PROGAT. The latter was a foundation that collected stray cats, controlled their health and left them in the street again. Thus, twenty-four cats attended at veterinary clinics were stray cats.

Variables registered were: date of collection, age, sex, municipality, living place, outdoor activities, demographic area (rural: <5,000 inhabitants; suburban 5,000 to 50,000; urban: > 50,000 inhabitants), healthy status, contact with animals, and ectoparasite infestation.

Blood samples were aseptically collected from the external jugular vein of each cat. One millilitre of blood was introduced in a serum-separating tube. Sera were obtained by centrifugation. Whole-blood samples were collected in sterile EDTA and heparin vacutainers. Samples were frozen at -80° C until used.

Eighty fleas were studied. Sixty-four were collected in kennels of the same areas. Most fleas were collected in the boxes and some were collected on animals. In addition, sixteen fleas were collected on cats attended at veterinary clinics. Fleas were identified according to taxonomic keys. Fleas were immersed in a solution of 70% ethanol, washed in sterile distilled water, dried, and individually transferred to tubes. Samples were stored at 4° C.

Indirect Immunofluorescence Assay

Antibodies titres against *Rickettsia typhi* were evaluated by indirect immunofluorescence assay (IFA) using a commercial antigen (*R. typhi-R. rickettsii* IgG, Focus Technologies, Inc., Herndon, VA). Briefly, 25µL of twofold dilutions of cat sera in phosphate-buffered saline (PBS) -3% non-fat dry milk were applied to the antigens. Sera of a seropositive cat and a seronegative cat, kindly obtained from Veterinary Faculty, were added to each experiment as positive and negative controls. The slides were incubated in a humidified chamber at 37° C for 30 min. Two washes (10 min) in PBS and one wash (5 min) in water were performed to remove unbound immunoglobulins. Slides were air dried. Binding sera were detected using a fluorescein isothiocyanate-labelled anti-cat IgG (Sigma-Aldrich Química, S.A., Madrid) diluted 1/128 in PBS -0.01% Evans Blue (bioMérieux, S.A., Madrid). The slides were incubated and washed as described above. The slides were examined with a fluorescence microscope at 400x. The highest dilution, at which distinct and specific fluorescence was seen, was scored as the end-point titer for the serum sample. Titres $\geq$ 1/64 were considered positive. The samples were subjectively evaluated and independently graded by two of the authors.

Molecular Detection

DNA was obtained from cats samples collected in an EDTA whole-blood vacutainer, and from fleas individually triturated in brain-heart infusion (BHI) (bioMérieux, Marcy-l'Etoile, France) [18]. Masterpure DNA purification kit (Epicentre, Madison, Wisconsin) was used according to the manufacturer's instructions. This kit included a Uracil DNA Glycosylase (UNG) and deoxyuracil triphosphates (dUTP) as well as a Hot start DNA polymerase. DNA from cultures of *Rickettsia typhi* (kindly obtained from the Unité de Rickettsies, France) was obtained and used as positive control. Measures to avoid contamination were performed using separate and dedicated rooms for DNA extraction and molecular detection.

DNA samples were tested by *R. typhi*-specific real-time PCR targeting the gene for *omp*B [19]. When there was enough DNA, a second real-time

PCR assay targeting the gene *glt*A was carried out [20]. PCR assays were performed and analysed using 7500 thermocycler (Applied Biosystems). Two negative controls (DNA-free water as template) and one positive control (purified *R. typhi* DNA) were included in all assays. PCRs were set up in a UV-sterilized workstation. Each sample was assayed three times. Amplification products were purified by Exosap-it (GE Healthcare, Bunkinghamshire, UK) and sequenced on 3130 Genetic Analyzer (Applied Biosystems) using a BigDye Terminator v 3.1 Cycle Sequencing Kit (Applied Biosystems, Foster City, CA). The sequences were compared with those in the GenBank by BLAST program (megablast algorithm).

Culture

Whole-blood samples collected in a heparin vacutainer were used. Three hundred microlitres of blood were added to two shell vials seeded with Vero cells (African green monkey epithelial cells). A shell vial is a tube with a coverslip that contains the cell monolayer. After adding blood, the shell vials were centrifuged at 700*g* for one hour. During centrifugation, microorganisms were in close proximity to the cells and, thus, infection efficiency was enhanced. The shell vials were incubated at 32° C. Every week, the medium was replaced. On day 20, cell monolayer was scraped with glass beads. Part of the culture collected was transferred to a confluent monolayer of Vero cells in a 25 cm^2 culture flask.

On day 40, monolayers were scraped with glass beads and cultures were collected. A drop was placed on a slide and evaluated by Giménez staining. DNA was obtained from each culture. Presence of *R. typhi* DNA was analysed using *omp*B *R. typhi*-specific PCR [19] and sequencing as described above.

Two slides were prepared for indirect immunofluorescence assay (IFA). Cultures collected were deposited onto the slides, air dried and fixed with acetone. They were incubated with murine typhus group (MT) positive control or spotted fever group (SFG) positive control of the commercial kit for *R. typhi-R. rickettsii* IgG detection (Focus Technologies, Inc., Herndon, VA). These controls contain antibodies against Typhus Fever group and Spotted Fever group rickettsiae, respectively. Three wells of a commercial slide, with antigen to *R. typhi-R. rickettsii*, were incubated with MT positive control, SFG positive control, and negative control of the same kit. IFA was performed according to the manufacturer's instructions of the kit.

Statistical analysis

To achieve an accuracy of 5.0% in the estimation of a confidence interval using a normal asymptotic finite population correction for the bilateral 95%, assuming that the expected proportion were the highest prevalences found in the literature (worst cases) and that the total size of the populations were 1000, the sample size was calculated.

Data analysis was carried out using the software application SPSS Statistics 18.0. Univariate analysis was performed using Chi-square test, Fisher exact test, and Mann–Whitney U test. A $p < 0.05$ was considered significant.

Ethics Statement

This study was approved by the Ethical Committee of Corporació Sanitària i Universitària Parc Taulí. This study was adhered to the Animal Protection Law (5/1995) of the Government of Catalonia, and RD1201/2005 of the Government of Spain, based on European Union directives 86/609/CEE and 2003/65/CE. Animal owners consented to have their cats involved in the study.

RESULTS

Cats' study

Study Population

The clinical and epidemiological characteristics are shown in Table 1. The mean age was 3.9 ± 4.2 years (0.4-17 years). Most samples were collected between February and July (74.6%). Samples of cats attended at veterinary clinics were collected throughout the year, whereas sample of stray cats controlled by municipalities were collected in May (88.9%) and June (11.1%).

Table 1. Demographic information from cats tested for antibodies to *Rickettsia typhi.*

Variables	Total N (%[1])	Positive N (%[1])
Sex		
Male	103 (46.6)	19 (54.3)

Female	114 (51.6)	16 (45.7)
Age		
Kitten	60 (27.1)	9 (25.7)
Adult	154 (69.7)	26 (74.3)
Living place		
Apartment	54 (24.4)	12 (34.3)
House	94 (42.5)	13 (37.1)
Stray	68 (30.8)[2]	10 (28.6)
Demographic area		
Urban	183 (82.8)	28 (80)
Suburban	26 (11.8)	6 (17.1)
Rural	12 (5.4)	1 (2.9)
Regions		
Barcelonés	44 (19.9)	9 (25.7)
Baix Llobregat	1 (0.5)	0 (0)
Maresme	19 (8.6)	3 (8.6)
Selva	1 (0.5)	1 (2.9)
Vallés Occidental	146 (66.2)	22 (62.9)
Vallés Oriental	10 (4.5)	0 (0)
On the coast		
Yes	59 (26.7)	11 (31.4)
No	162 (73.3)	24 (68.6)
Contact with animals		
Yes	172 (77.8)	27 (77.1)
No	37 (16.7)	8 (22.9)
Contact with cats	132 (59.7)	18 (51.4)
Contact with dogs	48 (21.7)	9 (25.7)
Contact with other animals	5 (2.3)[3]	1 (2.9)[4]
Illness		
Yes	29 (13.1)[5]	7 (24.1)[6]
No	160 (72.4)	22 (75.9)
TOTAL	**221**	**35**

[1] The percentage is calculated against the total number of animals that belong to the study population or the seropositive population. [2] 24 (10.9%) stray cats were attended at PROGAT foundation, and 44 (19.9%) belonged

to groups of stray cats subjected to control by municipalities (controlled cats). [3] Other animals (total population): birds, mice, sheep, wild animals and rabbits; [4] Other animals (seropositive population): bird and mice. [5] Diseases (total population): abscess, dehydration, abortion, sepsis, cystitis, conjunctivitis, anaemia, diabetes mellitus, diarrhoea, fever, gingivitis, hepatic diseases, respiratory diseases, Feline immunodeficiency, poisoning, feline leukaemia, breast lump, nasal lump, mouth infection, urinary tract infection, worms, and uterus infection. No association was observed between the presence of illness and any of the variables. [6] Diseases (seropositive population): abscess, cystitis, mouth infection, hepatic diseases, gingivitis, and respiratory diseases (two cats)

Fleas were found on 55 (24.9%) cats. Geographical distribution of infested cats is shown in Figure 2. A statistically significant association was found between infestation and stray cats ($p < 0.001$), contact with animals ($p < 0.001$), and month of collection ($p<0.001$, Figure 3).

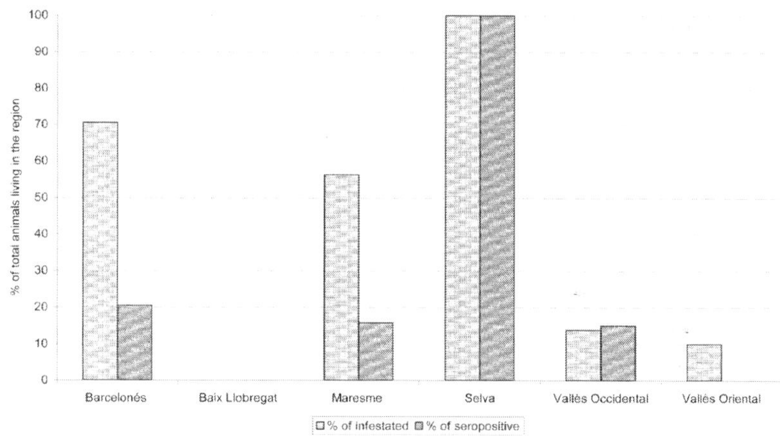

Figure 2. Geographical distribution: percentage of infested and seropositive cats compared to the entire population of each region.

https://doi.org/10.1371/journal.pone.0071386.g002

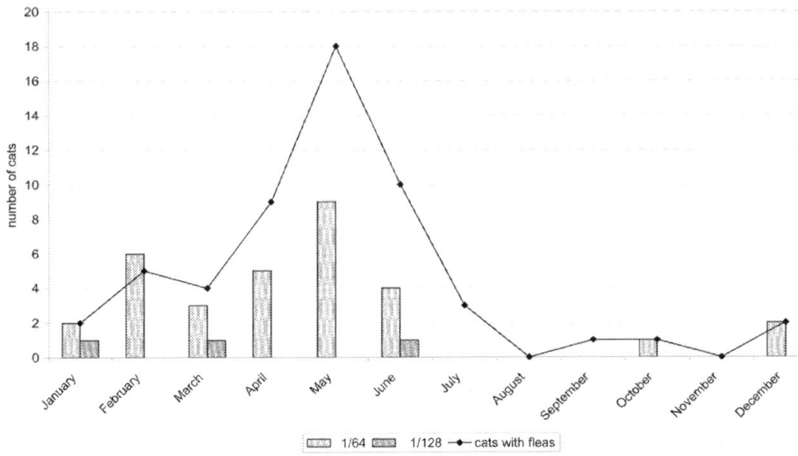

Figure 3. Annual distribution of infested animals and titres obtained by indirect immunofluorescence antibody assay.

https://doi.org/10.1371/journal.pone.0071386.g003

Twenty-nine (13.1%) cats showed some type of illness. The diseases observed were: abscess, dehydration, abortion, sepsis, cystitis, conjunctivitis, anemia, diabetes mellitus, diarrhea, fever, gingivitis, hepatic diseases, respiratory diseases, Feline immunodeficiency, poisoning, feline leukemia, breast lump, nasal lump, mouth infection, urinary tract infection, worms, and uterus infection. Fifty-five percent of stray cats ($p < 0.05$), and 55.6% ($p=0.001$) of infested cats were ill.

Sero-epidemiological study

Thirty-five (15.8%) sera had antibodies against *R. typhi* (1/64: 32 [14.5%], 1/128: 3 [1.4%]) (Figure 4). Distribution of titres throughout the year is shown in Figure 3. Cats with titres of 1/128 were healthy adults, from urban area near the coast, and they were not infested. Two of these lived in apartments, and the other one lived in a house.

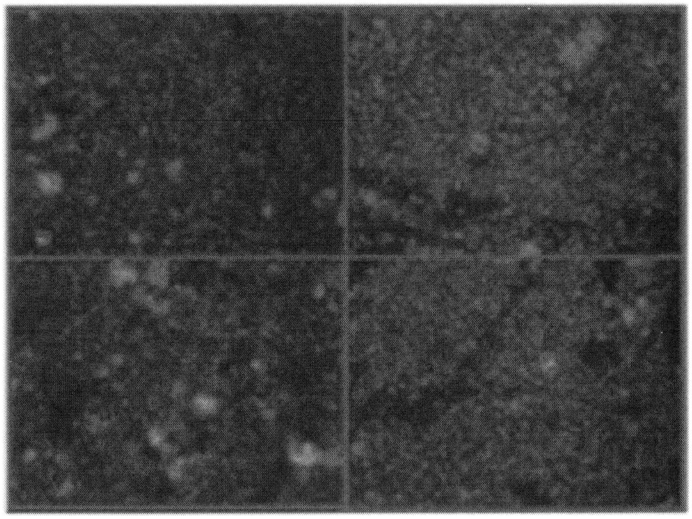

Figure 4. Images of positive indirect immunofluorescence assays.

https://doi.org/10.1371/journal.pone.0071386.g004

The mean age of seropositive cats was 3.44 ± 3.09 years (0.5-11 years) and the mean age of seronegative cats was 3.98 ± 4.37 years (0.4-17 years) (non-significant). The relationship between the seropositivity and the surveyed items is shown in Table 1. There was no statistically significant association between any of the items and seropositivity or titres.

Outdoors activities were reported in 43 pets. A new variable called outdoors/indoors was built. Stray cats and pets with outdoors activities were considered as "Outdoors cats". Fifteen cats with outdoors exposure were seropositive (non- significant).

Eleven seropositive cats had fleas (50% stray cats, 30.8% adults). The geographical distribution of seropositive cats is shown in Figure 2.

Molecular detection

Whole-blood sample in EDTA vacutainers was collected in 23 seropositive cats. In addition of these, samples from 23 seronegative cats were analysed. Five seropositive cats (numbers: 32, 39, 44, 170, 281) were positive by *R. typhi*-specific PCR assay (*omp*B). The sequences were identical among them and with those of *R. typhi* in the GenBank (CP003398, CP003397, AE017197, L04661). Table 2 shows information concerning these cats.

Cat number 32 was also positive by PCR targeting *gltA* gene. However, the amount of DNA amplified was not enough to analyse it by sequencing. No amplification was obtained in seronegative cats.

Table 2. Demographic information from cats with positive results by molecular detection.

Sample	32	39	44	170	281
Month of collection	March	April	April	March	December
Sex	Female	Female	Male	Male	Male
Age	2 years	2 years	6 months	> 1 year	8 years
Living place	Apartment	Apartment	Apart-ment	Street (Stray cat)	Apartment
Demo-graphic area	Urban	Urban	Urban	Urban	Urban
Ectopara-sites	No	No	No	Fleas	No
Illness	No	No	No	1. Respiratory disease, 2. feline im-munodefi-ciency	No
Contact with ani-mals	Cats	Dogs	No	Cats	No
Titres	1/128	1/64	1/64	1/64	1/64
Real-time PCR (blood):					
***omp*B**	+	+	+	+	+
***glt*A**	+	neg[1]	Neg	Neg	neg
Real-time PCR (cul-ture):		NP[2]	NP	NP	
***omp*B**	+				neg

Culture

Cultures were carried out in those cats in which rickettsial DNA was detected, and whose whole-blood samples were collected in heparin vacutainers (numbers: 32, 281). On day 40, a few rickettsiae were observed in Gimenez staining of cat 32. Fluorescence was observed inside few Vero cells of cat 32 incubated with antibodies against *R. typhi.* Presence of rickettsial DNA was detected in culture from cat 32. The sequence obtained was identical to those of *R. typhi* in the GenBank (CP003398, CP003397, AE017197, L04661). However, culture could not be established subsequently.

Fleas' study

All fleas were *Ctenocephalides felis. omp*B *R. typhi*-specific PCR yielded positive results in 44 (55%) fleas: 33/64 [51.6%] of fleas from kennels, 11/16 [68.8%] of those from Veterinary Clinics (non-significant). Nine amplicons were chosen at random and sequenced. The sequences were identical among them and to those of *R. typhi* in the GenBank (CP003398, CP003397, AE017197, L04661).

Presence of *Rickettsia felis* had been studied in 78 of the same fleas (previously published [21]). Twenty (25.6%) of these fleas contained DNA of both species (10/63 [15.8%] of fleas from kennels, and 10/55 [66.7%] of those from veterinary clinics).

DISCUSSION

Rickettsia typhi is the aetiological agent of murine typhus. This disease can be misdiagnosed or underreported due to its non-specific symptoms, or the absence of epidemiological criteria [2–5,12]. Seroepidemiological studies have demonstrated the presence of *R. typhi* infection in human populations from many Spanish regions [12–15]. However, while Southern Spain is considered as an endemic area [5,7,8], few clinical cases have been reported in other regions, most of them in the Canary Islands [9–11]. Interestingly, some areas with no or few clinical cases show higher prevalences than those found in Southern or Canary Islands. Therefore, murine typhus may be underreported. Catalonia is a Mediterranean region in Northeastern Spain. Many cases of murine typhus have been described in the Mediterranean area, which is considered a risk area [22]. Previous results showed a considerable proportion of people from our region with past *R. typhi* infection [12]. Thus, it was necessary to identify reservoirs and vectors involved in *R. typhi* cycle and in close contact with people. Even though a peridomestic cycle involving cats and their fleas has been described [1], there is not much information, especially focused on the natural infection in cats. In this study, the presence of *R. typhi* in cats and fleas is demonstrated.

Although human *R. typhi* infection directly from cats has not been described, association between high rates of infected cats has been related to human cases of murine typhus. For instance, 90% of seropositive cats were found in an area of Los Angeles where human cases were described, whereas all cats living in a control area, without human clinical cases, were seronegative [16]. Recently, an outbreak of murine typhus in Austin was described [4]. Cats, dogs and opossums analyzed during this outbreak were seropositive whereas raccoons and rats were negative. Cats cannot only act as transport host of infective fleas into the human environment [16], but also they are susceptible to infection, and produce antibodies against it [16,17]. In fact, 15.8% of cats studied in our region were seroreactive. Moreover, a direct detection of *R. typhi* in some of these cats was obtained by PCR. Even though isolation of *R. typhi* was not achieved, results obtained by culturing also seem to point to the presence of *R. typhi* infection in the cats. To our knowledge, this is the first time of the *R. typhi* detection in naturally infected cats. As it was suspected, the rate of molecular detection was lower than seroprevalence. *R. typhi* is an intracellular pathogen and it could be found in tissues, as a consequence, bacteraemia may be short or not detected with only one sample. Moreover, as titres found in our study are not very high, the higher seroprevalence could also be due to a past infection with persistent antibodies.

It is important to highlight that many seropositive cats, even cats in which *R. typhi* has been directly detected, lived in close contact with humans. Like other studies [23,24], although stray cats or cats with outdoor activities may be more exposed to *R. typhi* infection, there were no significant differences among seropositivity and either habitat or outdoor activities.

There was no association between seroprevalence and illness. This fits in with the fact that cats may present subclinical infection [16,17]. Besides, *R. typhi* was detected by PCR in four healthy cats. One of these was the cat no. 32 which also had the highest titres, and a positive culture.

Cats are often infested by fleas. In fact, almost half the seropositive cats were infested, and half of these were pets. Taking into account that more than half the fleas surveyed in our study presented *R. typhi* DNA, vector control is very important, even in pet cats. On the other hand, four of five cats in which molecular detection was positive, lived in apartments and did not have ectoparasites. One of these cats presented positive results in culture too. In this case, its infection could be recent. Consequently, if infected cats are in close contact to humans, they may not present clinical symptoms, and they may not be infested, more controls on pets are necessary.

It has been suggested that kittens have lower rates of seropositivity due to their shorter contact with the microorganism. However, seroprevalence observed in kittens were similar to that observed in adult cats. Other studies have described the same results [23].

It has been described that murine typhus tends to be seasonal. Although cases can be detected throughout the year, the highest prevalences are found in summer, whereas a few cases are described in winter [7,10,25]. These results are in accordance with the seasonality of infestation, which may tend to be higher during summer. According to this, the highest percentage of our infested cats was found between April and June. However, 44 of the 68 stray cats (with higher proportion of infestation) were surveyed in these months, and, thus, association between infestation and seasonality is not as accurately as it should be. On the other hand, although 19 seropositive cats were found between March and June; there were 11 seroreactive cats in winter and *R. typhi* was detected by PCR in one of these. Therefore, despite seroreactive cats in winter mainly reflect the persistence of antibodies over time, some cats may also be infected in this season.

The present study is the first molecular detection of *R. typhi* in *Ctenocephalides felis* from Spain. There are not many studies on the presence of *R. typhi* in *C. felis* [25–27]. Our elevated prevalence suggests that *C. felis* may play a role in the epidemiological cycle of *R. typhi*. Considering *C. felis* avidly feeds on humans, it could transmit *R. typhi* to them. *C. felis* has a broad host range and is frequent on pets. It should be pointed out that fleas analysed not only were collected in kennels but they were also collected on cats attended at veterinary clinics, and therefore, living in a household.

Eleven (13.8%) fleas contained DNA from both *R. felis* and *R. typhi*. Although there may be an interspecific competition between rickettsiae in arthropods [26], Noden demonstrated that cat fleas could be infected with *R. felis* and *R. typhi*, and could maintain both species [28]. On the other hand, other authors suggested that fleas containing *R. felis* may not acquire *R. typhi* as readily as uninfected fleas. Although our data could not analyse this fact, dual infection was demonstrated as it was observed in *X. cheopis* [29].

CONCLUSIONS

Evidences of presence of *R. typhi* in cats, and fleas have been shown in this study. Percentages observed were higher than those observed in the human population because animals usually present more exposure to vectors. Risk of infection in animals with close contact with humans is similar to that observed in stray animals or those living in kennels. Absence of ectoparasites or clinical symptoms does not guarantee that the animal is not infected. Although infection may be a seasonal, cases may appear throughout the year. Therefore, probably cases of murine typhus may be misdiagnosed in our area considering infected animals could be in close contact with people, past human infection has been detected in our region [12], and one clinical case was recently described near our area [30]. This is of great importance because delayed diagnosis and inappropriate therapy are related to severity and death.

REFERENCES

1. Gillespie JJ, Ammerman NC, Beier-Sexton M, Sobral BS, Azad AF (2009) Louse- and flea-borne rickettsioses: biological and genomic analyses. Vet Res 40: 12. doi:10.1051/vetres:2008050. PubMed: 19036234.
2. Centers for Disease Control and Prevention (CDC) (2009) Outbreak of *Rickettsia typhi* infection - Austin, Texas, 2008. MMWR Morb Mortal Wkly Rep 58: 1267-1270. PubMed: 19940832.
3. Restrepo MI, Vasquez EM, Echeverri C, Fiebelkorn KR, Anstead GM (2010) Fibrin ring granulomas in *Rickettsia typhi* infection. Diagn Microbiol Infect Dis 66: 322-325. doi:10.1016/j.diagmicrobio.2009.10.019. PubMed: 20071129.
4. Adjemian J, Parks S, McElroy K, Campbell J, Eremeeva ME et al. (2010) Murine typhus in Austin, Texas, USA, 2008. Emerg Infect Dis 16: 412-417. doi:10.3201/eid1603.091028. PubMed: 20202415.
5. Beltrán LM, García S, Vallejo AJ, Bernabeu-Wittel M (2011) [Bilateral anterior uveitis and *Rickettsia typhi* infection]. Enferm Infecc Microbiol Clin 29: 235-236. doi:10.1016/j.eimc.2010.05.009. PubMed: 21324556.
6. Kaabia N, Letaief A (2009) Characterization of rickettsial diseases in a hospital-based population in central Tunisia. Ann N Y Acad Sci 1166: 167-171. doi:10.1111/j.1749-6632.2009.04521.x. PubMed: 19538277.
7. Bernabeu-Wittel M, Pachón J, Alarcón A, López-Cortés LF, Viciana P et al. (1999) Murine typhus as a common cause of fever of intermediate duration: a 17-year study in the south of Spain. Arch Intern Med 159: 872-876. doi:10.1001/archinte.159.8.872. PubMed: 10219934.
8. Espinosa N, Cañas E, Lara R, León E, Santos J et al. (2010) [SAEI prospective multicenter study on fever syndrome of intermediate duration: preliminary results]. Enf Infec Microb Clin 28: 317.
9. Hernández Cabrera M, Angel-Moreno A, Santana E, Bolaños M, Francès A et al. (2004) Murine typhus with renal involvement in Canary Islands, Spain. Emerg Infect Dis 10: 740-743. doi:10.3201/eid1004.030532. PubMed: 15200874.
10. Miguélez M, Laynez P, Linares M, Hayek M, Abella L et al. (2003) Murine typhus in Tenerife. Clinico-epidemiological study and

differential clinical features with Q fever. Med Clin (Barc) 121: 613-615. doi:10.1016/S0025-7753(03)74033-0.

11. Basrai D, Pox C, Schmiegel W (2007) Fever of intermediate duration after return from the Canary Islands. Internist (Berl)48: 413-419.
12. Nogueras MM, Cardeñosa N, Sanfeliu I, Muñoz T, Font B et al. (2006) Serological evidence of infection with *Rickettsia typhi* and *Rickettsia felis* among the human population of Catalonia, in the Northeast of Spain. Am J Trop Med Hyg 74: 123-126. PubMed: 16407356.
13. Bernabeu-Wittel M, del Toro MD, Nogueras MM, Muniain MA, Cardeñosa N et al. (2006) Seroepidemiological study of *Rickettsia felis, Rickettsia typhi*, and *Rickettsia conorii* infection among the population of southern Spain. Eur J Clin Microbiol Infect Dis 25: 375-381. doi:10.1007/s10096-006-0147-6. PubMed: 16767485.
14. Lledó L, Gegúndez MI, Saz JV, Beltrán M (2001) Prevalence of antibodies to *Rickettsia typhi* in an area of the center of Spain. Eur J Epidemiol 17: 927-928. doi:10.1023/A:1016248630809. PubMed: 12188012.
15. Bolaños-Rivero M, Santana-Rodriguez E, Angel-Moreno A, Hernández-Cabrera M, Limiñana-Canal JM et al. (2011) Seroprevalence of *Rickettsia typhi* and *Rickettsia conorii* infections in the Canary Islands (Spain). Int J Infect Dis 15: e481-e485. doi:10.1016/j.ijid.2011.03.019. PubMed: 21612963.
16. Sorvillo FJ, Gondo B, Emmons R, Ryan P, Waterman SH et al. (1993) A suburban focus of endemic typhus in Los Angeles County: association with seropositive domestic cats and opossums. Am J Trop Med Hyg 48: 269-273. PubMed: 8447530.
17. Matthewman L, Kelly P, Hayter D, Downie S, Wray K et al. (1997) Domestic cats as indicators of the presence of spotted fever and typhus group rickettsiae. Eur J Epidemiol 13: 109-111. doi:10.1023/A:1007375718204. PubMed: 9062789.
18. Stevenson HL, Labruna MB, Montenieri JA, Kosoy MY, Gage KL et al. (2005) Detection of *Rickettsia felis* in a New World flea species, *Anomiopsyllus nudata* (Siphonaptera: *Ctenophthalmidae*). J MedEntomol42:163-167.doi:10.1603/0022-2585(2005)042[0163:DORFIA]2.0.CO;2. PubMed: 15799525.

19. Henry KM, Jiang J, Rozmajzl PJ, Azad AF, Macaluso KR et al. (2007) Development of quantitative real-time PCR assays to detect *Rickettsia typhi* and *Rickettsia felis*, the causative agents of murine typhus and flea-borne spotted fever. Mol Cell Probes 21: 17-23. doi:10.1016/j.mcp.2006.06.002. PubMed: 16893625.
20. Stenos J, Graves SR, Unsworth NB (2005) A highly sensitive and specific real-time PCR assay for the detection of spotted fever and typhus group Rickettsiae. Am J Trop Med Hyg 73: 1083-1085. PubMed: 16354816.
21. Nogueras MM, Pons I, Ortuño A, Lario S, Segura F (2011) *Rickettsia felis in fleas from Catalonia (Northeast Spain)* . Vector Borne Zoonotic Dis 11: 479-483. doi:10.1089/vbz.2010.0008. PubMed: 20586606.
22. Angelakis E, Botelho E, Socolovschi C, Sobas CR, Piketty C et al. (2010) Murine typhus as a cause of Fever in travelers from Tunisia and mediterranean areas. J Travel Med 17: 310-315. doi:10.1111/j.1708-8305.2010.00435.x. PubMed: 20920051.
23. Case JB, Chomel B, Nicholson W, Foley JE (2006) Serological survey of vector-borne zoonotic pathogens in pet cats and cats from animal shelters and feral colonies. J Feline Med Surg 8: 111-117. doi:10.1016/j.jfms.2005.10.004. PubMed: 16434226.
24. Luria BJ, Levy JK, Lappin MR, Breitschwerdt EB, Legendre AM et al. (2004) Prevalence of infectious diseases in feral cats in Northern Florida. J Feline Med Surg 6: 287-296. doi:10.1016/j.jfms.2003.11.005. PubMed: 15363760.
25. Williams SG, Sacci JB Jr, Schriefer ME, Andersen EM, Fujioka KK et al. (1992) Typhus and typhuslike rickettsiae associated with opossums and their fleas in Los Angeles County, California. J Clin Microbiol 30: 1758-1762. PubMed: 1629332.
26. Schriefer ME, Sacci JB Jr, Taylor JP, Higgins JA, Azad AF (1994) Murine typhus: Updated roles of multiple urban components and a second typhuslike rickettsia. J Med Entomol 31: 681-685. PubMed: 7966170.
27. Boostrom A, Beier MS, Macaluso JA, Macaluso KR, Sprenger D et al. (2002) Geographic association of *Rickettsia felis*-infected opossums with human murine typhus, Texas. Emerg Infect Dis 8: 549-554. doi:10.3201/eid0806.010350. PubMed: 12023908.
28. Noden BH, Radulovic S, Higgins JA, Azad AF (1998) Molecular identification of *Rickettsia typhi* and *R. felis* in co-

infected *Ctenocephalides felis* (Siphonaptera: *Pulicidae*). J Med Entomol 35: 410-414. PubMed: 9701920.

29. Eremeeva ME, Warashina WR, Sturgeon MM, Buchholz AE, Olmsted GK et al. (2008) *Rickettsia typhi* and *R. felis* in rat fleas (*Xenopsylla cheopis*), Oahu, Hawaii. Emerg Infect Dis 14: 1613-1615. doi:10.3201/eid1410.080571. PubMed: 18826827.

30. Nicolás Sánchez FJ, Ribes Amorós I, Cervera del Pino M, Sarrat Nuevo RM, Bestard Solivellas J et al. (2008) Cavited pneumonia caused by murine typhus. An Med Interna 25: 46-47. PubMed: 18751331.

3

Molecular Detection of Acinetobacter Species in Lice and Keds of Domestic Animals in Oromia Regional State, Ethiopia

Bersissa Kumsa[1,2], Cristina Socolovschi[2], Philippe Parola[2], Jean-Marc Rolain[2], Didier Raoult[2]

[1] Department of Parasitology, College of Veterinary Medicine and Agriculture, Addis Ababa University, Bishoftu, Ethiopia,
[2] Aix Marseille Universite´, URMITE, UM63, CNRS 7278, IRD 198, Inserm 1095, Marseille, France

ABSTRACT

This study was conducted to determine the presence of *Acinetobacter* and *Rickettsia* species DNA in lice and *Melophagus ovinus* (sheep ked) of animals from Oromia Regional State in Ethiopia. From September through November 2011, a total of 207 cattle, 85 sheep, 47 dogs and 16 cats were examined for ectoparasites. Results of morphological identification revealed several species of ectoparasites: *Linognathus vituli (L. vituli)*, *Bovicola*

Citation: Kumsa B, Socolovschi C, Parola P, Rolain J-M, Raoult D (2012) Molecular Detection of Acinetobacter Species in Lice and Keds of Domestic Animals in Oromia Regional State, Ethiopia. PLoS ONE 7(12): e52377. https://doi.org/10.1371/journal.pone.0052377

bovis (B. bovis) and *Solenopotes capillatus (S. capillatus)* on cattle; *B. ovis* and *Melophagus ovinus (M. ovinus)* on sheep; and *Heterodoxus spiniger (H. spiniger)* on dogs. There was a significantly ($p \leq 0.0001$) higher prevalence of *L. vituli* observed in cattle than both *S. capillatus* and *B. bovis*. Molecular identification of lice using an 18S rRNA gene analysis confirms the identified lice species by morphological methods. We detected different *Acinetobacter* species among lice (11.1%) and keds (86.4%) including *A. soli* in *L. vituli* of cattle, *A. lowffii* in *M. ovinus* of sheep, *A. pittii* in *H. spiniger* of dogs, 1 new *Acinetobacter* spp. in *M. ovinus* and 2 new *Acinetobacter* spp. in *H. spiniger* of dogs using partial *rpoB* gene sequence analysis. There was a significantly higher prevalence of *Acinetobacter* spp. in keds than in lice ($p \leq 0.00001$). Higher percentage of *Acinetobacter* spp. DNA was detected in *H. spiniger* than in both *B. ovis* and *L. vituli* ($p \leq 0.00001$). Carbapenemase resistance encoding genes for blaOXA-23, blaOXA-24, blaOXA-58, blaNDM-1 and blaOXA-51 were not found in any lice and keds. These findings suggest that synanthropic animals and their ectoparasites might increase the risk of human exposure to zoonotic pathogens and could be a source for *Acinetobacter* spp. infections in humans. However, additional epidemiological data are required to determine whether ectoparasites of animals can act as environmental reservoirs and play a role in spreading these bacteria to both animal and human hosts.

INTRODUCTION

Lice and sheep keds (*Melophagus ovinus*) are two of the most common and economically important ectoparasites of domestic animals worldwide. They are responsible for a wide range of health problems in domestic animals [1]. In infested animals they cause losses in productivity; due to anemia, loss of wool or hair due to scratching, biting and rubbing, disruption in feeding, hide and skin damage, secondary skin infections and damage, reduced newborn birth weights, abortion in pregnant animals, and damage to fences, equipment and buildings due to excessive rubbing and scratching [2], [3].

Several species of lice have been found on Ethiopian cattle; one species of chewing lice, *Bovicola* (*Damalinia*) *bovis*, and three species of sucking lice, including *Linognathus vituli* (the long-nosed cattle louse), *Solenopotes capillatus* (the little blue cattle louse) and *Haematopinus quadripertusus* (the cattle tail louse) [4]. On sheep, the biting lice *Bovicola (Damalinia) ovis*; one species of sucking lice, *Linognathus ovillus* and one species of the fly, *Melophagus ovinus*, are common ectoparasites [5], [6]. On goats, one species of biting lice, *Bovicola (Damalinia) caprae*, and the sucking lice *Linognathus stenopsis* have been reported [6], [7]. On dogs, the biting lice *Heterodoxus spiniger* and *Trichodectes canis* as well as one species of sucking lice, *Linognathus steosus*, were reported. In contrast, there are no reports available on the lice from cats in Ethiopia [8].

All the previous reports on the lice and sheep keds of animals in Ethiopia resulted from studies of other ectoparasites, specifically ticks and mange mites [6], [7], [9], [10], [11]. The only lice-specific study in Ethiopia was carried out by Kumsa and Bekele (2008) on the lice of cattle in the Endegagn district. Earlier studies focus on the epidemiology, species composition, distribution and impact of ectoparasites on the skin and hides of food animals. Studies on the role of lice and keds of domestic animals as vectors of pathogens of veterinary and medical importance have not yet been conducted in Ethiopia.

Acinetobacter species are Gram-negative coccobacilli bacteria commonly found in water, soil, mud, living organisms, vegetables, as well as in the feces, urine and skin of humans and animals [12], [13], [14], [15]. Currently, the genus comprises 23 validly named species and 12 genomic species. Owing to difficulty of precise identification of all members to the species level with advancement and development of new techniques in molecular methods, the taxonomy of the genus *Acinetobacter* has been continually revised [16], [17], [18]. Recently, formerly *Acinetobacter* genomic sp. 3 was renamed as *A. pittii, Acinetobacter* genomic sp. 13TU as *A. nosocomialis* and *Acinetobacter* genomic sp. G13 as *A.lwoffii* [16].

In humans, members of the genus *Acinetobacter* have emerged as opportunistic pathogens, and are frequently implicated in various types of infections, especially in immunocompromised individuals and intensive health care units [14] throughout the world. In tropical countries, they are associated with severe community-acquired infections [13]. They have the capability to survive for prolonged periods under a wide range of environmental conditions [13]. *A. baumannii* is described as the most common species in this genus frequently associated with outbreaks and has been repeatedly reported to develop a high level of resistance against all available classes of antimicrobial drugs in many parts of the world [12]. *A. pittii* is reported as the second most commonly isolated *Acinetobacter* species after *A. baumannii* in human patients [16], [17]. More recently other less known species such as *A. lwoffii* and *A. soli* have been associated with serious infections and considered as emerging pathogens [14], [19], [20]. For instance *A. lwoffii* has been associated with acute gastroenteritis in USA [21], multidrug resistant *A. lwoffii* in southern Thailand [22], *A. soli* outbreak as a cause of infection in neonatal intensive health care unit in Korea [23] and nosocomial bloodstream infections due to *A. pittii* in United States were reported.

In animals infection due to *Acinetobacter* species is considered as an emerging problem due to escalating in the number of reports from many countries of the world. Nosocomial infection by *A. baumannii* in dogs and cats in intensive care unit in Switzerland [24], infection due to *A. baumannii* in pets and horses in Switzerland [25], multidrug resistant *Acinetobacter* spp. in veterinary clinics in Germany [12], detection of *A. baumannii* in samples from cattle and pigs slaughtered for human consumption in major Scottish abattoirs [26], carbapenemase producing *Acinetobacter* spp. in cattle from France [27] and OXA-23 producing *Acinetobacter* spp. from faeces of horses in Belgium [28] are some of the current reports that notify the growing importance of *Acinetobacter* spp. in veterinary medicine mirroring the situation happening in human medicine.

In arthropods *Acinetobacter* spp. have been detected in different species including tsetse flies, sand flies, mosquitoes, fleas and ticks in many countries of the world. Details on *Acinetobacter* spp. detected in arthropods, species of vectors and methods of detection is presented (Table 1). Also, recently several investigators detected *A. baumannii* in both body and head lice of humans from some countries [15], [29], [30], [31]. Despite the worldwide distribution and great economic significance of these ectoparasites, information is not available on the occurrence of *Acinetobacter* species in the lice and flies found on domestic animals. In addition, there is little known about *Rickettsia* species in the lice and flies of domestic animals in Ethiopia. Therefore, the current study investigated the presence of these bacteria in arthropods of domestic animals in six districts in Oromia Regional State, Ethiopia.

Table 1. Summary of *Acinetobacter* **spp. detected in various species of arthropods from different countries of the world.**

Acinetobacter sp.	Arthropod sp.	Detection method	Country	Reference
A. radioresistans	*Ctneocephalides felis*	Culture and PCR	Australia	[53]
A. johnsonii	*Ct. felis*	Culture and PCR	Australia	[53]
A. junii	*Ct. felis*	Culture and PCR	Australia	[53]
A. lwoffii	*Ixodes holocyclus*	Culture and PCR	Australia	[53]
A. lwoffii	*Boophilus microplus*	Culture and PCR	Australia	[53]
A. johnsonii	*Boophilus microplus*	Culture and PCR	Australia	[53]
A. junii	*Boophilus microplus*	Culture and PCR	Australia	[53]
A. radioresistans	*Boophilus microplus*	Culture and PCR	Australia	[53]
A. baumannii	*Lutzomyia longipalpis*	Culture and PCR	Brazil	[39]
A. juni	*Culex quinquefasciatus*	PCR	India	[54]
A. calcoaceticus	*Culex quinquefasciatus*	PCR	India	[54]
A. calcoaceticus	*Anopheles stephensis*	PCR and biochemical	Iran	[55]
A. calcoaceticus	*Anopheles maculipennis*	PCR and biochemical	Iran	[55]
Acinetobacter sp.	*Lutzomyia longipalpis*	PCR	Brazil	[56]
Acinetobacter sp.	*Bemisia tabaci*	PCR	Iran	[44]
Acinetobacter sp.	*Glossina palpalis palpalis*	Culture and PCR	Cameroon	[57]
Acinetobacter sp.	*Bactericera cockerelli*	PCR	USA	[42]
A. genomosp. 3	*Aedes albopictus*	PCR	Madagascar	[58]
A. genomosp. 13U	*Aedes aegypti*	PCR	Madagascar	[58]
A. baumannii	*Glossina palpalis palpalis*	Culture and PCR	Angola	[40]

https://doi.org/10.1371/journal.pone.0052377.t001

MATERIALS AND METHODS

Study areas and Animals

Lice and *Melophagus ovinus* were collected from indigenous cattle, sheep and dogs in six different districts in Oromia Regional State, Ethiopia: Asalla, Walmara, Shano, Ada›a, Bedele and Gachi. The districts are located in 5 zones in the central, southeast and southwestern parts of the country, with various climates and agroecology (Table 2). Ectoparasite collections were performed from September through November of 2011. A long rainy season from July to September, a short rainy season from March to May and a dry season from November to April prevail in all the study districts. In addition, October is a post rainy month with only few days of raining while November is a month during which the dry season commences hence is without raining or rarely for some days. The farming system in all the districts is characterized by a mixed crop-livestock production system. The livestock in the study areas are traditionally managed under extensive production systems [32].

Table 2. Description of study districts and number of study animals in Oromia Regional State.

District	Zone	Km from Addis Ababa	Agroecol-ogy	Altitude in m.a.s.l	Coordinates	Annual rainfall in mm	Av. Annual Temp in °C	No. Animals examined
Asalla	Arsi	175 south-east	Highland	2500	7°56′58′.57″N 39°8′23.42″ E	2000–4000	20–30	Sheep=39
Shano	North Showa	70 North-east	Highland	2861	9°20.00.00″N 39°18′00.00″E	945.4	6.1–18.5	Cattle=38 Sheep=19
Walmara	West Showa	45 west	Highland	2500	9°7′50.64″N 38°28′38.46″E	1060	4.6–23.3	Cattle=42
Ada'a	East Showa	47 east	Midland	1911	8°44′37.69″N 38°59′19.28″E	1911	13–26.5	Cattle=53 Dogs=47 Cats=16
Bedele	Illubabora	483 south-west	Midland	1974	8°,27′1.76″N 36°21′5.08″E	1400	12.5–27.5	Cattle=32 Sheep=21
Gachi	Illubabora	460 south-west	Midland	1751	9°14′2.57″N 35° 4′48.46″E	1300	13–28	Cattle=42 Sheep=6

m.a.s.l=meters above sea level; Av.=average; °C= degrees Celsius.

https://doi.org/10.1371/journal.pone.0052377.t002

Collection and Morphological Identification of Ectoparasites

Lice and flies were carefully removed manually, using forceps or by hand, to avoid any damage to the body and were then placed in vials containing 70% ethanol for subsequent identification. In the case of lice collected from dogs, the dog's body was brushed for ten minutes with a flea comb as previously described [8]. All lice and flies from the same animal were put in the same vial and transported to the Laboratory of the World Health Organization Collaborative Center for Rickettsial Diseases and Arthropod-borne Bacterial Diseases located in Marseille, France. Morphological identification of ectoparasites and molecular studies were performed from the end of 2011 through 2012. All of the ectoparasites were identified to the species level using a microscope and the morphological identification keys described [1], [3]. Photographs of the dorsal and ventral body parts of each ectoparasite were captured (Fig. 1), and the number and sex of each louse and fly was determined.

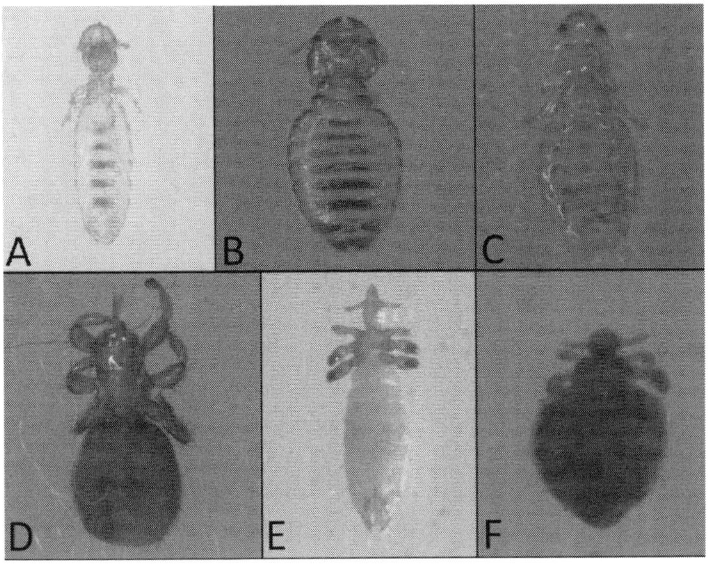

Figure 1. Photographs of morphologically identified 5 species of lice and *Melophagaus ovinus* **collected from domestic animals**. A, *Bovicola ovis* of sheep; B, *Bovicola bovis* of cattle; C, *Heterodoxus spiniger* of dog; D, *Linognathus vituli* of cattle; E, *Solenopotes capillatus* of cattle; F, Melophagus ovinus (sheep ked) of sheep.

https://doi.org/10.1371/journal.pone.0052377.g001

Molecular identification of ectoparasites

Prior to DNA extraction, each specimen was rinsed twice in sterile water for 15 minutes and then dried on sterile filter paper. Each specimen was longitudinally cut into two equal halves. Genomic DNA was extracted from each specimen using the QIAamp DNA tissue extraction kit (Qiagen, Hilden, Germany) as per the instructions of the manufacturer. DNA from each ectoparasite was eluted in 200 μl of TE buffer and stored at −20°C under sterile conditions to preclude any contamination until the sample was used for PCR. The second half of each louse and fly was kept at −80°C as a backup sample.

For molecular species identification, DNA samples of 3 to 8 randomly selected individual lice per species were subjected to standard PCR in an automated DNA thermal cycler to amplify a fragment of the 18S rRNA gene as described [33]. The PCR was carried out in a Peltier PTC-200 model thermal cycler (MJ Research Inc, Watertown, Mass.). The amplified products were detected by electrophoresis on 2% agarose gels in TBE 0.5× buffer, stained with ethidium bromide and visualized using ultraviolet (UV) transillumination. A DNA molecular weight marker (Boehringer-Mannheim VI, Germany) was used to estimate the size of the products. The positive controls consisted of one sample from *Pediculus humanus capitis* collected in Mali and one sample of *P. humanus humanus* collected in Algeria. These samples were included in the PCR assay, and sterile water was used as negative control.

The PCR products were cleaned of excess primers and nucleotides using a QIAquick Spin PCR Purification Kit (Qiagen) as per instructions of the manufacturer. Purified DNA was sequenced using the BigDye Terminator Cycle Sequencing Ready Reaction Kit (ABI PRISM, PE Applied Biosystems, Foster City, CA). All obtained sequences were assembled and edited using Chromas Pro1.34 (Technelyium Pty. Ltd., Tewantin). The sequences of the 18S rRNA genes were then subjected to BLAST analysis to determine similarities to those available in GenBank and to construct a phylogenetic tree using Mega 5 software (Molecular Evolution Genetic Analysis; The Biodesign Institute, Tempe, AZ).

Molecular detection of *Acinetobacter* and *Rickettsia* species

All the specimens were individually tested for the presence of *Acinetobacter* species DNA targeting the *rpoB* gene [30] by real-time quantitative (q) PCR as per the instructions of the manufacturer (Applied Biosystems, Foster City, CA). In addition, the ectoparasite DNA was tested for spotted fever group *Rickettsia* with primers targeting the *glt*A gene specific for this group [34] and for typhus group *Rickettsia* with primers targeting the Rpr274P gene, which encodes a hypothetical protein [35]. Sterile water was used as negative control; *A. baumannii*, *R. montanensis* and *R. typhi* DNA were used as positive controls. The samples were considered positive when cycle thresholds (Ct) were <35.

Molecular detection of Carbapenemase Encoding Genes

All the DNA of lice (n=82) and flies (n=19) positive for *Acinetobacter* species were tested for the presence of carbapenemase encoding genes by qPCR targeting blaOXA-23, blaOXA-24, blaOXA-58 and blaNDM-1 using primers, probes and all conditions as has been described before [15], [36]. In addition, a total of 32 lice and 10 flies with sufficient amount of DNA were evaluated for carbapenemase encoding genes by standard PCR targeting blaOXA-51 and blaOXA-23 with primers and all conditions as described before [15].

Molecular identification of *Acinetobacter* spp. by partial *rpoB* gene

A total of 32 lice and 10 keds DNA of sufficient amount and positive for *Acinetobacter* spp. by qPCR were further subjected to standard PCR targeting partial *rpoB* gene (zone 1) to identify *Acinetobacter* spp. using the primers and all conditions as described before [17]. Sterile water and *A. genomosp* DNA were used as negative and positive controls, respectively. Detection of amplified products, cleaning of excess primers and nucleotides from DNA, sequencing, assembling and edition of sequences, BLAST analysis and *rpoB* gene phylogenetic tree construction with Maximum likelihood statistics of Mega 5 were all performed using similar methods as described for 18S rRNA gene for lice above.

Ethical Statement

Ethical approval for the collection of lice and flies from domestic animals was obtained from the animal research ethics board (Agreement # 14/160/550/2011) of the College of Veterinary Medicine and Agriculture of Addis Ababa University. All necessary oral permits were obtained for the described field studies, including permission of administration and agricultural office of each Ethiopian district and from each animal owner. Ectoparasite collections are not harmful and are not against the welfare of animals. No collection had been done from privately-owned, wildlife, national park or other protected areas and endangered or protected species.

Data Analysis

Microsoft Excel was used for data management. Descriptive statistics such as percentages and means were employed to summarize the proportions of infestations with lice and keds. Statistical analysis was performed with EpiInfo™7 and a P-value of <0.05 was considered significant.

RESULTS

Morphological Identification of Ectoparasites

A total of 207 cattle, 85 sheep, 47 dogs and 16 cats in six districts were examined for the presence of lice and sheep keds (Table 2). The results of the morphological identification of the collected lice and flies from infested animals revealed a total of 408 *Linognathus vituli*, 4 *Bovicola bovis* and 3 *Solenopotes capillatus* from cattle; 22 *Heterodoxus spiniger* from dogs; 298 *Bovicola ovis* and 22 *Melophagus ovinus* from sheep (Fig. 1). Furthermore, the study showed that out of all of the examined cattle, 19.3% (40/207) were infested with *L. vituli*, 0.5% (1/207) with *S. capillatus* and 0.5% (1/207) with *B. bovis*. In cattle, there was a significantly higher prevalence of *L. vituli* than both *S. capillatus* and *B. bovis* ($p \leq 0.0001$). Of the total examined sheep, 48.2% (41/85) were infested with *B. ovis* and 21.05% (4/19) were positive for *M. ovinus*. In addition, 19.1% (9/47) of the examined dogs were infested with *H. spiniger*. Alternatively, lice were not detected on any of the cats studied.

MOLECULAR identification of Ectoparasites

Molecular identification of the lice based on the 18S rRNA gene analysis was used to confirm the species of lice identified by morphological methods. A BLAST analysis of 18S rRNA gene sequences of lice from dogs morphologically identified as *H. spiniger* (n=5) showed 100% (509/509) similarity to the GenBank reference of *H. spiniger* collected from Japan (GU569166) (Fig. 2). Likewise, a BLAST analysis of 18S rRNA gene sequences of lice that were collected from cattle and were morphologically identified as *L. vituli* (n=7) showed 99.3% (553/557) similarity to *Linognathus vituli* collected in Australia (GenBank Access. No. AY077774). Four mutations were detected between our sequence and the reference (AY077774) at 283 bp (T-G), 327 bp (T-A), 352 bp (T-C), and 420 bp (T-C). This sequence was submitted to GenBank under accession number JX401573.

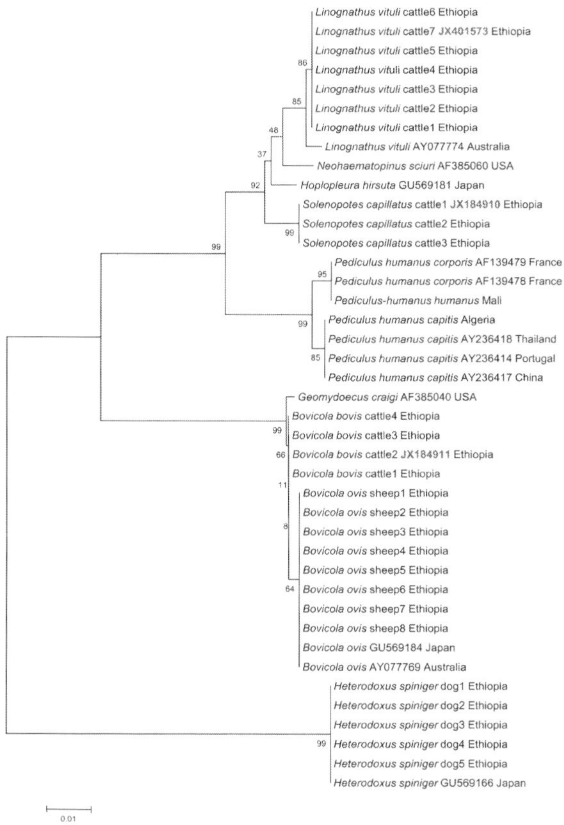

Figure 2. Phylogenetic tree based on 18S gene sequences of lice species collected from domestic animals. Accession numbers in red color are 18S gene sequence of lice of animals from Ethiopia recently deposited in the GenBank. Minimum evolution method was used to build the phylogentic tree. Bootstrap values are indicated at the nodes.

https://doi.org/10.1371/journal.pone.0052377.g002

Interestingly, a sequence analysis of the 18S rRNA gene of 3 lice from cattle morphologically identified as *Solenopotes capillatus* showed 92.2% (536/581) homology with the GenBank sequence of *Neohaematopinus sciuri* (AF423798) and 92% (520/565) homology with *Hoplopleura hirsute* collected in Japan from the cotton rat *Sigmodon hispidus* (GU569181).

A BLAST analysis of the 18S rRNA gene of 4 *Bovicola bovis* lice from cattle showed 99.8% (494/495) similarity to *B. ovis* (GenBank Access No. GU569184) and 99.4% (510/513) similarity to *Geomydoecus craigi*

(GenBank Access No. AF385040) (Fig. 2). There are no 18S rRNA gene sequences available in GenBank for *S. capillatus* and *Bovicola bovis*. Thus, these two sequences were submitted to GenBank under accession numbers JX184910 and JX184911, respectively. An analysis of the 18S rRNA gene of *Bovicola ovis* (n=8) from sheep revealed 100% (495/495) similarity with *Bovicola ovis* detected in Japan and Australia with the sequences GU569184 and AY077769, respectively.

An analysis of the 18S rRNA gene from *Pediculus humanus humanus* (n=1) from Mali revealed 99.4% (496/499) similarity with *Pediculus humanus corporis* detected in France with Accessions Nos. AF139479 and AF139478, respectively. An analysis of the 18S rRNA gene from *Pediculus humanus capitis* (n=1) from Algeria revealed 99.8% (518/520) similarity with the *Pediculus humanus capitis* detected in Thailand, Portugal and China with Accessions Nos. AY236418, AY236414 and AY236417, respectively.

A phylogenetic tree was constructed from the 18S rRNA gene sequences of lice collected in this study and the most similar sequences in the GenBank using the Neighbor-joining statistics of Mega 5. *H. spiniger* reference lice and our lice sample from dogs that belong to the family Boophidae in the suborder Amblycera formed one group on the phylogenetic tree (Fig. 2). Reference *Linognathus vituli* and our lice sample from cattle that belong to the family Linognathidae under the suborder Anoplura clustered together. The 3 individual *S. capillatus* lice from cattle, which shared the same family and suborder with *L. vituli*, clustered near the cattle louse. Both the control and the reference *P. humanus capitis* and *P. humanus humanus* isolated from human head and body lice, belonging to the family Pediculidae under the suborder Anoplura, formed a separate group on the phylogenetic tree that was close to *L. vituli* from cattle. The *Bovicola ovis* reference lice and our lice sample from sheep that belong to the family Trichodectidae in the suborder Ischnocera clustered in a separate group close to the four *B. bovis* cattle lice, which belong to the same genus, family and suborder (Fig. 2).

Detection of *Acinetobacter* and *Rickettsia* species in lice and *M. ovinus*

We detected *Acinetobacter* spp. in a total of 82 lice (11.1%) and 19 flies (86.4%) (Table 3). In our study, there was a significantly higher prevalence of *Acinetobacter* spp. in flies than in lice (82/735 vs. 19/22; $p \leq 0.00001$). The study showed that a higher percentage of *Acinetobacter* spp. DNA was detected in *H. spiniger* of dogs than in *B. ovis* and *L. vituli* (15/22 vs. 19/298; 47/408; p≤0.00001). The prevalence of *Acinetobacter* spp. in *L. vituli* collected from cattle was significantly higher than in *B. ovis* from sheep (47/408 vs 19/298; p=0.02). *Acinetobacter* spp. was not detected in any of the 4 *Bovicola bovis* collected from cattle (Table 3).

Table 3. Percentage of *Acinetobacter* **spp. detected by qPCR in lice and flies collected from domestic animals** in six districts in Oromia.

District (Number of positive samples for *Acinetobacter* **spp.**/Number of tested samples) (%)							
Lice spp.	Asalla (%)	Shano (%)	Walmara (%)	Ada'a (%)	Bedele (%)	Gachi (%)	Total (%)
B. ovis	13/234(5.5)	5/57(8.8)	-	-	1/7(14.3)	-	19/298(6.4)
L. vituli	-	7/51(13.7)	10/93(10.7)	4/84(4.8)	1/12(8.3)	25/168 (14.9)	47/408(11.5)
S. capillatus	-	-	-	1/3(33.3)	-	-	1/3(33.3)
B. bovis	-	0/4(0)	-	0/0 (0)	-	-	0/4(0)
H. spiniger	-	-	-	15/22(68.2)	-	-	15/22(68.2)
Total lice							**82/735(11.1)**
Fly species							
M. ovinus	-	19/22(86.4)	-	-	-	-	19/22(86.4)
Total flies							**19/22(86.4)**

https://doi.org/10.1371/journal.pone.0052377.t003

Results of our study revealed that out of the total lice infested animals, 88.9% (8/9) of dogs were infested with *H. spiniger*, 45% (18/40) of cattle were infested with *L. vituli* and 34.4% (14/41) of sheep were infested with *B. ovis* and harbored at least 1 louse positive for *Acinetobacter* spp. (Table 4). Similarly, *Acinetobacter* spp. was detected at least in one *M. ovinus* in all the 4/4 (100%) infested sheep. Highest proportion of *Acinetobacter* spp. was observed in cattle infested with *L. vituli* in Walmara and Gachi districts whereas highest percentage of *Acinetobacter* spp. was noted in sheep infested with *B. ovis* in Shano and Asalla districts (Table 4).

Table 4. Proportion of infested animals positive for *Acinetobacter* **spp. in lice and keds by qPCR in six districts in Oromia.**

District	No. animals positive for *Acinetobacter* **spp./No. animals infested with lice or fly**					
	B. ovis **(Sheep)**	*L. vituli* **(Cattle)**	*S. capillatus* **(Cattle)**	*B. ovis* **(Cattle)**	*H. spiniger* **(Dog)**	*M. vinus* **(Sheep)**
Asalla	9/29 (31.0%)	-	-	-	-	-
Walmara	-	6/9 (66.7%)	-	-	-	-
Shano	4/10 (40%)	2/9 (22.2%)	-	0/1	-	4/4 (100%)
Ada'a	-	3/9 (33.3%)	1/1 (100%)	-	8/9 (88.9%)	-
Bedele	1/2(50%)	1/4 (25%)	-	-	-	-
Gachi	-	6/9 (66.7%)	-	-	-	-
Total	**14/41(34.4%)**	**18/40 (45%)**	**1/1 (100%)**	**0/1**	**8/9 (88.9%)**	**4/4 (100%)**

https://doi.org/10.1371/journal.pone.0052377.t004

A molecular investigation of the 735 lice collected from domestic animals and 22 *Melophagus ovinus* collected from sheep using qPCR did not produced any positive results for either spotted fever or typhus group *Rickettsia* species.

Molecular identification of *Acinetobacter* spp

We succeeded in the amplification of 350 bp fragment of partial *rpoB* gene from a total of 10 samples including one from *L. vituli* of cattle, 3 from *M. ovinus* of sheep and 6 from *H. spiniger* of dogs (Table 5). BLAST analysis of partial *rpoB* gene sequence and *rpoB* phylogenetic tree showed the presence of *A. soli* in *L. vituli* of cattle, *A. lwoffii* in *M. ovinus* of sheep, 1 new *Acinetobacter* sp. in *M. ovinus* of sheep, *A. pittii* in *H. spiniger* of dogs and 2 new *Acinetobacter* sp. in *H. spiniger* of dogs (Table 5 and Fig. 3). Five of these 10 sequences were submitted to GenBank under accession numbers KC130085-89, respectively.

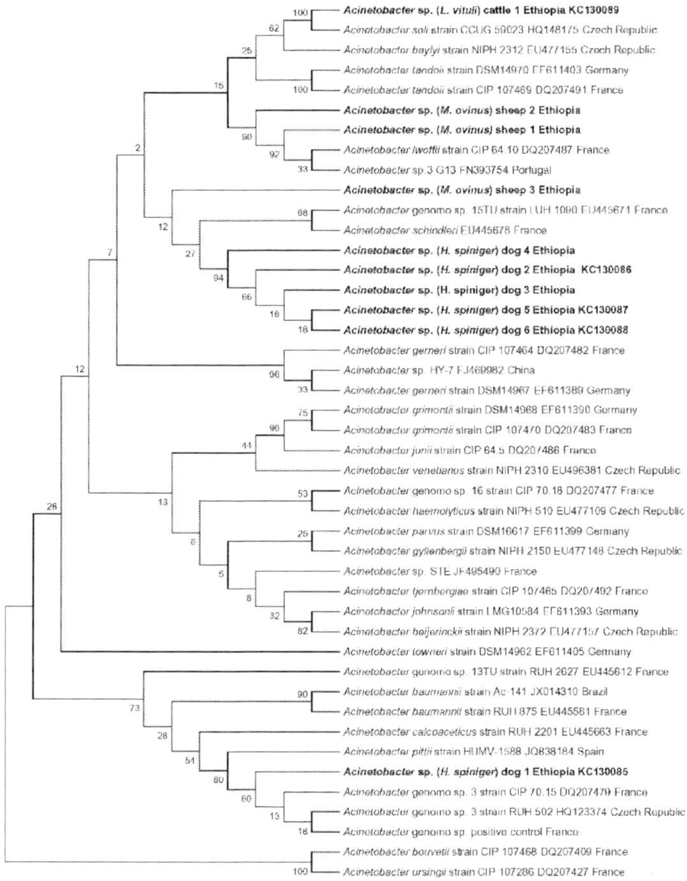

Figure 3. Phylogenetic tree based on partial *rpoB* **gene sequences of** *Acinetobacter* **species.**

Maximum Likelihood method was used to build the phylogentic tree. Bootstrap values are indicated at the nodes. Bold indicates the taxonomic position of Acinetobacter species identified in this study.

https://doi.org/10.1371/journal.pone.0052377.g003

Table 5. Summary of BLAST analysis of partial *rpoB* **gene sequences obtained from lice and keds of domestic animals** in six districts in Oromia, Ethiopia.

Lice/fly spp	Host spp	Length (bp)	Nearest match in Gen-Bank	% similarity
H. spiniger	Dog	383	*Acinetobacter* **genomosp. 3 (DQ207479)**	99.2%
H. spiniger	Dog	351	*Acinetobacter* **sp.Hy-7 (FJ469982)**	90.8%
H. spiniger	Dog	391	*Acinetobacter* **sp.Hy-7 (FJ469982)**	90.7%
H. spiniger	Dog	310	*Acinetobacter* **sp. Hy-7 (FJ469982)**	86.5%
H. spiniger	Dog	387	*Acinetobacter* **sp. Hy-7 (FJ469982)**	90.5%
H. spiniger	Dog	388	*Acinetobacter* **sp. Hy-7 (FJ469982)**	90.5%
L. vituli	Cattle	381	*Acinetobacter soli* **(HQ148175)**	98.9%
M. ovinus	Sheep	365	*Acinetobacter* **sp.G13 (FN393754)**	98.9%
M. ovinus	Sheep	293	*Acinetobacter* **sp.G13 (FN393754)**	94.39%
M. ovinus	Sheep	328	*Acinetobacter* **sp-Hy-7 (FJ469982)**	90.2%
P. h. humanus **(Control)**	Man	383	*Acinetobacter* **genomosp.3 (DQ207479)**	99.2%

https://doi.org/10.1371/journal.pone.0052377.t005

Detection of carbapenemase encoding genes in *Acinetobacter* species

A molecular investigation of 82 lice and 19 keds positive for *Acinetobacter* spp. by qPCR did not produced any positive results for blaOXA-23, blaOXA-24, blaOXA-58 and blaNDM-1 genes encoding for carbapenemase resistance. Likewise, investigation of 32 lice and 10 keds DNA by standard PCR never produced any positive results for both blaOXA-23 and blaOXA-51 genes encoding for carbapenemase resistance.

DISCUSSION

In this study, we detected, for the first time, *Acinetobacter* spp. in the lice and flies of animals from Ethiopia. In addition, we obtained two new sequences of 18S rRNA genes from *Bovicola bovis* and *Solenopotes capillatus* collected from cattle. We believe that the results of this study are valid, as all of the negative control samples produced negative results and all the positive control samples tested positive for both *Acinetobacter* spp. and *Rickettsia* spp. detection in the lice and flies of these animals. These findings confirmed that our molecular study conditions precluded any accidental cross-contamination of samples during the study period. Similarly, in our 18S rRNA gene study of lice, we obtained sequences from both positive controls, *Pediculus humanus capitis* and *Pediculus humanus humanus*, confirming the appropriateness of our working conditions.

In the study, the species of lice morphologically identified as *L. vituli*, *B. bovis* and *S. capillatus* from cattle; *H. spiniger* from dogs; and *B.ovis* from sheep (Fig. 1) were confirmed molecularly using 18S rRNA gene sequences. The 18S rRNA gene has been used previously as an important tool to investigate human lice phylogeny [33], to study the evolution of sucking lice [37] and to study the phylogeny of lice [38]. The phylogenetic tree constructed from the 18S rRNA gene sequences of the collected lice and reference lice in the GenBank presented here supports the information from previous studies [37], [38].

In this study, we sequenced the 18S rRNA genes of *B. bovis* and *S. capillatus* from cattle for the first time. *B. bovis* is the only biting lice species in cattle, and it has important morphological features such as a rounded head, reddish-brown color and dark transverse bands on the abdomen [1] (Fig. 1). *S. capillatus* from cattle are the smallest sucking lice, with morphological characteristics that include a hexagonal sterna plate on the thorax, a weak front pair of legs and prominent abdominal tubercles [3]. In Ethiopia, both species are typically reported with low prevalence [4]. The finding in this study that there was a significantly ($p \leq 0.0001$) higher prevalence of *L. vituli* than both *S. capillatus* and *B. bovis* in cattle is in agreement with previous work [4]. Our findings in sheep, dogs and cats of this study are also in line with previous reports [5], [6], [8].

The observation that 11.1% of the lice in the current study were found to contain *Acinetobacter* spp. is lower than the earlier report of 47% in human head lice and 71% human body lice from Ethiopia [31], 21% in human body lice [29] and 33% in human head lice from Paris [30]. Recently, a low prevalence (4%) of *A. baumannii* in head lice was reported from Senegal [15]. However, the overall percentage of 86.4% (19/22) of *Acinetobacter* spp. detected in *M. ovinus* of sheep in our study is greater than that found in the aforementioned reports. The variation among study districts in the percentage of *Acinetobacter* spp. in lice and keds in infested animals is most probably attributed to differences in agroecology, animal management and factors like age, sex, physiological status or presence of other concurrent diseases that may favor infection of lice or the animal by the bacteria. This observation coincides with the findings of differences in the prevalence of infection in lice of humans by *A. baumannii* among different study areas with different altitudes in south western Ethiopia [31].

In line with our findings, *Acinetobacter* spp. were detected in many species of arthropods in different parts of the world (Table 1). For instance, *Acinetobacter* spp. have been detected in 13% of *Lutzomyia longipalpis* (sand flies) in Brazil [39], 7.7% of *Glossina palpalis palpalis* (tsetse fly) in Angola [40], 8.7% in the gut of the *Prionoplus reticularis* larvae (wood feeding beetle) in New Zealand [41], up to 75% in *Bacterria cockerelli* (potato psyllid) in the USA [42], 18.9% in chewing lice of pocket gophers in the USA [43] and 1% in *Bemisia tabaci* (tobacco whitefly) in India [44]. Moreover, *A. baumannii* have been detected with a prevalence of 5.1% in the feces of domestic animals in Senegal [15].

Most earlier investigators suggested that it is still yet not exactly determined how both body and head human lice acquired *A. baumannii* infection [30], [31]. However, some authors argued undiagnosed transient *A. baumannii* bacteremia in infested patients as a source of infection for body lice but they stated it is not possible to rule out the possibility of acquiring *A. baumannii* infection in human body lice from external environmental contamination [29]. Other investigators pointed out that bacterial spp. such as *Acinetobacter* abundant in the environment reside in the gut of several species of arthropods as transient or natural flora [39] acquired commonly by vertical transmission [40] and also are maintained and spread by several mechanisms of horizontal transmission including mating, cofeeding, or contact with contaminated faeces [41], [42], [44]. Having all these facts in mind and due to the ubiquitous nature of *Acinetobacter* spp. in the environment including on the skin and in the faces of animals [15] lice and keds possibly had acquired *Acinetobacter* spp. infection from the skin, faces or transient *Acinetobacter* spp. bacteremia of their host animals. Moreover, it is not possible to rule out infection of lice and keds by vertical route and also probably animals may play a role as reservoirs for these bacteria.

We did not detect *Rickettsia* DNA in the lice and flies in our study. This finding contrasts with previous work that reported detecting *R. helvetica* in *M. ovinus* from sheep, in *Linognathus stenopsis* from goats, and *Rickettsia* spp. in *Haematopinus eurysternus* from cattle in Hungary [45], [46].

The absence of any positive results for blaOXA-23, blaOXA-24, blaOXA-58, blaNDM-1 and blaOXA-51 genes encoding for carbapenemase resistance by both qPCR and standard PCR in *Acinetobacter* spp. in the lice and keds of domestic animals in our study is in line with the previous finding of absence of resistance to several antimicrobials in *A. soli* in intensive health care units of neonates in Brazil [23], full susceptibility to several antibiotics of *A. lwoffii* from acute gastroenteritis in USA [21] and absence of blaOXA-like genes encoding for carbapenemase resistance in *A. baumannii* from faeces of domestic animals in Senegal [15]. On the other hand our findings contrast the previous multidrug resistance in *A. baumannii* reported from several countries of the world [12], [36], [47]. This finding also contradicts the observations of multidrug-resistant isolates of *Acinetobacter* spp. from a range of environmental sources in South Korea [20] and the presence of resistance to antibiotics in *A. baumannii* and other *Acinetobacter* spp. from blood cultures in Norway [48]. This variation is most probably attributed to differences in strains of the bacteria among various studies and other many factors contributing for emergence of resistance.

Acinetobacter spp. identification study from DNA of lice and keds of animals in Oromia uncovered the occurrence of 3 previously described species and 3 new *Acinetobacter* spp. (Table 5 and Fig. 3). All the 3 previously described species, we detected: *A. soli* from *L. vituli* of cattle, *A. lowffii* from keds of sheep and *A. pittii* from *H. spiniger* of dogs with high nucleotide sequence identities of 98–100% with their respective reference species in the GenBank (Table 5). This finding supports the criteria established for *Acinetobacter* spp. identification using partial *rpoB* gene sequence analysis [17], [18]. In line with our observation higher predominance of other *Acinetoabacter* spp. (24.8%) than *A. baumannii* (only 8.8%) from human blood culture isolates was recently reported from Norway [48]. Furthermore, these species are nowadays reported to cause various types of human infections worldwide [21], [23], [49], [50], [51] and are implicated as emerging *Acinetobacter* spp.

We also identified two new *Acinetobacter* species from *H. spiniger* of dogs and one from ked of sheep (Table 5 and Fig. 3). All these 3 new *Acinetobacter* spp. demonstrated low nucleotide homology with reference *rpoB* sequence in the GenBank and low bootstrap value in the *rpoB* phylogenetic tree. Results of our study suggest the presence of specific *Acinetobacter* species in lice and keds of domestic animals unlike *A. baumannii* in human lice. We believe that additional in depth epidemiological studies involving other species of lice and ectoparasites of different animal species from vast areas in the world are needed as has been done during the last decade for *Bartonella* species, ectoparasites and their animal hosts [52]. Further studies are also required to isolate and determine the human health significance of the new *Acinetobacter* spp. detected in the lice and keds from animals.

To our knowledge, this study is the first to report the presence of DNAs from different *Acinetobacter* spp. in various species of lice collected from domestic animals and in flies collected from sheep. Our study demonstrates that *Acinetobacter* spp. are not only common as hospital pathogens and in human lice but they can also be detected in the ectoparasites of animals. Our findings suggest that synanthropic animals and their ectoparasites might play a role to increase the risk of human exposure to zoonotic pathogens and could be a source for *Acinetobacter* spp. infections in humans. However, additional epidemiological data are required to justify the significance of this finding and to determine whether ectoparasites from animals can act as environmental reservoirs and play a role in spreading these bacteria to both animal and human hosts.

ACKNOWLEDGMENTS

The authors greatly acknowledge all animal owners in Oromia and Yonas Abiyi, Teferi Banti, Fikadu and Gurara Megerssa Veterinary professionals of Oromia Regional laboratories for their help during sample collection. Also Amina Boutellis and Seydina M. Diene PhD students at the Faculty of Medicine of Aix Marseille University, are gratefully acknowledged for their professional help during laboratory work of the study.

REFERENCES

1. Wall R, Shearer D (1997) Veterinary Entomology; 1, editor. London: Chapman and Hall. 438 p.
2. Small RW (2005) A review of *Melophagus ovinus* (L.), the sheep ked. Vet Parasitol 130: 141–155.
3. Taylor MA, Coop RL, Wall R (2007) Veterinary Parasitology: Wiley-Blackwell. 847 p.
4. Kumsa B, Bekele S (2008) Lice Infestation On Cattle In Endegagn District, Southern Ethiopia: Species Composition, Prevalence And Seasonal Pattern. Bull Anim Hlth Prod Afr 56: 213–222.
5. Kumsa B, Beyecha K, Geloye M (2012) Ectoparasites of sheep in three agro-ecological zones in central Oromia, Ethiopia Onderstepoort J Vet Res 79(1): 7 pages.
6. Sertse T, Wossene A (2007) A study on ectoparasites of sheep and goats in eastern part of Amhara region, Northeast Ethiopia. Small Rumin Res 69: 55–61.
7. Beyecha K, Beyene D, Kumsa B (2012) Ectoparasites of goats in three agroecologies in central Oromia, Ethiopia. Comp Clin Pathol DOI 10.1007/s00580-012-1563-x.
8. Kumsa B, Mekonnen S (2011) Ixodid ticks, fleas and lice infesting dogs and cats in Hawassa, southern Ethiopia. Onderstepoort J Vet Res 78: 4 pages.
9. Yacob HT, ATAKLTY H, Kumsa B (2008) Major ectoparasites of cattle in and around Mekelle, northern Ethiopia. Entom Res 28: 26–30.
10. Chanie M, Negash T, Sirak A (2010) Ectoparasites are the major causes of various types of skin lesions in small ruminants in Ethiopia. Trop Anim Health Prod 42: 1103–1109.
11. Abebayehu T, Kibrom M (2010) Study on ectoparasitic defects of processed skins at Sheba Tannery, Tigray, Northern Ethiopia. Trop Anim Health Prod 42: 1719–1722.
12. Zordan S, Prenger-Berninghoff E, Weiss R, van der Reijden T, van den Broek P, et al. (2011) Multidrug-resistant *Acinetobacter baumannii* in veterinary clinics, Germany. Emerg Infect Dis 17: 1751–1754.
13. Peleg AY, Seifert H, Paterson DL (2008) *Acinetobacter baumannii*: emergence of a successful pathogen. Clin Microbiol Rev 21: 538–582.
14. Turton JF, Shah J, Ozongwu C, Pike R (2010) Incidence of

Acinetobacter species other than *A. baumannii* among clinical isolates of *Acinetobacter*: evidence for emerging species. J Clin Microbiol 48: 1445–1449.

15. Kempf M, Rolain JM, Diatta G, Azza S, Samb B, et al. (2012) Carbapenem Resistance and *Acinetobacter baumannii* in Senegal: The Paradigm of a Common Phenomenon in Natural Reservoirs. PLoS One 7: e39495.
16. Nemec A, Krizova L, Maixnerova M, van der Reijden TJ, Deschaght P, et al. (2011) Genotypic and phenotypic characterization of the *Acinetobacter calcoaceticus-Acinetobacter baumannii* complex with the proposal of *Acinetobacter pittii* sp. nov. (formerly *Acinetobacter* genomic species 3) and *Acinetobacter nosocomialis* sp. nov. (formerly *Acinetobacter* genomic species 13TU). Res Microbiol 162: 393–404.
17. Gundi VA, Dijkshoorn L, Burignat S, Raoult D, La Scola B (2009) Validation of partial *rpoB* gene sequence analysis for the identification of clinically important and emerging *Acinetobacter* species. Microbiology 155: 2333–2341.
18. La Scola B, Gundi VA, Khamis A, Raoult D (2006) Sequencing of the *rpoB* gene and flanking spacers for molecular identification of *Acinetobacter* species. J Clin Microbiol 44: 827–832.
19. Narciso-da-Rocha C, Vaz-Moreira I, Svensson-Stadler L, Moore ER, Manaia CM (2012) Diversity and antibiotic resistance of *Acinetobacter* spp. in water from the source to the tap. Appl Microbiol Biotechnol.
20. Choi JY, Kim Y, Ko EA, Park YK, Jheong WH, et al. (2012) *Acinetobacter* species isolates from a range of environments: species survey and observations of antimicrobial resistance. Diagn Microbiol Infect Dis 74: 177–180.
21. Regalado NG, Martin G, Antony SJ (2009) *Acinetobacter lwoffii*: bacteremia associated with acute gastroenteritis. Travel Med Infect Dis 7: 316–317.
22. Nakwan N, Wannaro J, Nakwan N (2011) Multidrug-resistant *Acinetobacter lwoffii* infection in neonatal intensive care units. Res and Reports in Neonatatology.
23. Pellegrino FL, Vieira VV, Baio PV, dos Santos RM, dos Santos AL, et al. (2011) *Acinetobacter soli* as a cause of bloodstream infection in a neonatal intensive care unit. J Clin Microbiol 49: 2283–2285.
24. Francey T, Gaschen F, Nicolet J, Burnens AP (2000) The role of

Acinetobacter baumannii as a nosocomial pathogen for dogs and cats in an intensive care unit. J Vet Intern Med 14: 177–183.

25. Endimiani A, Hujer KM, Hujer AM, Bertschy I, Rossano A, et al. (2011) *Acinetobacter baumannii* isolates from pets and horses in Switzerland: molecular characterization and clinical data. J Antimicrob Chemother 66: 2248–2254.
26. Hamouda A, Findlay J, Al Hassan L, Amyes SG (2011) Epidemiology of *Acinetobacter baumannii* of animal origin. Int J Antimicrob Agents 38: 314–318.
27. Poirel L, Bercot B, Millemann Y, Bonnin RA, Pannaux G, et al. (2012) Carbapenemase-producing *Acinetobacter* spp. in Cattle, France. Emerg Infect Dis 18: 523–525.
28. Smet A, Boyen F, Pasmans F, Butaye P, Martens A, et al.. (2012) OXA-23-producing *Acinetobacter* species from horses: a public health hazard? J Antimicrob Chemother.
29. La Scola B, Raoult D (2004) *Acinetobacter baumannii* in human body louse. Emerg Infect Dis 10: 1671–1673.
30. Bouvresse S, Socolovshi C, Berdjane Z, Durand R, Izri A, et al. (2011) No evidence of *Bartonella quintana* but detection of *Acinetobacter baumannii* in head lice from elementary schoolchildren in Paris. Comp Immunol Microbiol Infect Dis 34: 475–477.
31. Kempf M, Abdissa A, Diatta G, Trape JF, Angelakis E, et al.. (2012) Detection of *Acinetobacter baumannii* in human head and body lice from Ethiopia and identification of new genotypes. Int J Infect Dis.
32. Central Statistics Agency (2008) Ethiopian agricultural sample survey report on livestock and livestock characteristics, Vol II, 2007/08, Addis Ababa, Ethiopia.
33. Yong Z, Fournier PE, Rydkina E, Raoult D (2003) The geographical segregation of human lice preceded that of *Pediculus humanus capitis* and *Pediculus humanus humanus* . C R Biol 326: 565–574.
34. Socolovschi C, Barbarot S, Lefebvre M, Parola P, Raoult D (2010) *Rickettsia sibirica mongolitimonae* in traveler from Egypt. Emerg Infect Dis 16: 1495–1496.
35. Walter G, Botelho-Nevers E, Socolovschi C, Raoult D, Parola P (2012) Murine typhus in returned travelers: a report of thirty-two cases. Am J Trop Med Hyg 86: 1049–1053.
36. Kusradze I, Diene SM, Goderdzishvili M, Rolain JM (2011)

Molecular detection of OXA carbapenemase genes in multidrug-resistant *Acinetobacter baumannii* isolates from Iraq and Georgia. Int J Antimicrob Agents 38: 164–168.

37. Light JE, Smith VS, Allen JM, Durden LA, Reed DL (2010) Evolutionary history of mammalian sucking lice (Phthiraptera: Anoplura). BMC Evol Biol 10: 292.
38. Barker SC, Whiting M, Johnson KP, Murrell A (2003) Phylogeny of the lice (Insecta, Phthiraptera) inferred from small subunit rRNA. Zoologica Scripta 32: 407–414.
39. Gouveia C, Asensi MD, Zahner V, Rangel EF, Oliveira SM (2008) Study on the bacterial midgut microbiota associated to different Brazilian populations of *Lutzomyia longipalpis* (Lutz & Neiva) (Diptera: Psychodidae). Neotrop Entomol 37: 597–601.
40. Geiger A, Fardeau ML, Grebaut P, Vatunga G, Josenando T, et al. (2009) First isolation of *Enterobacter*, *Enterococcus*, and *Acinetobacter* spp. as inhabitants of the tsetse fly (*Glossina palpalis palpalis*) midgut. Infect Genet Evol 9: 1364–1370.
41. Reid NM, Addison SL, Macdonald LJ, Lloyd-Jones G (2011) Biodiversity of active and inactive bacteria in the gut flora of wood-feeding huhu beetle larvae (Prionoplus reticularis). Appl Environ Microbiol 77: 7000–7006.
42. Nachappa P, Levy J, Pierson E, Tamborindeguy C (2011) Diversity of endosymbionts in the potato psyllid, *Bactericera cockerelli* (Triozidae), vector of zebra chip disease of potato. Curr Microbiol 62: 1510–1520.
43. Reed DL, Hafner MS (2002) Phylogenetic analysis of bacterial communities associated with ectoparasitic chewing lice of pocket gophers: a culture-independent approach. Microb Ecol 44: 78–93.
44. Singh ST, Priya NG, Kumar J, Rana VS, Ellango R, et al. (2012) Diversity and phylogenetic analysis of endosymbiotic bacteria from field caught *Bemisia tabaci* from different locations of North India based on 16S rDNA library screening. Infect Genet Evol 12: 411–419.
45. Hornok S, de la Fuente J, Biro N, Fernandez de Mera IG, Meli ML, et al. (2011) First molecular evidence of *Anaplasma ovis* and *Rickettsia* spp. in keds (Diptera: Hippoboscidae) of sheep and wild ruminants. Vector Borne Zoonotic Dis 11: 1319–1321.
46. Hornok S, Hofmann-Lehmann R, de Mera IG, Meli ML, Elek V, et al. (2010) Survey on blood-sucking lice (Phthiraptera: Anoplura)

of ruminants and pigs with molecular detection of *Anaplasma* and *Rickettsia* spp. Vet Parasitol 174: 355–358.

47. Higgins PG, Dammhayn C, Hackel M, Seifert H (2010) Global spread of carbapenem-resistant *Acinetobacter baumannii* . J Antimicrob Chemother 65: 233–238.
48. Karah N, Haldorsen B, Hegstad K, Simonsen GS, Sundsfjord A, et al. (2011) Species identification and molecular characterization of *Acinetobacter* spp. blood culture isolates from Norway. J Antimicrob Chemother 66: 738–744.
49. Wisplinghoff H, Paulus T, Lugenheim M, Stefanik D, Higgins PG, et al. (2012) Nosocomial bloodstream infections due to *Acinetobacter baumannii*, *Acinetobacter pittii* and *Acinetobacter nosocomialis* in the United States. J Infect 64: 282–290.
50. Vaz-Moreira I, Novo A, Hantsis-Zacharov E, Lopes AR, Gomila M, et al. (2011) *Acinetobacter rudis* sp. nov., isolated from raw milk and raw wastewater. Int J Syst Evol Microbiol 61: 2837–2843.
51. Debarry J, Garn H, Hanuszkiewicz A, Dickgreber N, Blumer N, et al. (2007) *Acinetobacter lwoffii* and *Lactococcus lactis* strains isolated from farm cowsheds possess strong allergy-protective properties. J Allergy Clin Immunol 119: 1514–1521.
52. Saisongkorh W, Rolain JM, Suputtamongkol Y, Raoult D (2009) Emerging *Bartonella* in humans and animals in Asia and Australia. J Med Assoc Thai 92: 707–731.
53. Murrell A, Dobson SJ, Yang X, Lacey E, Barker SC (2003) A survey of bacterial diversity in ticks, lice and fleas from Australia. Parasitol Res 89: 326–334.
54. Pidiyar VJ, Jangid K, Patole MS, Shouche YS (2004) Studies on cultured and uncultured microbiota of wild culex quinquefasciatus mosquito midgut based on 16s ribosomal RNA gene analysis. Am J Trop Med Hyg 70: 597–603.
55. Dinparast Djadid N, Jazayeri H, Raz A, Favia G, Ricci I, et al. (2011) Identification of the midgut microbiota of *An. stephensi* and *An. maculipennis* for their application as a paratransgenic tool against malaria. PLoS One 6: e28484.
56. Sant'Anna MR, Darby AC, Brazil RP, Montoya-Lerma J, Dillon VM, et al. (2012) Investigation of the bacterial communities associated with females of *Lutzomyia* sand fly species from South America. PLoS One

7: e42531.

57. Geiger A, Fardeau ML, Njiokou F, Joseph M, Asonganyi T, et al. (2011) Bacterial diversity associated with populations of Glossina spp. from Cameroon and distribution within the Campo sleeping sickness focus. Microb Ecol 62: 632–643.

58. Zouache K, Raharimalala FN, Raquin V, Tran-Van V, Raveloson LH, et al. (2011) Bacterial diversity of field-caught mosquitoes, *Aedes albopictus* and *Aedes aegypti*, from different geographic regions of Madagascar. FEMS Microbiol Ecol 75: 377–389.

4

Ticks Infesting Domestic Dogs in the UK: A Large-scale Surveillance Programme

Swaid Abdullah[1], Chris Helps[2], Severine Tasker[2], Hannah Newbury[3] and Richard Wall[1]

[1]Veterinary Parasitology and Ecology Group, School of Biological Sciences, Life Sciences Building, University of Bristol, Bristol, UK.
[2] Molecular Diagnostic Unit, Langford Veterinary Services and School of Veterinary Sciences, University of Bristol, Bristol, UK.
[3] MSD Animal Health, Walton Manor, Walton, Milton Keynes, UK.

ABSTRACT

Background

Recent changes in the distribution of tick vectors and the incidence of tick-borne disease, driven variously by factors such as climate change, habitat modification, increasing host abundance and the increased movement of people and animals, highlight the importance of ongoing, active surveillance.

Citation: Abdullah, S., Helps, C., Tasker, S. et al. Ticks infesting domestic dogs in the UK: a large-scale surveillance programme. Parasites Vectors 9, 391 (2016). https://doi.org/10.1186/s13071-016-1673-4

This paper documents the results of a large-scale survey of tick abundance on dogs presented to veterinary practices in the UK, using a participatory approach that allows relatively cost- and time-effective extensive data collection.

Methods

Over a period of 16 weeks (April–July 2015), 1094 veterinary practices were recruited to monitor tick attachment to dogs and provided with a tick collection and submission protocol. Recruitment was encouraged through a national publicity and communication initiative. Participating practices were asked to select five dogs at random each week and undertake a thorough, standardized examination of each dog for ticks. The clinical history and any ticks were then sent to the investigators for identification.

Results

A total of 12,000 and 96 dogs were examined and 6555 tick samples from infested dogs were received. *Ixodes ricinus* (Linnaeus) was identified on 5265 dogs (89 %), *Ixodes hexagonus* Leach on 577 (9.8 %) and *Ixodes canisuga* Johnston on 46 (0.8 %). Ten dogs had *Dermacentor reticulatus* (Fabricius), one had *Dermacentor variabilis* (Say), three had *Haemaphysalis punctata* Canesteini & Fanzago and 13 had *Rhipicephalus sanguineus* Latreille. 640 ticks were too damaged for identification. All the *R. sanguineus* and the single *D. variabilis* were on dogs with a recent history of travel outside the UK. The overall prevalence of tick attachment was 30 % (range 28–32 %). The relatively high prevalence recorded is likely to have been inflated by the method of participant recruitment.

Conclusion

The data presented provide a comprehensive spatial understanding of tick distribution and species abundance in the UK against which future changes can be compared. Relative prevalence maps show the highest rates in Scotland and south west England providing a valuable guide to tick-bite risk in the UK.

BACKGROUND

Ticks are globally important arthropod vectors of disease that transmit an extensive range of viral, bacterial and protozoan pathogens to vertebrate hosts [1]. Tick survival, phenology and biting activity is highly dependent on environmental conditions [2] and is highly responsive to changes in factors such as climate and habitat modification. Long-term increases in the abundance of ticks, such as *Ixodes ricinus* (Linneaus), have been recorded in temperate habitats over recent decades [3], along with evidence of altitudinal and latitudinal expansion in central and northern Europe [4–8]. Data collected by questionnaire from 20 districts in Sweden showed a significant increase in tick abundance in west-central regions where previously ticks had been rare. Blanket dragging in 54 regions along the perceived latitudinal boundary for ticks supported this observation [9, 10]. Similarly, in the UK, over recent decades there has been an estimated 17 % expansion in the distribution of *I. ricinus* and an increase in abundance at 73 % of locations surveyed [11]. The tick *Dermacentor reticulatus* (Fabricius), an important vector of canine babesiosis in Europe, is also believed to have extended its distribution and populations have become established in Poland [12], Belgium [13], Germany [14], the Netherlands [15] and in southern England [16, 17]. Changes in tick-borne disease prevalence are also reported, in association with the changes in vector distribution [18, 19].

Climate change is considered likely to be responsibe in part for these observed changes [20] and further climate-related impacts might be expected if predicted increases in global temperatures of up to 4.8 °C in the next hundred years occur [21]. Longer summer seasons, with a warmer and wetter spring or autumn, might be expected to promote higher tick challenge and longer exposure. Tick mortality may be lower given milder winters, but higher in hotter drier summers. In addition, vector potential may be enhanced by biological changes stimulated by temperature, as is reported in other arthropod vectors [22]. However, ticks may also adapt their seasonal activity and some species may aestivate during very hot conditions and perhaps adopt a more bimodal pattern of activity, pushing the period of feeding to earlier and later in the year. Nevertheless, along with climate, changes in habitat management, land use by people and animals, host movement patterns and changes in host abundance, particularly deer, may also be equally important in explaining changes in patterns of tick abundance and activity [23].

A central problem associated with understanding changing arthropod-borne disease patterns is that systematic surveillance in animals is not routinely undertaken [24]. The cost and complexity of this monitoring is high because vector and pathogen prevalence are often relatively low and large samples sizes are required for meaningful results. Surveillance programmes often rely on passive reporting, which may render them subject to significant levels of spatial and temporal bias. Nevertheless, systematic surveillance is essential to allow detection of changes in the distribution of arthropod vectors and arthropod-borne diseases, particularly because subtle changes at the early stages of an epidemic curve are hard to spot. Routine surveillance is also needed to allow informed risk analysis and the evaluation of the potential spread to new areas or the new introduction of exotic species or diseases [24–26]. This necessitates clear and exhaustive knowledge of the distribution of arthropod vectors and associated vector-borne diseases in different areas [1].

The aim of the work described here therefore was to undertake a national survey of tick abundance on dogs presented to veterinary practices in the UK and to provide a comprehensive spatial understanding of the distribution and species abundance against which future changes could be compared. This study also aimed to evaluate an approach to large-scale surveillance that allows relatively cost- and time-effective extensive data collection.

METHODS

Veterinary practices from throughout the UK were recruited largely through an intensive media and communication initiative, designed to raise awareness of ticks and tick-borne disease amongst veterinarians and the general public. The campaign, launched in March 2015, used an intense period of radio, television, print and social media to highlight the role of ticks as vectors of canine disease in the UK. Interested veterinary practices were then able to register their interest by e-mailing a contact name and their practice details. Once enrolled, they were sent a pack which contained 40 questionnaires, stamped addressed envelopes, a tick removal hook, specimen tubes and protocols. The protocol asked registered veterinary practices to examine five different dogs for ticks each week, for a total of 8 weeks, using a specified standard grooming procedure and then complete a questionnaire relating to the clinical and travel history of each dog. Veterinarians were asked to ensure that the dogs selected for inspection were a random cross-section of animals brought into the surgery for routine procedures such as vaccination and were not those known to be carrying ticks when selected, but no formal randomisation procedure was included in the protocol.

The examination protocol for included dogs required first checking the head for ticks. Special attention was given to the ears, carefully checking the pinnae and inside the external ear canal. The dog was then checked on the neck and chest area, legs, armpits and interdigital spaces. After that the dog's hair, from head to tail, was checked manually using sufficient pressure to detect small lumps. Finally, a louse or flea comb was used to part the hair along the length of the body. The examination was estimated to take an average of 5 min per dog. All attached ticks were removed using a tick hook or forceps, placed in a tube labelled with the dog's name and inspection date and stored in a freezer at -20 °C. Each week stored ticks along with the completed questionnaires were posted to the investigators. The instructions sent to each registered veterinary practice stressed the importance of completing and sending questionnaires for dogs that were found not to have ticks, to allow a true prevalence figure to be calculated.

Each tick sample received by the investigators was given a unique identification number and placed in freezer at -20 °C pending analysis. Subsequently, ticks were identified to species, life-cycle stage and sex using a range of keys [27–29] and a sub-sample of identifications cross-checked by an independent investigator. The questionnaire data were entered into an Excel (Microsoft) worksheet.

The history and sex, breed and age of each dog were recorded. To determine any effect of sex on tick attachment, four categories were considered: male, female, male neutered and female neutered. For the effect of breed, the dogs were categorised following Kennel Club classifications: gundogs, hounds, pastoral, terriers, toy, utility and working, plus mongrels and crossbreds. To assess any effects of age, dogs were divided into five classes: less than 1 year-of-age, 1 to 3, 4 to 6, 7 to 10 and any of greater than 10 years-of-age. Each dog's travel history and any history of visits to kennels were also recorded. The data were subjected to binary logistic regression using SPSS (version 23). The distribution of the participating veterinary practices, dogs and various tick species were mapped using QGIS (version 2.8.1) using the practice or owner's postcodes.

RESULTS

A total of 1000 and 94 veterinary practices from across the UK participated in the 16 week study (Fig. 1); in this time 12,096 dogs were examined for each of which a completed questionnaires was submitted. 6555 tick samples were also submitted, of which 5915 were identified, while 640 were too damaged to allow identification. The median number of ticks per sample was 1, although the maximum number reported on an individual dog was about 200. Of the tick samples received, 98.7 % were adults alone, 0.83 % were nymphs alone and 0.11 % were larvae alone. All the remaining samples were mixed stage. The nymphs and adults were identified to species, the very small numbers of larvae received were not identified. Almost all ticks were semi- or fully-engorged.

Figure 1. Distribution of the 1094 veterinary practices that participated in the UK survey for ticks on dogs over 16 weeks (March to July) in 2015

Species Abundance and Distributions

Amongst dogs that had not travelled outside the UK, 89.2 % of those that had ticks were infected by *I. ricinus* and 9.8 % by *Ixodes hexagonus* Leach (Table 1). Both species were widely distributed throughout the UK (Fig. 2a, b). Smaller numbers of *Ixodes canisuga* Johnston were recorded, less than 0.78 % of the dogs were infected by this species, and these tended to have a more southerly distribution in England and Wales with none submitted from the northern half of Scotland (Fig. 2c). Ten dogs were found to be infected with *D. reticulatus*, largely from populations in western Wales and south-west England (Fig. 2d). The one case recorded in north-east England, was from a dog known to have travelled to Wales in the previous week. Three dogs were infected by *Haemaphysalis punctata* Canestrini & Fanzago;

all were from one specific location in south east England (not plotted). Five dogs with mixed species infestations were detected, in three of these cases *I. ricinus* was found along with *I. hexagonus*, whereas in other two *I. ricinus* was found with *I. canisuga* and *D. reticulatus*.

Table 1 The number and percentage of dogs that had not travelled outside the UK in the previous 2 weeks, infested by each species of tick as submitted by veterinary practices that participated in the UK survey

Tick species	**Number of dogs**	**Percentage**
Ixodes ricinus	5236	89.2
Ixodes hexagonus	577	9.8
Ixodes canisuga	46	0.78
Haemaphysalis punctata	3	0.05
Dermacentor reticulatus	10	0.17
Total number of dogs infested by endemic ticks	5872	

Figure. 2. Distributions of the ticks *Ixodes ricinus* (**a**), *Ixodes hexagonus* (**b**), *Ixodes canisuga* (**c**) and *Dermacentor reticulatus* (**d**) in samples submitted by the veterinary practices in the UK survey

Fifty-six dogs had travelled outside the UK in the previous 2 weeks (Table 2); 29 were infected with *I. ricinus*. Thirteen were infected by *Rhipicephalus sanguineus* Latrille all of which were found on animals that had entered from the Mediterranean region (predominantly Cyprus and Spain). One dog with *Dermacentor variabilis* (Say) was detected, on an animal imported from the USA.

Table 2. Amongst the samples submitted by veterinary practices, 56 dogs had travelled outside the UK in the previous 2 weeks, 43 of which were infested by one of three species of tick

Tick species	Number of travelled dogs infested (%)
Ixodes ricinus	29 (67.4)
Dermacentor variabilis	1 (2.3)
Rhipicephalus sanguineus	13 (30.2)
Total number with ticks	43

Host Associations and Risk Factors

Logistic regression was used to consider the effects of various risk factors on the likelihood that dogs may get bitten by ticks. This analysis showed that breed, neutered status and age significantly predicted the likelihood of dogs having ticks, but visits to kennels and exposure to different habitats did not (Table 3). Pastoral and Gundogs were the breeds most likely to have ticks ($R^2=0.037$, $df_{9, 17}$, $P<0.001$); neutered male and female dogs were at lower risk of tick infestation than male or female unneutered dogs ($R^2=0.037$, $df_{3, 17}$, $P<0.001$). All age groups were significantly more likely to have ticks than dogs of 1 year-old or below ($R^2=0.037$, $df_{5, 17}$, $P<0.001$).

Table 3. Significance, odds ratios and 95 % confidence intervals (CIs) of the logistic regression between presence and absence of ticks and an array of significant tick risk factors

	Significance	Odds ratio	95 % CI
Breed type			
Pastoral	0.009	1.809	1.162–2.817
Gundogs	0.026	1.619	1.060–2.471
Dog sex			

Male neutered	< 0.0001	0.638	0.553–0.736
Female neutered	< 0.0001	0.570	0.494–0.659
Dog age			
1 to 3 years	< 0.0001	1.400	1.184–1.655
3 to 6 years	0.001	1.331	1.128–1.571
6 to 10 years	0.036	1.202	1.012–1.427
Above 10 years	0.014	1.288	1.052–1.577

Hosmer-Lemeshow test: $\chi^2=2.117$, $df=8$, $P=0.977$

Prevalence of Tick Attachment

Prevalence estimation requires random selection of dogs without bias towards dogs known to have attached ticks. However, while many participating veterinary practices, as expected, sent more negative questionnaire reports than positive, some veterinary practices sent questionnaire reports only from infested dogs. It is therefore likely that these practices misunderstood the study protocol and only submitted reports when ticks were found. All data submitted from practices that sent only positive samples from any single week in which it participated in the survey were not included in the prevalence analysis. Data from practices that submitted reports from three or fewer dogs any week were also removed, since it was considered unlikely that any practice would see fewer than three dogs in a week. This removal of inspection records from practices where miss- or over-reporting was suspected, resulted in the removal of 4994 dogs. Following this, the total number of dog records remaining was 7102 and of these 2182 of the dogs had ticks. The prevalence of ticks on dogs over the entire 16-week period was thus 30.7 % (95 % exact binomial confidence interval ± 0.011) (Table 4). Calculation of the prevalence over each four-week period of the study showed little variation, with a range of 28–32 % (Table 4). The prevalence data were mapped by county to give a visual indication of geographical differences (Fig. 3). Where there were small sample sizes from any one county, counties were merged to give a minimum sample of 200 cases per reporting area and a pooled prevalence was plotted. Prevalence was then mapped on a relative scale of 1 to 5. The data show that the highest prevalence of tick infestation are in south west England, East Anglia and Scotland, but are also high throughout most of central and northern England.

Table 4. The number of dogs inspected and the number found to have at least one tick attached in each 4 week period of the 16 week study, in samples submitted by veterinary practices in the UK survey

Time period (weeks)	Number of dogs	Number of tick infested dogs	Prevalence (%)	95 % confidence interval
1–4	287	81	28.2	0.052
5–8	1784	503	28.2	0.021
9–12	2938	959	32.6	0.017
13–16	2093	639	30.5	0.019
1–16	7102	2182	30.7	0.011

The estimated percentage prevalence with exact binomial 95 % confidence intervals are also presented

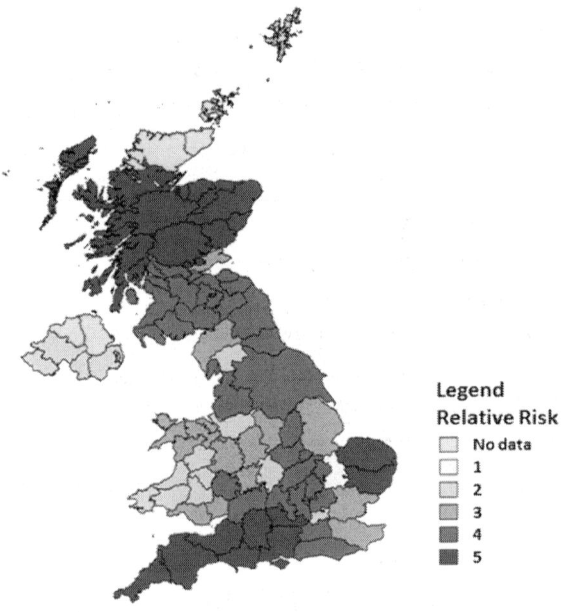

Figure 3. The relative risk of tick attachment on dogs on a scale of 1 to 5, based on the prevalence of ticks found in different regions of the UK in samples submitted by the veterinary practices in the UK survey

DISCUSSION

The three tick species most commonly encountered on humans, livestock and companion animals in the UK are *Ixodes ricinus*, *I. hexagonus* and *I.*

canisuga; of these, *I. ricinus* is the most widely distributed and poses the greatest biting risk [16]. This species is the known principal vector for Lyme borreliosis and *Anaplasma*, louping ill virus (LIV) and various species of *Babesia* [2, 6, 30]. In mainland Europe *I. ricinus* also acts as a vector of tick-borne encephalitis (TBE) in humans and *Ehrlichia canis*; the latter has been recorded recently for the first time in a non-travelled dog in the UK [31]. In a previous study, based on 56 veterinary practices and 280 animals in the UK, 52 % of dogs and cats with ticks carried *I. ricinus*, 39 % carried *I. hexagonus*, and 11 % were infested with *I. canisuga* [32]. More recently, a study involving 180 veterinary practices found that 72.1 % of tick infestations on domestic dogs were due to *I. ricinus*, 22 % due to *I. hexagonus* and 6 % due to *I. canisuga* [17]. In the present study, data collected from over 1000 veterinary practices and 12,000 dogs showed that the proportion of *I. ricinus* was considerably higher than in earlier studies at 89 %, while *I. hexagonus* was the second most abundant species (10 %) and *I. canisuga* was relatively rare (1 %). Despite differences from study to study, the pattern of abundance of the three species appears relatively consistent; variation may be due to differences in time of year, geographic focus or the sample population examined.

The abundance of each tick species is strongly determined by differences in climate, host availability and vegetation cover, which affects microclimate. *Ixodes ricinus* is most commonly associated with woodland and moorland habitats, although high densities may also be found in urban recreational spaces [33] as with *I. hexagonus* [32]. In the present study, both species were found throughout the UK as far north as Scotland. In contrast, *I. canisuga* has been reported as being more commonly found in boarding kennel environments [27], although infestation by *I. canisuga* was seen here in dogs that had no previous exposure to kennels, as has been reported previously [32]. Hence this species evidently lives in association with wildlife hosts, such as mustelids [27]. As seen in previous studies [17], *I. canisuga* appeared to have a more strongly southerly distribution in the current study.

Risk factors associated with tick attachment are highly inconsistent between studies. Dog breed, sex, age and neutered status were not found to be significant predictors of tick infestation [34]. Earlier work [17] found that hound, toy and utility breeds and neutered dogs had a lower probability of tick attachment, but reported no significant effect of sex. In the current study, older dogs were more likely to have ticks than dogs of 1 year-of-age or below, pastoral and gundogs were at higher risk than other breeds while neutered dogs were at lower risk than unneutered dogs. No effect of sex was detected. It is possible that the highly variable results between studies are due to the fact that tick attachment rate has been shown to be most strongly correlated with exposure rather than any dog physiological or phenotypic characteristics, as demonstrated by Jennet et al. [34]. Surprisingly, dogs that were restricted to urban habitats were no less likely to have ticks that dogs exposed to more rural habitats. This corresponds with the growing number of reports of high numbers of ticks in urban environments [32, 33]. Almost all the ticks found were adults; only 76 were immatures. It is relatively likely that dogs are bitten by a greater proportion of immatures than is apparent from the data presented here, but their absence in the samples received strongly suggests that this life-cycle stage is undetected in clinical examination.

The results of this study show that a large-scale, cost-effective national tick prevalence assessment can be conducted using voluntary enrolment. Publicity and media interest were pivotal to the approach, generating enthusiasm amongst the participating veterinary surgeons. However, there were evident limitations; veterinarians who signed up agreed to inspect a given number of dogs per week and these were to be a random selection of otherwise healthy animals brought to the surgery. In the study reported by Smith et al. [17] which used a similar protocol, 60 veterinary practices participated in tick collection at any one time and incoming samples were monitored closely and veterinary practices were contacted individually by telephone when unexpected patterns were detected to ensure that the protocols were rigorously followed. Given the number of veterinary practices that participated in the current study, this approach could not be adopted and more than half of participating practices sent too few or only positive samples for at least 1 week during their participation. Furthermore, over 200 ticks were sent from cats (plus samples from humans and birds). On the other hand, the very large sample size meant that a more rigorous *post-hoc* approach could be taken to exclude specific categories of return.

After exclusion of returns where only cases from dogs positive for tick infestation were submitted, an overall attachment prevalence on dogs of 30 % was recorded over the 16-week sample period. There was little change in this prevalence figure when sub-divided into 4-week periods (range 28–32 %). This attachment rate is still relatively high; in the study by Smith et al. [30] a median frequency of infestation of 14.9 % was reported, but 19 % of veterinary practices found no ticks and 14.6 % reported that more than 50 % of the dogs inspected carried ticks. In the present study, it is likely that the veterinarians who enrolled were those with the greatest interest in ticks and tick-borne disease or were in practices with a known history of tick problems amongst their clients resulting in over-reporting. Nevertheless, if it is assumed that over-reporting has no regional bias, the data can be used on a relative scale to compare geographical differences in relative risk. Scotland, south-west England and East Anglia showed the highest regional prevalence, although prevalence was also high throughout central and northern England. The relatively low prevalence in Wales was perhaps surprising, given the vegetation and climatic requirements, particularly of *I. ricinus*.

The distribution of *D. reticulatus* is of particular current interest because it is the primary vector of canine babesiosis. Much like *I. ricinus*, *D. reticulatus* is also known to have extended its distribution northwards through mainland Europe [13, 15, 35]. Historical records show that this tick has been found in the UK for over 100 years (http://data.nbn.org.uk). However, in recent years it has become more widely established in southern England and Wales [17, 36]. There are now known to be at least four established predominantly coastal populations: west Wales, south Devon, north Devon and Essex [16, 36]. Other populations may as yet be undiscovered. Here, samples were found in Wales and south-west England, confirming established distribution pattern of this species in these areas, but no cases were submitted from the Essex population. In the UK, there have been an increasing number of cases of babesiosis in dogs that have travelled to Europe [37], with other cases probably unreported [38]. In March 2016, a cluster of cases of canine babesiosis, due to *B. canis*, was reported in Essex in non-travelled dogs confirming that this pathogen is now well established in the UK [39]. This outbreak highlights the urgent need for an improved understanding of the ecology and behaviour of the vector, and in particular, an understanding of its distribution and mechanism(s) of dispersal.

Fifty-six dogs were known to have travelled outside the UK in the 2 weeks prior to their inclusion in the study and 43 of these were found to be carrying attached ticks, predominantly *I. ricinus*. The 13 cases of the tick, *R. sanguineus*, that were detected in travelled dogs, highlight the concern regarding the import and potential establishment of this species following changes to the pet passport scheme in 2012 which had previously required dogs entering the UK to be treated against ticks [26]. This species has now been shown to have overwintered in the UK in at least two locations [26]. The data clearly emphasise the importance of appropriate treatment against ticks for dogs that are travelling and the persistent threat of introduction and establishment of non-endemic ticks and their pathogens into the UK.

CONCLUSIONS

This study has shown how very large samples can be generated through voluntary participation of veterinary surgeries following a high profile media and communication campaign. However, despite a clear protocol for participants, this approach resulted in a prevalence of tick infestation that is considerably higher than seen in previous studies, probably as a result of over-reporting. Nevertheless, the data presented provide a comprehensive spatial understanding of tick distribution and species abundance against which future changes can be compared while the relative prevalence maps show the highest rates in Scotland and south-west England, providing a valuable guide to tick-bite risk in the UK.

REFERENCES

1. Otranto D, Wall R. New strategies for the control of arthropod vectors of disease in dogs and cats. Med Vet Entomol. 2008;22:219–302.
2. Gray JS. The development and seasonal activity of the tick *Ixodes ricinus*: a vector of Lyme borreliosis. Rev Med Vet Entomol. 1991;79:323–333.
3. Korotkov Y, Kozlova T, Kozlovskaya L. Observations on changes in abundance of questing *Ixodes ricinus*, castor bean tick, over a 35-year period in the eastern part of its range (Russia, Tula region) Med Vet Entomol. 2015;29:129–36. doi: 10.1111/mve.12101.
4. Daniel M, Danielová V, Kríz B, Jirsa A, Nozicka J. Shift of the tick *Ixodes ricinus* and tick-borne encephalitis to higher altitudes in central Europe. Eur J Clin Microbiol Inf Dis. 2003;22:327–328.
5. Kirby AD, Smith AA, Benton TG, Hudson PJ. Rising burden of immature sheep ticks (*Ixodes ricinus*) on red grouse (*Lagopus lagopus scoticus*) chicks in the Scottish uplands. Med Vet Entomol. 2004;18:67–70. doi: 10.1111/j.0269-283X.2004.0479.x.
6. Pietzsch ME, Medlock JM, Jones L, Avenell D, Abbott J, Harding P, Leach S. Distribution of *Ixodes ricinus* in the British Isles: investigation of historical records. Med Vet Entomol. 2005;19:306–314. doi: 10.1111/j.1365-2915.2005.00575.x.
7. Jore S, Viljugrein H, Hofshagen M, Brun-Hansen H, Kristoffersen AB, Nygard K, et al. Multi-source analysis reveals latitudinal and altitudinal shifts in range of *Ixodes ricinus* at its northern distribution limit. Parasit Vectors. 2011;4:84–95. doi: 10.1186/1756-3305-4-84.
8. Danielová V, Rudenko N, Daniel M, Holubová J, Materna J, Golovchenko M, Schwarzová L. Extension of *Ixodes ricinus* ticks and agents of tick-borne diseases to mountain areas in the Czech Republic. I J Med Microbiol. 2006;296:48–53. doi: 10.1016/j.ijmm.2006.02.007.
9. Tälleklint L, Jaenson TGT. Increasing geographical distribution and density of *Ixodes ricinus* (Acari: Ixodidae) in Central and Northern Sweden. J Med Entomol. 1998;35:521–526. doi: 10.1093/jmedent/35.4.521.
10. Lindgren E, Tälleklint L, Polfeldt T. Impact of climatic change on the northern latitude limit and population density of the disease-transmitting European tick *Ixodes ricinus*. Env Health Persp. 2000;108:119–23. doi: 10.1289/ehp.00108119.
11. Scharlemann JPW, Johnson PJ, Smith AA, Macdonald DW, Randolph

SE. Trends in ixodid tick abundance and distribution in Great Britain. Med Vet Entomol. 2008;22:238–247. doi: 10.1111/j.1365-2915.2008.00734.x.

12. Zygner W, Górski P, Wedrychowicz H. New localities of *Dermacentor reticulatus* tick (vector of *Babesia canis canis*) in central and eastern Poland. Pol J Vet Sci. 2009;12:549–555.
13. Beugnet F, Marié JL. Emerging arthropod-borne diseases of companion animals in Europe. Vet Parasitol. 2009;163:298–305. doi: 10.1016/j.vetpar.2009.03.028.
14. Dautel H, Dippel C, Kämmer D, Werkhausen A, Kahl O. Winter activity of *Ixodes ricinus* in a Berlin forest. I J Med Microbiol. 2008;298:50–54. doi: 10.1016/j.ijmm.2008.01.010.
15. Matjila TP, Nijhof AM, Taoufik A, Houwers D, Teske E, Penzhorn BL, et al. Autochthonous canine babesiosis in The Netherlands. Vet Parasitol. 2005;131:23–29. doi: 10.1016/j.vetpar.2005.04.020.
16. Jameson LJ, Medlock JM. Tick surveillance in Great Britain. Vector-borne Zoonotic Dis. 2011;11:403–412. doi: 10.1089/vbz.2010.0079.
17. Smith F, Ballantyne R, Morgan E, Wall R. The incidence of tick infestation on domestic dogs in the UK. Med Vet Entomol. 2011;25:377–384. doi: 10.1111/j.1365-2915.2011.00954.x.
18. Lindgren E, Gustafson R. Tick-borne encephalitis in Sweden and climate change. Lancet. 2001;358:16–8. doi: 10.1016/S0140-6736(00)05250-8.
19. Daniel M, Kříž B, Danielová V, Beneš Č. Sudden increase in tick-borne encephalitis cases in the Czech Republic, 2006. I J Med Microbiol. 2008;298:81–87. doi: 10.1016/j.ijmm.2008.02.006.
20. Parmesan C, Yohe G. A globally coherent fingerprint of climate change impacts across natural systems. Nature. 2003;21:37–42. doi: 10.1038/nature01286.
21. IPCC Annex I. Atlas of global and regional climate projections et al. Climate change 2013: the physical science basis. In: van Oldenborgh GJ, Collins M, Arblaster J, Christensen JH, Marotzke J, Power SB, et al., editors. Contribution of working group I to the fifth assessment report of the intergovernmental panel on climate change. Cambridge: Cambridge University Press; 2013.
22. Carpenter S, Wilson A, Barber J, Veronesi E, Mellor P, Venter G, Gubbins S. Temperature dependence of the extrinsic incubation period

of orbiviruses in *Culicoides* biting midges. PLoS One. 2011;6:e27987. doi: 10.1371/journal.pone.0027987.

23. Randolph SE. Evidence that climate change has caused "emergence" of tick-borne diseases in Europe? I J Med Microbiol. 2004;293:5–15.
24. Wall R. Ectoparasites: future challenges in a changing world. Vet Parasitol. 2007;148:62–74. doi: 10.1016/j.vetpar.2007.05.011.
25. Randolph SE, Rogers DJ. The arrival, establishment and spread of exotic diseases: patterns and predictions. Nature Rev Microbiol. 2010;8:361–371. doi: 10.1038/nrmicro2336.
26. Hansford KM, Pietzsch ME, Cull B, Medlock JM, Wall R. Overwintering of the brown dog tick in residential properties in England – raising awareness. Vet Rec. 2015;177:156. doi: 10.1136/vr.h4227.
27. Arthur DR. British ticks. Wallingford: CABI; 1963. p. 218.
28. Hillyard, PD, Ticks of North-west Europe. Synopses of the British Fauna 52;1996. Field Studies Council, Shrewsbury.
29. Walker A, Bouattour A, Camicas J-L, Estrada-Peña A, Horak IG, Latif AA, et al. Ticks of domestic animals in Africa: a guide to identification of species. UK: Bioscience Reports Edinburgh; 2014.
30. Smith FD, Ellse L, Wall R. Prevalence of *Babesia* and *Anaplasma* in ticks infesting dogs in Great Britain. Vet Parasitol. 2013;198:18–23. doi: 10.1016/j.vetpar.2013.08.026.
31. Wilson HE, Mugford AR, Humm KR, Kellett-Gregory LM. *Ehrlichia canis* infection in a dog with no history of travel outside the United Kingdom. J Small Anim Pract. 2013;54:425–7. doi: 10.1111/jsap.12088.
32. Ogden NH, Cripps P, Davison CC, Owen G, Parry JM, Timms BJ, Forbes AB. The ixodid tick species attaching to domestic dogs and cats in Great Britain and Ireland. Med Vet Entomol. 2000;14:332–338. doi: 10.1046/j.1365-2915.2000.00244.x.
33. Nelson C, Banks S, Jeffries CL, Walker T, Logan JG. Tick abundances in South London parks and the potential risk for Lyme borreliosis to the general public. Med Vet Entomol. 2015;29:448–452. doi: 10.1111/mve.12137.
34. Jennett AL, Smith FD, Wall R. Tick infestation risk for dogs in a peri-urban park. Parasit Vectors. 2013;6:358. doi: 10.1186/1756-3305-6-358.

35. Sréter T, Széll Z, Varga I. Spatial distribution of *Dermacentor reticulatus* and *Ixodes ricinus* in Hungary: evidence for change? Vet Parasitol. 2005;128:347–51. doi: 10.1016/j.vetpar.2004.11.025.
36. Medlock JM, Jameson LJ, Phipps LP. Status of *Dermacentor reticulatus* in the UK. Vet Rec. 2011;168:386–387. doi: 10.1136/vr.d2186.
37. Holm LP, Kerr MG, Trees AJ, McGarry JW, Munro ER, Shaw SE. Fatal babesiosis in an untravelled British dog. Vet Rec. 2006;159:179–180. doi: 10.1136/vr.159.6.179.
38. Tasker S. Exotic diseases - a growing concern? J Small Anim Pract. 2013;54:393–394. doi: 10.1111/jsap.12108.
39. Phipps LP, Fernandez De Marco MDM, Hernández-Triana LM, Johnson N, et al. *Babesia canis* detected in dogs and associated ticks from Essex. Vet Rec. 2016;178:243–244. doi: 10.1136/vr.i1265.

5

Use of Big Data in the Surveillance of Veterinary Diseases: Early Detection of Tick Paralysis in Companion Animals

Vanina Guernier[1], Gabriel J. Milinovich[1], Marcos Antonio Bezerra Santos[1,2], Mark Haworth[1], Glen Coleman[1] and Ricardo J. Soares Magalhaes[1,3]

[1]School of Veterinary Science, The University of Queensland, Gatton Campus, Gatton 4343, QLD, Australia.

[2] Federal Rural University of Pernambuco, Academic Unit of Garanhuns, Garanhuns 55292-270, PE, Brazil.

[3] Child Health Research Centre, The University of Queensland, Herston 4006, QLD, Australia.

ABSTRACT

Background

Tick paralysis, resultant from envenomation by the scrub-tick *Ixodes holocyclus*, is a serious threat for small companion animals in the eastern

Citation: Guernier, V., Milinovich, G.J., Bezerra Santos, M.A. et al. (2016) Use of big data in the surveillance of veterinary diseases: early detection of tick paralysis in companion animals. Parasites Vectors 9, 303. https://doi.org/10.1186/s13071-016-1590-6

coast of Australia. We hypothesise that surveillance systems that are built on Internet search queries may provide a more timely indication of high-risk periods more effectively than current approaches.

Methods

Monthly tick paralysis notifications in dogs and cats across Australia and the states of Queensland (QLD) and New South Wales (NSW) were retrieved from *Disease WatchDog* surveillance system for the period 2011–2013. Internet search terms related to tick paralysis in small companion animals were identified using Google Correlate, and corresponding search frequency metrics were downloaded from Google Trends. Spearman's rank correlations and time series cross correlations were performed to assess which Google search terms lead or are synchronous with tick paralysis notifications.

Results

Metrics data were available for 24 relevant search terms at national level, 16 for QLD and 18 for NSW, and they were all significantly correlated with tick paralysis notifications ($P<0.05$). Among those terms, 70.8, 56.3 and 50 % showed strong Spearman's correlations, at national level, for QLD, and for NSW respectively, and cross correlation analyses identified searches which lead notifications at national or state levels.

Conclusion

This study demonstrates that Internet search metrics can be used to monitor the occurrence of tick paralysis in companion animals, which would facilitate early detection of high-risk periods for tick paralysis cases. This study constitutes the first application of the rapidly emerging field of Internet-based surveillance to veterinary science.

BACKGROUND

Tick paralysis is a potentially fatal neurological condition affecting companion animals, livestock and humans. It occurs globally, but the majority of cases are restricted to the Pacific Northwest regions of North America and the eastern coast of Australia [1]. The disease is resultant from exposure to a toxin present in the saliva of paralysis tick species [2–4]. About 69 species of ticks from around the world are capable of inducing paralysis [4], and *Ixodes holocyclus* is the main species in Australia [1, 2, 5].

Annually, *Ix. holocyclus* causes tick paralysis in about 10,000 companion animals [6, 7]. It is considered highly toxic, with one female able to kill a cat or a dog [8]. In Australia, the majority of tick paralysis cases occur between July and December [2], with adult ticks number peaking during spring and early summer [5, 6, 9]. In companion animals, tick paralysis may exhibit a progressive generalised paresis leading to respiratory failure in conjunction with pulmonary oedema and aspiration pneumonia [10, 11]. If left untreated, death can occur within 24 to 48 h of onset of clinical signs [11–13].

Given the acute clinical presentation of the disease and the potentially fatal outcome, reliable surveillance data for the early warning of the occurrence of high-risk periods is essential for the timely preparedness of primary care services. Sentinel surveillance approaches have been used for surveillance of a number of companion animal diseases in Australia [14]. Tick paralysis is among the diseases monitored by the *Disease WatchDog* surveillance system, a prospective national disease surveillance project held in Australia, focusing on dogs and cats [14]. Data are collected from practicing veterinarians as cases of diseases are diagnosed, providing monthly online updates.

In recent years, an emerging field of research has focused on the development of surveillance approaches that utilise various Internet-based data streams to analyse queries from the Internet search engines to predict high-risk periods [15–17]. This approach works on the premise that people who contract a disease are likely to actively seek for information on the symptoms and/or disease on the Internet, and that reliable and timely estimates of diseases activity in the community can be produced by monitoring the frequency of specific Internet search terms [16]. This approach has been documented by numerous studies and produced useful tools for human communicable disease surveillance that work in near real-time and assist the allocation of public health interventions [15, 16, 18, 19].

Recent research has suggested that Internet-based surveillance systems have wider epidemiological applicability for human infectious disease surveillance than has previously been recognised [16]. However, there is no research reporting the use of this approach in veterinary medicine. The present study aims to examine the utility of digital epidemiological approaches based on Internet searches as a means of surveillance of tick paralysis in companion animals in Australia. The specific objectives are twofold: (i) to investigate the statistical correlation between tick paralysis notifications from endemic regions in Australia and search term metrics extracted from Google Trends; and (ii) to demonstrate the utility of digital surveillance systems, specifically systems built on Internet search metrics, as tools for use in veterinary science.

METHODS

Tick Paralysis Data

Data on tick paralysis cases used in this study were extracted from *Disease WatchDog* surveillance website (http://www.diseasewatchdog.org). Information available on the *Disease WatchDog* site includes suburb level disease incidence data and monthly updates for registered users [14]. Monthly reports of tick paralysis cases were obtained for a three-year period, spanning 01 January 2011 to 31 December 2013; we focused our analysis on data from this period as it represented the time period with more reliable data series. The reasons behind this decision are two-fold. First, from 01 January 2011 Google Trends data have undergone an improvement to geographical assignment. As the disease watchdog started reporting from September 2010, we decided to discard the 2010 data (only 4 months' worth of data) to ensure we only used the best from Google. Secondly, upon downloading data from *Disease WatchDog* website we noted that reporting of tick paralysis from January 2014 to the database became inconsistent compared to previous years and we decided to utilise data up until December 2013 to ensure we only used the best from *Disease WatchDog*. Monthly tick paralysis cases were considered only for Queensland (QLD) and New South Wales (NSW) as these two states account for the vast majority of cases reported in Australia (98 %).

Google Search Term Selection and Collection of Search Metric Data

We analysed the words (i.e. search terms) that people used in their searches of the World Wide Web when seeking information on tick paralysis. Counts (i.e. metrics) of these search terms were taken from the Google Trends website (www.google.com/trends). Search metrics are provided as normalised time series; all data are presented as values ranging from 0 to 100, with 100 representing the time point with the highest value (during which the most searches were performed) and other points scaled accordingly. Methods for the identification of relevant search terms, collection of data and the use of these data have previously been described [16]. Briefly, search terms of interest were identified using a two-step approach. Firstly, time series of paralysis tick notifications downloaded from the *Disease WatchDog* website were used to query Google Correlate (www.google.com/trends/correlate). This allows identifying up to 100 search terms with the highest correlation to the three queried time series (i.e. time series related to Australia, QLD, and NSW). Second, search metrics for to the final list of terms were downloaded from the Google Trends website; metrics were collected at national level and at state level for QLD and NSW. Whilst Google Trends provides a function to download.csv text files of search metrics, the level of data aggregation (daily, weekly, monthly) cannot be specified by the user [16]. As such, the required data was "scraped" from Google Trends website using a previously developed script, allowing the problems associated with the level of aggregation to be overcome [16].

Statistical Analyses

Spearman's rank correlations were performed between monthly notifications of tick paralysis cases submitted to the *Disease WatchDog* website and Google Trends search metrics for all terms with available data; these were performed at both state and national level, allowing the search metrics with the highest degree of correlation with notification data to be identified. A strong correlation was defined as $0.60 \leq$ Spearman's rho < 0.80 and a very strong correlation was defined as rho ≥ 0.80. Bonferroni correction was applied to significance levels to account for multiple testing. All reported *P*-values in this document correspond to one-tailed tests.

Linear associations between Google search metrics and tick paralysis notifications were assessed using cross correlations. Cross-correlations were conducted for Google search terms that exhibited strong or very strong correlations (rho ≥ 0.60) with tick paralysis notifications for Australia, QLD and NSW separately. Cross-correlations were performed with lag values ranging from −7 to +7 using IBM SPSS version 21 software [20]. Cross-correlations should be interpreted as product-moment correlations between search metrics and tick paralysis notifications. This approach allows dependence to be identified between the two time series with respect to temporal offsets, referred to as lags, and to visually judge whether one series tends to lead or lag the other. A lag value of +1 describes correlations that have been performed using Google search data that has been shifted forward (i.e. occurred after tick paralysis notifications) one unit (i.e. one month); conversely, a negative lag indicates that the Google search data has been shifted backwards by the number of units denoted, and thereby occurred before tick paralysis notifications data. Within the context of this work, cross-correlations with significant negative lag values are of most interest as they demonstrate that Google search metrics data are (i) significantly associated with the tick paralysis notifications and (ii) lead the clinical notification by a number of months. Synchronous cross-correlations (i.e. significant positive cross-correlation at lag 0) are also of interest as they demonstrate Google search terms that are coincident with tick paralysis reports.

RESULTS

Tick Paralysis Notifications and Google Trends Data

Between 1st January 2011 and 31st December 2013, a total number of 8,414 tick paralysis cases were registered nationally in the *Disease WatchDog* website in both canines and felines; a total of 4,891 (58 %) cases were reported as originating in NSW and 3,523 (42 %) in QLD (Fig. 1). The figure indicates that each year the rise and peak of tick paralysis notifications in QLD occurs between September-October while in NSW it occurs in October-November. This difference is statistically significant, cross-correlations analyses between the two states time series showing that the peak of notifications in QLD leads the peak in NSW by one month. In total, 31 search terms were identified as having potential relevance to tick paralysis (Table 1). Search metrics associated to these terms were downloaded from the Google Trends website; they were available for 24 of the 31 identified terms for Australia, 16 for QLD and 18 for NSW.

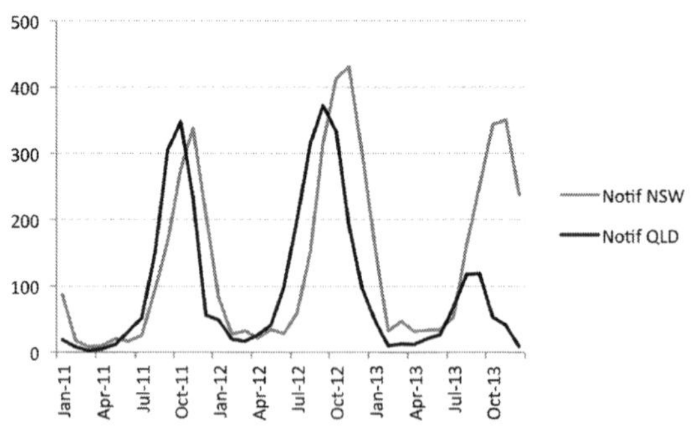

Figure 1. Tick paralysis cases notified on the *Disease WatchDog* website. Monthly notifications originating in the states of New South Wales or Queensland, 2011–2013

Table 1. Spearman's rho rank for the search terms related to tick paralysis in Google Trends

	Australia		QLD		NSW	
Search term	**rho**	***P*-value**	**rho**	***P*-value**	**rho**	***P*-value**
Paralysis tick	0.95[a]	< 0.01	0.81[a]	< 0.01	0.89[a]	< 0.01
Tick paralysis	0.91[a]	< 0.01	0.92[a]	< 0.01	0.89[a]	< 0.01
Ticks	0.83[a]	< 0.01	0.82[a]	< 0.01	0.83[a]	< 0.01
Tick	0.81[a]	< 0.01	0.83[a]	< 0.01	0.87[a]	< 0.01
Paralysis ticks	0.85[a]	< 0.01	0.68[b]	< 0.01	0.66[b]	< 0.01
Dog tick	0.76[b]	< 0.01	0.59	< 0.01	0.80[a]	< 0.01
Dogs ticks	0.72[b]	< 0.01	0.88[a]	< 0.01	0.70[b]	< 0.01
Removing ticks	0.70[b]	< 0.01	0.65[b]	< 0.01	0.65[b]	< 0.01
Dog ticks	0.67[b]	< 0.01	0.40	0.016	0.60[b]	< 0.01
Remove tick	0.78[b]	< 0.01	0.78[b]	< 0.01	0.52	< 0.01
Cat tick	0.75[b]	< 0.01	0.72[b]	< 0.01	0.42	< 0.01
Paralysis tick dogs	0.79[b]	< 0.01	0.40	0.015	0.37	0.020
Dog paralysis tick	0.79[b]	< 0.01	0.45	< 0.01	0.43	< 0.01
Paralysis tick dog	0.71[b]	< 0.01	0.37	0.026	0.44	< 0.01
Cats ticks	0.70[b]	< 0.01	0.40	0.015	0.38	0.020
Tick on dog	0.61[b]	< 0.01	0.36	0.026	0.34	0.037
Paralysis tick symptoms	0.77[b]	< 0.01	No data		0.44	< 0.01
Tick in dog	0.58	< 0.01	No data		0.43	< 0.01
Paralysis tick treatment	0.48	0.030	No data		No data	
Tick paralysis in dogs	0.46	< 0.01	No data		No data	
Tick symp-toms dog	0.43	< 0.01	No data		No data	
Paralysis tick humans	0.39	0.018	No data		No data	
Dog tick treat-ment	0.36	< 0.01	No data		No data	
Paralysis tick Australia	0.36	0.030	No data		No data	

Brown dog tick	No data		No data		No data	
Dog tick removal	No data		No data		No data	
Paralysis tick cat	No data		No data		No data	
Paralysis tick cats	No data		No data		No data	
Tick paralysis in humans	No data		No data		No data	
Tick paralysis in cats	No data		No data		No data	
Ticks and dog paralysis	No data		No data		No data	

[a]Top ranked terms that have very strong Spearman's correlation with the *Disease WatchDog* data (rho$\geq$0.8)

[b]Terms that have strong Spearman's correlation (rho$\geq$0.6); other terms are correlated in a lesser degree. *P*-values for all search terms with data available were $P < 0.05$, which are represented beside each column

No data, the metrics for that term were not available

Correlations between Google search terms and Tick Paralysis Cases

Our results demonstrate that all Google search terms analysed were significantly correlated to tick paralysis notifications in Australia, QLD and NSW (P<0.05; Bonferroni corrected) (Table 1). Of the 24 terms analysed at the national level, 12 terms exhibited a strong correlation and five exhibited a very strong correlation with the number of tick paralysis notifications. In QLD, out of the 16 Google search terms, four presented strong correlation and five very strong correlations. In NSW, out of 18 Google search terms, four exhibited strong correlations and five exhibited very strong correlations (Table 1). The temporal relationships between Google search metrics and tick paralysis notifications for all search terms that exhibited strong or very strong correlations (rho$\geq$0.60) are presented in Figs. 2, 3 and 4.

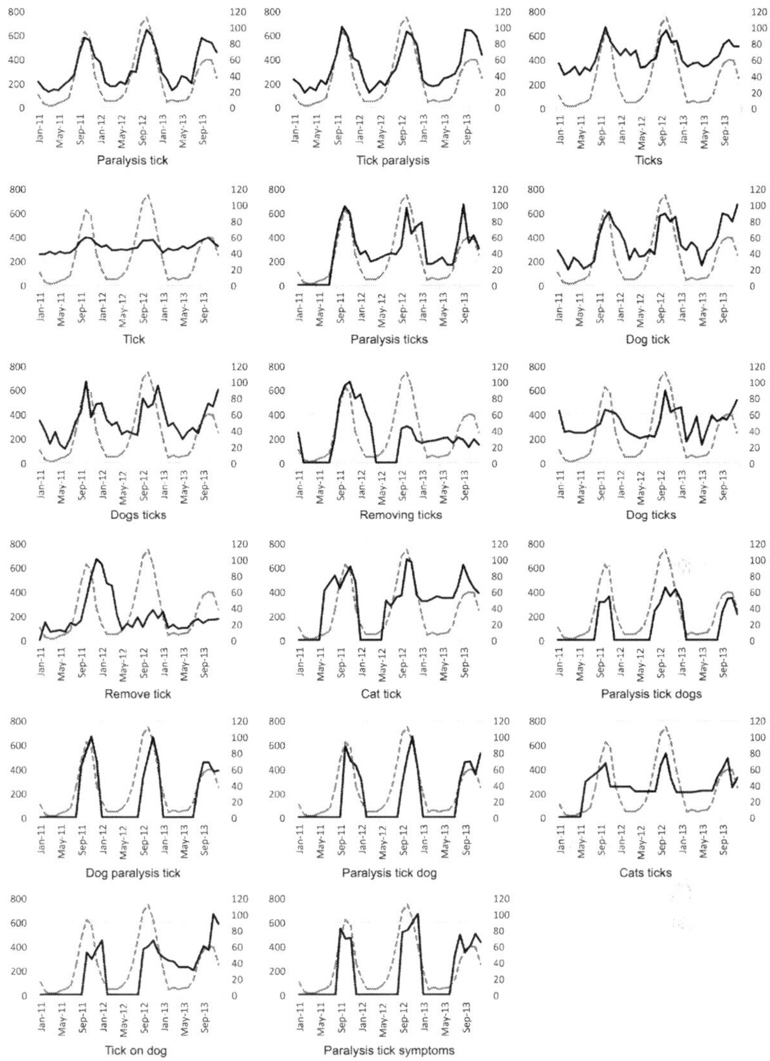

Figure 2. Spearman's top ranked search terms at the national level. Gray dotted lines show the monthly notifications of tick paralysis cases in Australia, while black lines show the metrics for each of the 17 search terms (on 24) that were ranked as having strongly and very strongly significant Spearman's correlations (rho≥0.6)

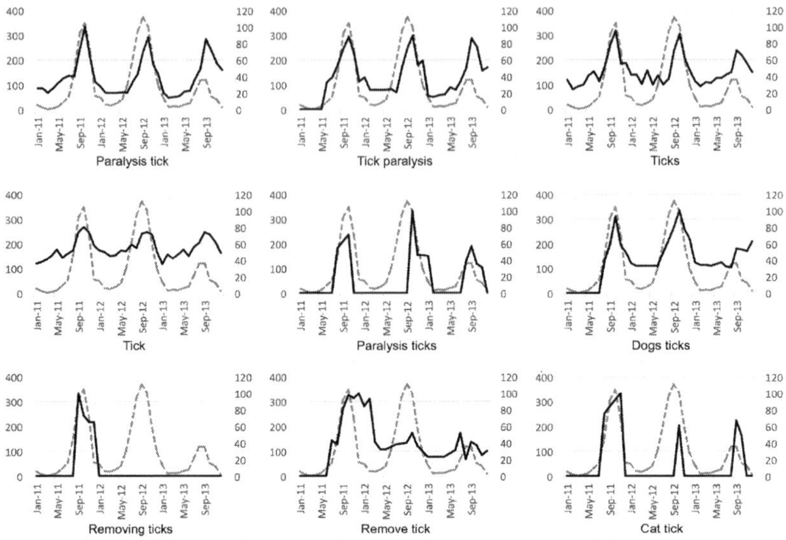

Figure 3. Spearman's top ranked search terms at the state level, QLD. Gray dotted lines show the monthly notifications of tick paralysis cases in QLD, while black lines show the metrics for each of the nine search terms (on 16) that were ranked as having strongly and very strongly significant Spearman's correlations (rho≥0.6)

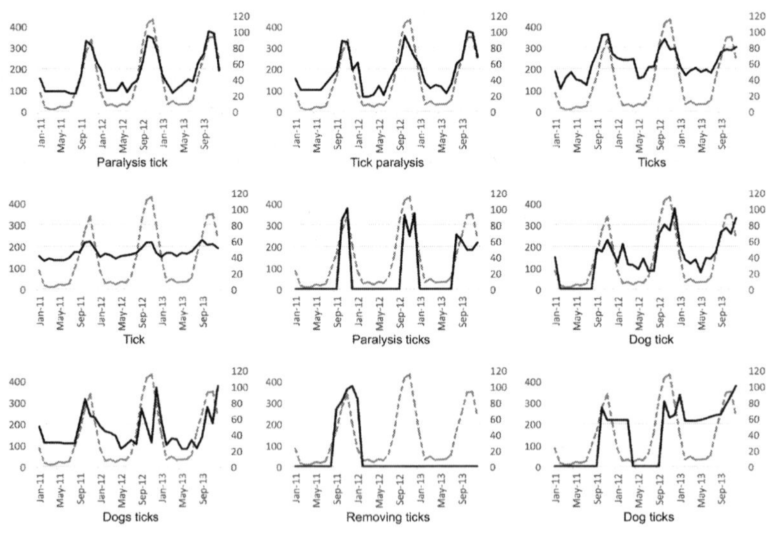

Figure 4. Spearman's top ranked search terms at the state level, NSW. Gray dotted lines show the monthly notifications of tick paralysis cases in NSW, while black lines show the metrics for each of the nine search terms (on 18)

that were ranked as having strongly and very strongly significant Spearman's correlations (rho≥0.6)

Cross-correlations between Google search terms and Tick Paralysis Cases

The highest cross-correlation values, i.e. for which the two times series are best aligned, occur for lag −1, 0 or +1 (in bold, Tables 2 and 3), meaning that the peak of Google searches occur from one month before to one month after the disease notifications. At national level (Table 2), 8/17 terms peak at lag 0 (peak of Google searches coincident with notifications), 8/17 terms peak at lag +1 (Google searches occur one month after notifications) and 1/17 terms, i.e. "cats ticks", peak at lag −1 (Google search occur one month before notifications). In QLD (Table 3), 6/9 terms peak at lag 0, while the other three peak at lag +1. In NSW (Table 3), 6/9 terms peak at lag 0, two at lag +1 and one, i.e. "removing ticks", is not significant (for positive cross-correlations).

Table 2. Cross-correlation results for Australia

	(−) Values: Google searches lag notifications					**(+) Values: notifications lag Google searches**			
Search terms	**–4**	**–3**	**–2**	**–1**	**0**	**1**	**2**	**3**	**4**
Paralysis tick[b]	–0.388	–0.116	0.264	*0.652*	**0.899**	*0.86*	*0.56*	0.132	–0.261
Tick paralysis[b]	–0.381	–0.126	0.24	*0.632*	**0.871**	*0.82*	*0.511*	0.077	–0.313
Ticks[b]	–0.428	–0.207	0.18	*0.616*	**0.847**	*0.795*	*0.507*	0.187	–0.095
Tick[b]	–0.322	–0.084	0.276	*0.649*	**0.814**	*0.685*	*0.36*	0.006	–0.267
Paralysis ticks[b]	–0.363	–0.121	0.239	*0.575*	**0.786**	*0.729*	*0.49*	0.209	–0.065
Dog tick[a]	–0.474	–0.299	0.052	*0.459*	0.756	**0.829**	*0.637*	0.326	–0.021
Dogs ticks[a]	–0.554	–0.375	–0.055	*0.357*	0.641	**0.785**	*0.724*	*0.493*	0.177
Removing ticks[a]	–0.577	–0.437	–0.129	0.212	0.475	**0.547**	*0.476*	0.322	0.128
Dog ticks[a]	–0.527	–0.362	–0.07	0.286	0.607	**0.758**	*0.624*	0.346	–0.011
Remove tick[a]	–0.394	–0.284	–0.082	0.186	0.426	**0.545**	*0.473*	0.279	0.008
Cat tick[a]	–0.125	0.183	*0.44*	*0.62*	**0.692**	*0.582*	0.268	–0.087	–0.355

Paralysis tick dogs[a]	–0.405	–0.184	0.194	*0.589*	**0.864**	*0.837*	*0.579*	0.23	–0.109
Dog paralysis tick[a]	–0.44	–0.261	0.092	*0.523*	*0.836*	**0.864**	*0.562*	0.138	–0.234
Paralysis tick dog[a]	–0.429	–0.282	–0.002	*0.414*	*0.746*	**0.844**	*0.637*	0.239	–0.145
Cats ticks[a]	–0.053	0.229	*0.497*	**0.694**	*0.677*	0.472	0.177	–0.088	–0.253
Tick on dog[a]	–0.436	–0.332	–0.117	0.198	*0.515*	**0.681**	*0.649*	*0.419*	0.121
Paralysis tick symptoms[a]	–0.354	–0.134	0.222	*0.598*	**0.807**	*0.721*	*0.425*	0.065	–0.253
Confidence Limits	±0.354	±0.348	±0.342	±0.338	±0.334	±0.338	±0.342	±0.348	±0.354

The cross-correlation values are shown only for the search terms that were strongly ([a]) or very strongly ([b]) significant (Spearman's rho$\geq$0.6). Significant positive cross-correlation values, i.e. above the upper confidence limit, are in italic, and highest values (i.e. for which the two times series are best aligned) are in bold

Table 3. Cross-correlation results for the states of QLD and NSW

	(–) Values: Google searches lag notifications					**(+) Values: notifications lag Google searches**			
Search terms	**–4**	**–3**	**–2**	**–1**	**0**	**1**	**2**	**3**	**4**
QLD									
Paralysis tick[b]	–0.304	–0.173	0.073	*0.445*	**0.738**	*0.705*	*0.391*	–0.018	–0.351
Tick paralysis[b]	–0.299	–0.106	0.161	*0.492*	**0.733**	*0.698*	*0.445*	0.091	–0.22
Ticks[b]	–0.282	–0.206	0.014	*0.411*	*0.719*	**0.723**	*0.453*	0.097	–0.233
Tick[b]	–0.235	–0.121	0.129	*0.470*	**0.758**	*0.724*	*0.411*	–0.007	–0.332
Paralysis ticks[a]	–0.376	–0.254	–0.017	0.3	**0.542**	*0.523*	*0.386*	0.195	–0.018
Dogs ticks[b]	–0.39	–0.205	0.1	*0.485*	*0.749*	**0.781**	*0.597*	0.32	0.047
Removing ticks[a]	–0.241	–0.208	0.009	0.292	**0.464**	*0.46*	0.286	0.054	–0.148
Remove tick[a]	–0.267	–0.149	0.067	0.293	*0.467*	**0.52**	*0.443*	0.316	0.125
Cat tick[a]	–0.268	–0.151	0.087	*0.355*	**0.581**	*0.511*	0.219	–0.111	–0.292
NSW									
Paralysis tick[b]	–0.312	–0.076	0.324	*0.752*	**0.941**	*0.787*	*0.409*	–0.02	–0.325
Tick paralysis[b]	–0.31	–0.019	*0.38*	*0.768*	**0.932**	*0.776*	*0.428*	–0.003	–0.326
Ticks[b]	–0.401	–0.103	0.328	*0.7*	**0.844**	*0.682*	*0.355*	0.056	–0.172
Tick[b]	–0.174	0.143	*0.512*	*0.813*	**0.853**	*0.588*	0.212	–0.098	–0.299
Paralysis ticks[a]	–0.293	–0.091	0.224	*0.621*	**0.805**	*0.664*	0.331	–0.008	–0.268

Dog tick[b]	–0.301	–0.092	0.242	*0.577*	**0.798**	*0.797*	*0.559*	–.277	–0.016
Dogs ticks[a]	–0.439	–0.25	0.049	*0.378*	*0.619*	**0.697**	*0.54*	0.258	–0.023
Removing ticks[a]	–0.335	–0.233	–0.045	0.128	0.25	0.284	0.186	–0.002	–0.157
Dog ticks[a]	–0.329	–0.254	–0.051	0.253	*0.514*	**0.632**	*0.594*	*0.434*	0.236
Confidence Limits	±0.354	±0.348	±0.342	±0.338	±0.334	±0.338	±0.342	±0.348	±0.354

The cross-correlation values are shown only for the search terms that were strongly ([a]) or very strongly ([b]) significant (Spearman's rho$\geq$0.6). Significant positive cross-correlation values, i.e. above the upper confidence limit, are in italic, and highest values (i.e. for which the two times series are best aligned) are in bold

If the peak of Google searches and the peak of notifications occur mostly simultaneously, a positive correlation between Google searches and notifications (in italic, Table 2) generally appear earlier, i.e. before the peak of notifications. Google searches can anticipate the notifications of tick paralysis up to two months (lag −2), depending on the Google search term (Tables 2 and 3). Our results indicate that Google searches significantly lead tick paralysis notifications for 13/17 significant terms for Australia (Table 2) (two months before for two terms, i.e. "cat tick" and "cats (cat's) ticks"; one month before for the others), 6/9 for QLD and 7/9 for NSW (Table 3). Significant search terms common to Australia and both states are: "paralysis tick", "tick paralysis", "ticks", "tick" and "dogs (dog's) ticks". For all five terms, Google search terms lead tick paralysis notifications by one month at national and state levels, except in NSW where "tick paralysis" and "tick" lead tick paralysis notifications by two months. The other search terms that significantly lead tick paralysis notifications but are specific to national or state levels are listed in Tables 2 and 3. Our results also indicate that in Australia (Table 2), the Google search terms "removing ticks", "dog ticks", "remove tick" and "tick on dog" are synchronous with tick paralysis notifications (significant positive cross-correlation at lag 0). In QLD, the Google search terms "paralysis ticks", "removing ticks" and "remove tick" are synchronous with tick paralysis notifications, whereas in NSW, the Google search term "dog ticks" only is synchronous with notifications (Table 3).

DISCUSSION

This study investigated the potential of Internet-based surveillance systems for the monitoring and early detection of tick paralysis in companion animals across Australia by comparing Internet search metrics with three years of notification data from the passive surveillance system *Disease WatchDog*. The analysis of these data has enabled us to identify a limited set of Google search terms of significant epidemiological importance for the monitoring and early detection of tick paralysis cases in companion animals in Australia.

Our findings are significant for three main reasons. First, previous studies in eastern Australia reported a seasonal abundance of young adult ticks during spring and early summer in QLD and NSW [6, 21, 22], coincident with a peak in tick paralysis notifications in Australia in spring, with 65.5 % of all cases being reported from September to November [2]. Our analysis of the time-series of monthly notifications of tick paralysis in dogs and cats extends current knowledge in that it suggests that the seasonal pattern differs between the eastern states of QLD and NSW; our results show that tick paralysis notifications in QLD rise and peak one month earlier compared to NSW (i.e. September-October in QLD *vs* October-November in NSW). The gap of one month that we report here between QLD and NSW could partly be explained by a temporal lag in optimal weather and environmental conditions for the proliferation of ticks between the two states [2, 23, 24].

Secondly, our study demonstrates that a tick paralysis surveillance system based on Internet search data may have great potential for the development of prevention and control strategies for the disease. We could highlight five search terms common to Australia, QLD and NSW which lag tick paralysis notifications by one or two months. Other terms were shown to be more specific to a geographic level but strongly significant, and anticipate notifications by one or two months. Five search terms were significant at the national level only, which could be related to small sample sizes at state level. Four search terms were best correlated with notifications at lag 0 at national and state levels, referring to synchronicity between Google searches and notifications. Two terms, i.e. "removing ticks" and "remove tick", were best correlated with notifications at lag +1 for Australia and QLD but were already significant at lag 0. The use of those search terms exhibits particular importance as they can be used to develop monitoring systems, and pinpoint the need for educating pet owners about how to remove ticks and/or for the distribution of tick removal tools.

Finally, the results of this study have important public health implications in that early detection of the seasonal increase of tick paralysis in states of Australia (potentially at different times) would allow veterinarians to anticipate high-risk periods, enabling them to prepare in a timely manner as case load escalates. Access to this sort of information would also allow clinics to provide improved, more accurate, evidence-based advice/reminders to owners on risk of tick envenomation and vigilance of prophylaxis. Indeed, vigilance of prophylaxis is even more important as client compliance in prophylaxis application has been demonstrated to be poor [10]. Moreover, as this surveillance is based on automated data acquisition, it is inexpensive and may avoid the reporting bias associated to conventional passive surveillance systems. Furthermore, despite affecting mostly dogs and cats, tick paralysis has been reported in humans and livestock, and is a potentially fatal illness of importance to both human and veterinary medicine [1, 4, 6, 9, 25]. Thus, the monitoring of tick paralysis in companion animals may present positive effects not only in the control of the disease in animals, but also in humans.

Some limitations need to be considered in this study. Firstly, we only had access to monthly data on tick paralysis through the *Disease WatchDog* system, when daily or weekly reports, which are commonly used in traditional surveillance systems, could potentially reveal more fine-grained, clinically relevant, trends [16]. Secondly, we aggregated tick paralysis cases from dogs and cats due to insufficient number of cases in cats; however, in future studies data from dogs and cats should be disaggregated to enable the identification of species-specific epidemiological characteristics of tick paralysis cases. Lastly, the study analysed notifications only for the three years that had relevant available data (2011–2013) totalizing 36 months, which is a low number of observations. In our study, the time-series of *Disease WatchDog* tick paralysis notifications showed a strong decrease of online notifications within and after 2013 compared to previous years which is very likely an artefact brought about by less compliance by users of the system [26]. It is therefore important to find complimentary data sources that can be used to monitor long-term trends in the number of tick paralysis cases. Despite these shortcomings, the results of this study demonstrate clear potential for digital surveillance approaches in monitoring not just tick paralysis, but diseases of veterinary significance in general.

CONCLUSION

In summary, the present study demonstrates that the monitoring and early detection of high-risk periods for tick paralysis cases in companion animals in the community could be performed through Internet-based surveillance using a limited set of Google search terms. Moreover, to our knowledge, this is the first study that demonstrates the applicability of an Internet-based surveillance approach to a disease of veterinary significance. Future work towards building of a relevant and effective real-time model of tick paralysis cases in companion animals should consider combining Google Search data identified in the present paper to other data on environmental factors such as climate time series on rainfall and temperature and physical environment time series such as landcover and landuse.

REFERENCES

1. Edlow JA. Tick paralysis. Curr Treat Options Neurol. 2010;12:167–77.
2. Eppleston KR, Kelman M, Ward MP. Distribution, seasonality and risk factors for tick paralysis in Australian dogs and cats. Vet Parasitol. 2013;196:460–8.
3. Gothe R, Neitz AWH. Tick paralysis: pathogenesis and aetiology. Adv Vector Res. 1991;8:177–204.
4. Hill CA, MacDonald JF. The biology and medical importance of ticks in Indiana. Purdue Ext Entomol. 2013;243:1–7.
5. Masina S, Broady KW. Tick paralysis: development of a vaccine. Int J Parasitol. 1999;29:535–41.
6. Barker SC, Walker AR. Ticks of Australia. The species that infest domestic animals and humans. Zootaxa. 2014;3816:1–144.
7. Stone BF, Aylward JH. Tick toxicoses and the causal toxins: tick paralysis. Proceedings of the First Asia–Pacific Congress on Animal, Plant and Microbial Toxins. 1987. p. 594–602.
8. Hall-Mendelin S, O'Donoghue P, Atwell RB, Lee R, Hall RA. An ELISA to detect serum antibodies to the salivary gland toxin of *Ixodes holocyclus* Neumann in dogs and rodents. J Parasitol Res. 2011;2011:e283416.
9. Ottaway S, Newby K. Paralysis ticks. NSW Agric Agnote. 2003.
10. Campbell FE, Atwell RB. Heart failure in dogs with tick paralysis caused by the Australian paralysis tick, *Ixodes holocyclus*. Int J Appl Res Vet Med. 2003;1:148–62.
11. Ilkiw JE, Turner DM, Howlett CR. Infestation in the dog by the paralysis tick *Ixodes holocyclus*. 1. Clinical and histological findings. Aust Vet J. 1987;64:137–9.
12. Atwell RB, Campbell FE, Evans EA. Prospective survey of tick paralysis in dogs. Aust Vet J. 2001;79:412–8.
13. Webster RA, Mackie JT, Haskins SC. Histopathological changes in the lungs from dogs with tick paralysis: 25 cases (2010–2012). Aust Vet J. 2013;91:306–11.
14. Ward MP, Kelman M. Companion animal disease surveillance: a new solution to an old problem? Spat Spatiotemporal Epidemiol. 2011;2:147–57.

15. Eysenbach, G. Infodemiology: Tracking flu-related searches on the Web for syndromic surveillance. AMIA Annual Symposium Proceedings. 2006; pp. 244–248
16. Milinovich GJ, Avril SMR, Clements ACA, Brownstein JS, Tong S, Hu W. Using internet search queries for infectious disease surveillance: screening diseases for suitability. BMC Infect Dis. 2014;14:690.
17. Wilson K, Brownstein JS. Early detection of disease outbreaks using the Internet. Can Med Assoc J. 2009;180:829–31.
18. Chan EH, Sahai V, Conrad C, Brownstein JS. Using web search query data to monitor dengue epidemics: a new model for neglected tropical disease surveillance. PLoS Negl Trop Dis. 2011;5:e1206.
19. Hulth A, Rydevik G, Linde A. Web queries as a source for syndromic surveillance. PLoS One. 2009;4:e4378.
20. IBM. SPSS Statistics for Windows. In: IBM Corp. NY: IBM Corp; 2012.
21. Doube BM. Seasonal patterns of abundance and host relationships of the Australian paralysis tick, *Ixodes holocyclus* Neumann (Acarina: Ixodidae), in southeastern Queensland. Aust J Ecol. 1979;4:345–60.
22. Jackson J, Beveridge I, Chilton NB, Andrews RH. Distributions of the paralysis ticks *Ixodes cornuatus* and *Ixodes holocyclus* in south-eastern Australia. Aust Vet J. 2007;85:420–4.
23. Brazier I, Kelman M, Ward MP. The association between landscape and climate and reported tick paralysis cases in dogs and cats in Australia. Vet Parasitol. 2014;204:339–45.
24. Rika-Heke T, Kelman M, Ward MP. The relationship between the Southern Oscillation Index, rainfall and the occurrence of canine tick paralysis, feline tick paralysis and canine parvovirus in Australia. Vet J. 2015;205(1):87–92.
25. Hall-Mendelin S, Craig SB, Hall RA, O'Donoghue P, Atwell RB, Tulsiani SM, Graham GC. Tick paralysis in Australia caused by *Ixodes holocyclus* Neumann. Ann Trop Med Parasitol. 2011;105:95–106.
26. Moore GE, Lund E. Disease reporting and surveillance. Where do companion animal diseases fit in? Vet Clin North Am Small Anim Pract. 2009;39:225–40.

6

The Microbiota of Hematophagous Ectoparasites Collected from Migratory Birds

Francesco Cerutti[1], Paola Modesto[2], Francesca Rizzo[3], Alessandra Cravero[4], Irena Jurman[5], Stefano Costa[6], Mauro Giammarino[7], Maria Lucia Mandola[3], Mariella Goria[4], Slobodanka Radovic[5], Federica Cattonaro[5] , Pier Luigi Acuti , Simone Peletto[1]

[1]S.S. Genetica e Immunobiochimica, Istituto Zooprofilattico Sperimentale del Piemonte, Liguria e Valle d'Aosta, Torino, Italy,
[2]S.S. Sezione di Genova, Istituto Zooprofilattico Sperimentale del Piemonte, Liguria e Valle d'Aosta, Genova, Italy,
[3]S.S. Laboratorio Specialistico Diagnostica Molecolare Virologica e Ovocoltura, Istituto Zooprofilattico Sperimentale del Piemonte, Liguria e Valle d'Aosta, Torino, Italy,
[4]S.S. Microbiologia Molecolare e Analisi Genomiche, Istituto Zooprofilattico Sperimentale del Piemonte, Liguria e Valle d'Aosta, Torino, Italy,
[5]IGA Technology Services, Udine, Italy,
[6]Laboratorio Chimico Camera Commercio Torino, Torino, Italy,
[7]Department of Prevention, ASL CN 1, Racconigi (CN), Italy

Citation: Cerutti F, Modesto P, Rizzo F, Cravero A, Jurman I, Costa S, et al. (2018) The microbiota of hematophagous ectoparasites collected from migratory birds. PLoS ONE 13(8): e0202270. https://doi.org/10.1371/journal.pone.0202270

ABSTRACT

Arthropod vectors are responsible for the transmission of human pathogens worldwide. Several arthropod species are bird ectoparasites, however, no study to date has characterized their microbiota as a whole. We sampled hematophagous ectoparasites that feed on migratory birds and performed 16S rRNA gene metabarcoding to characterize their microbial community. A total of 194 ectoparasites were collected from 115 avian hosts and classified into three groups: a) *Hippoboscidae* diptera; b) ticks; c) other arthropods. Metabarcoding showed that endosymbionts were the most abundant genera of the microbial community, including *Wolbachia* for *Hippoboscidae* diptera, Candidatus *Midichloria* for ticks, *Wolbachia* and *Arsenophonus* for the other arthropod group. Genera including pathogenic species were: *Rickettsia*, *Borrelia*, *Coxiella*, *Francisella*, *Bartonella*, *Anaplasma*. Co-infection with *Borrelia-Rickettsia* and *Anaplasma-Rickettsia* was also observed. A global overview of the microbiota of ectoparasites sampled from migratory birds was obtained with the use of 16S rRNA gene metabarcoding. A novel finding is the first identification of *Rickettsia* in the common swift louse fly, *Crataerina pallida*. Given their possible interaction with pathogenic viruses and bacteria, the presence of endosymbionts in arthropods merits attention. Finally, molecular characterization of genera, including both pathogenic and symbiont species, plays a pivotal role in the design of targeted molecular diagnostics.

INTRODUCTION

Arthropod vectors are responsible of numerous diseases (named vector-borne diseases) worldwide [1]. Mosquitoes, ticks, Phlebotominae and Simuliidae flies are ectoparasites that can transmit viruses (e.g., Dengue virus, Yellow fever virus, West Nile virus (WNV), and Zika virus), bacteria (e.g., *Borrelia* spp., *Rickettsia* spp., *Francisella tularensis*, *Coxiella burnetii*), and parasites (e.g., malaria *Plasmodium spp.*, trypanosomes, *Leishmania spp.*) [1]. The 2016 Zika virus pandemic is only the most recent example of a global vector-borne disease emergency among the many other pathogens for which there is an epidemic trend [2]. For example, the hard tick *Ixodes ricinus*, present throughout Europe, is involved in the transmission of a variety of pathogens of medical and veterinary importance including *Borrelia burgdorferi* s.l., tick-borne encephalitis virus, *Anaplasma phagocytophilum*, *Francisella tularensis*, *Rickettsia helvetica* and *Rickettsia monacensis*, *Babesia divergens* and *Babesia microti*, Louping ill virus, and Tribec virus [3].

Some of the arthropods responsible for disease transmission share their environment with birds. Mosquitoes belonging to *Culex* are, in fact, mainly ornithophilic and are the main vectors of WNV and Usutu virus. Moreover, birds physically carrying arthropods (such as ticks or mites) feeding on them can introduce novel species to Europe, as recently recorded for the U.K. [4–6].

Owing to its geographical location, the Italian peninsula is crossed by migratory routes from North and sub-Saharan Africa. To our knowledge, no data have been published on the whole microbiota of ectoparasites collected directly from migratory birds, though a few studies have described the presence and prevalence of specific genera of bacteria in ticks collected from birds or their nests [7, 8]. Since these ectoparasitic arthropods may carry pathogens, it may be relevant to study their microbial communities. Other than bacteria of public health interest, the microbiota of arthropods is complex. It has been described in ticks and mosquitoes [9–12] and the role of symbionts in influencing the microbial composition has been highlighted mainly in its interaction with pathogens. Symbionts like *Wolbachia* can influence arthropod reproduction, including male-killing, parthenogenesis, feminization, and embryonic mortality [13]. Furthermore, they may evolve the necessary adaptations to parasitize vertebrate cells, as recently demonstrated that the intracellular bacterium *Coxiella burnetii* evolved from a maternally-inherited endosymbiont of ticks [14]. Adaptation may also occur in the opposite direction, as in the case of the *Francisella*-like endosymbiont that evolved from *Francisella tularensis* [15].

For this study, we collected ectoparasites feeding on migratory birds during ringing sessions and then processed the arthropod samples for 16S rRNA gene metabarcoding to characterize their microbial community. Special care was paid to identify the genera commonly associated with pathogens. The samples reporting these bacteria were further tested with genus-specific or species-specific molecular assays.

MATERIALS AND METHODS

Sample Collection

The ectoparasites were collected from birds during ringing sessions from November 2012 to October 2014. A total of 35 sessions were carried out at 14 different sites in five regions (Piedmont, n = 105; Lombardy, n = 5;

Sicily, n = 4; Latium, n = 1; Liguria, n = 1;). A total of 194 ectoparasites were collected from 115 birds, divided into 120 pools by parasite type [a) *Hippoboscidae* diptera; b) ticks; c) other arthropods], host species, sampling site, date, and location. The species included in the other-arthropods group were: *Anatoecus dentatus*, *Anaticola*, *Lucilia caesar*, *Colpocephalum turbinatum*, *Anystis*, and *Aphidiinae* spp. Details on host species are reported in Table.

The birds were caught with mist nets according to the Euring Ringing System and retrieved by authorized personnel. After capture the birds were ringed with a metal ring on the right leg. In Italy, bird leg rings are supplied by the Institute for Environmental Protection and Research (ISPRA) and they bear a unique, permanent code identifying any ringed bird for life. The birds were then identified by species, sexed, and assigned to age categories according to plumage. They were released after ectoparasite collection by veterinarians with ISPRA authorization. Being a non-invasive procedure, no special permission was needed for collection. Parasites from common swifts, mainly *Hippoboscidae* diptera, were collected either directly from the birds or from their nests in dedicated stations.

To preserve nucleic acids and obtain good quality material for metabarcoding, live parasites were stocked in RNAlater™ stabilization solution (Invitrogen, Carlsbad, CA, USA) and stored at -80°C until processed. The parasites collected from each bird were pooled together in a single vial, except for two birds (*Apus apus*), for which the parasites were stored separately for preliminary evaluation of RNA integrity. Data on sampling site location, bird age, sex, and health status were collected and entered in a database.

RNA Extraction

As the rationale of the study was to describe the living bacteria (i.e., synthesizing RNA), we analyzed the total RNA to characterize only the active microbiota and to remove bias from the DNA carried over from dead prokaryotic cells. RNA purification was performed with TRIzol™ (Invitrogen) in combination with a Nucleospin miRNA kit (Macherey-Nagel, Düren, Germany) following the manufacturer's protocol for RNA purification of small and large RNA in two fractions. The large and small RNA fractions were stored at -80°C for further analysis.

Total RNA concentration and purity was estimated using a spectrophotometer for small volumes (Vivaspec, Sartorius, Göttingen,

Germany) and a fluorometer (Qubit 2.0, Thermo Fisher Scientific, Waltham, MA, USA). The quality of total RNA was evaluated using a 2100 BioAnalyzer and an RNA 6000 Nano Kit (Agilent Technologies, Santa Clara, CA, USA). Though it was not possible to calculate the RIN (RNA Integrity Number) values [16], since the 28S rRNA subunit of many arthropods contains two hydrogen-bonded fragments that dissociate and co-migrate with the 18S subunit [17], the graph showed a 28S/18S sharp peak associated with a flat baseline that indicated the absence of degradation.

Reverse Transcription and Arthropod Species Identification

Total RNA was reverse transcribed using a High-Capacity cDNA Reverse Transcription Kit (Thermo Fisher Scientific) with 7 μl RNA as input and then stored at -20°C until processed. Vector species was determined by partial amplification and sequencing of the cytochrome c oxidase I (COI) gene, as described by Hebert and colleagues [18]. Briefly, the reaction mix was composed of 12.5 μl SuperMix PCR-UDG 2X (qPCR ProbesMaster, Jena Bioscience, Jena, Germany), 0.38 μl primer LC01490 20 μM, 0.38 μl primer HC02198 20 μM, 11 μl H2O, 0.75 μl cDNA, for a total volume of 20 μl. The thermal profile was: 50°C x 2 m; 95°C x 2 m; 40 cycles {94°C x 30 s, 49°C x 30 s, 72°C x 1 m}; 72°C x 5 m.

Successful amplification was verified using E-Gel® precast agarose gels at 2% (Thermo Fisher Scientific). Amplicons were then purified with a EUROGOLD Cycle-Pure kit (Euroclone, Pero, MI, Italy). The cycling reaction was performed with a BigDye® Terminator v1.1 cycle sequencing kit (Thermo Fisher Scientific): 2 μl BigDye® Terminator v1.1 ready reaction mix, 1 μl 5X sequencing buffer, 0.32 μl primer 100 μM, 4.68 μl H2O, 2 μl purified amplicon. The thermal profile was: 96°C x 1 m; 25 cycles at 96°C x 1 m, 50°C x 5 s, 60°C x 4 m. The reaction was purified with a GE Healthcare Illustra™ AutoSeq G-50 columns kit (GE Healthcare, Chicago, IL, USA) to remove dye terminators, and then submitted to sequencing on an Applied Biosystems AbiPrism 3130 (Foster City, CA, USA).

Chromatograms were analyzed with Sequencing Analysis v5.2 software (Thermo Fisher Scientific) for the base call and with FinchTV (Geospiza, Inc, Seattle, WA, USA) for hand editing. The obtained nucleotide sequences were used as query in a Blastn search on the GenBank nt database and in the BOLD database (Barcoding Of Life Database, www.boldsystems.org).

16S Metabarcoding

The 16S rRNA gene metabarcoding of 116 out of 120 samples was performed following the protocol suggested by Illumina (four samples were discarded as they were not of adequate quality for sequencing). Briefly, 22.5 ng of cDNA was used as input for the first PCR using 16S amplicon PCR forward and reverse primers, amplifying V3-V4 regions of the 16S rDNA. After purification and second (index) PCR with a Nextera XT Index kit (Illumina, San Diego, CA, USA), the libraries were normalized according to fragment length and dsDNA molarity. The samples were pooled and processed in four sessions on a MiSeq platform (Illumina) using a MiSeq reagent kit v3-600 for 2x300 paired-end sequencing at the IGA Technology Services facility. The datasets generated and analyzed for this study are available in the BioProject database, with SubmissionID: SUB2898018 and BioProject ID: PRJNA396024.

Bioinformatic and Statistical Analyses

A first level analysis for all samples was achieved with MiSeq Reporter Metagenomics Workflow (MSR, Illumina) to gain an overview of the microbial community for each pool. The dataset was then analyzed following DADA2 workflow within the R framework, including quality check, error rate estimation, forward/reverse reads merge, chimera removal, ribosomal sequence variants (RSVs, equivalent to OTUs) determination, and taxa assignment to the GreenGenes gg_13_8_train_set_97, RDP Training Set 14 and SILVA version 128 reference databases for comparison [19–23]. Since SILVA performed better than the other two in classifying arthropod bacterial symbionts to appropriate taxa, only these results are presented. Data were also analyzed with the QIIME v.1.9.1 pipeline, and the results were largely consistent with those obtained with DADA2.

Alpha and beta diversities were estimated with the *pyhloseq* and *vegan* packages and visualized with the *ggplot2* package, and differential expression was assessed with *DeSeq2* [24–27]. Alpha diversity was estimated based on the observed species and Shannon index using the whole dataset after removing RSVs unassigned or assigned to Eukaryota. Beta diversity was estimated based on evenly sampled Bray-Curtis distance after filtering the low-frequency RSVs, pruning the samples with a sample size less than 1,800 and rarefaction to sample size of 1,800, where sample size was the number of individuals observed for each sample. Only data from DADA2 analysis using the SILVA database are presented in Results.

Molecular Diagnostics for Potential Pathogen Confirmation

To confirm the presence of the prokariotic genera identified by 16S rRNA metabarcoding, the pools were tested using a genus-specific or species-specific PCR for each of the potential pathogens detected by the MiSeq Reporter Metagenomics Workflow. The molecular tests included the following genera: *Rickettsia* spp., *Anaplasma* spp., *Borrelia* spp., *Coxiella* spp., *Francisella* spp., and *Bartonella* spp. For all PCR assays, amplicons were purified and submitted to Sanger sequencing as described earlier for the PCR targeting the COI gene.

Two different PCR assays to confirm and characterize *Rickettsia* were used: one targeting the citrate synthase gene, according to Regnery and colleagues [28], and the other targeting the 16S rRNA gene, according to Sprong and colleagues [29]. A PCR amplifying the partial 16S rRNA gene was used to identify *A. phagocytophilum* according to Stuen and colleagues [30]. Two different approaches were applied to identify *Borrelia burgdorferi* s. l. Real-time PCR targeting a tract of the 23S rRNA gene, highly conserved in all *Borrelia* species, was used to confirm positivity, as described in Courtney and colleagues [31]. An end-point PCR targeting a fragment of the flagellin gene, specific for the *Borrelia burgdorferi* s. l. group, was then used to determine whether the detected strain belonged to the causative agents of Lyme borreliosis [32]. The genomic group of samples positive for *Borrelia burgdorferi* s. l. was then identified by sequencing. A qualitative PCR targeting the I 111 repetitive transposon-like region of *Coxiella burnetii* was performed to confirm *Coxiella* spp., as recommended by the Manual of Diagnostic Tests and Vaccines for Terrestrial Animals [33, 34]. *Francisella* spp. was investigated using a real-time PCR TaqMan® *Francisella tularensis* detection kit (Applied Biosystems) that targets the Tul4 and fopA genes. *Francisella* characterization was performed by targeting the 16S rRNA gene, according to Forsman and colleagues [35].

RESULTS

Host and Parasite Species

A total of 115 bird hosts were included Table presents the species and the number of samples collected. Hosts were classified based on their migratory behavior as: resident, short-distance, mid-distance (North Africa) or long-distance (trans-Sahara) migration. A total of 194 parasites were collected and categorized into three groups based on the type of arthropod: *Hippoboscidae* diptera (n = 51, pool (n) = 49), ticks (n = 114, pool (n) = 60), other arthropods (OA, n = 29, pool (n) = 7). *Hippoboscidae* diptera were collected from common swifts (*Apus apus*) and classified by barcoding as *Crataerina pallida*. One sample of *Hippoboscidae* was collected from a goldcrest (*Regulus regulus*) and was classified as *Ornithomya fringillina*. The tick group included species of the genera *Ixodes*, *Hyalomma*, *Ambylomma*, and *Haemaphysalis*. The OA included lice (Mallophaga: *Colpocephalum turbinatum*, *Anatoecus dentatus*, and unidentified species), blowflies (Diptera: Calliphoridae: *Lucilia caesar*), mites (Trombidiformes: *Anystis*), the parasitoid wasps (Hymenoptera: Braconidae: Aphidiinae). A detailed list of ectoparasite species and number of sampled individuals and pools is given in Table.

16S Metabarcoding

A total of 116 pools were processed for V3-V4 16S rRNA gene amplification and massive parallel sequencing on an Illumina MiSeq platform, generating also a taxonomic report with MiSeq Reporter Software. The total amount of reads generated was 45,286,016 (median = 323,508, Q1 = 203,740, Q3 = 510,136); after quality filtering 24,012,404 reads per pair (median = 82,992, Q1 = 48,206, Q3 = 143,460, with read length uniformed to 230 bp) were obtained. Details on raw and filtered read numbers for each sample are reported in Table. The resulting 257,298 dereplicated non-chimeric sequences were assigned to 2,257 RSVs, belonging to Bacteria and Eukaryota domains. Taxa assigned to Eukaryota or unassigned were removed, reducing the final number of RSVs to 2,184, classified in 23 different phyla. The most abundant phyla in the whole dataset were the Proteobacteria (90.72%) and the Firmicutes (5.44%). The most common genera for the whole dataset are reported in Table 1, and the taxa with relative abundance >1% are reported by group in Table 2. Fig 1 presents a graphical overview of the genera identified in each sample. The most abundant taxonomic groups of the bacterial composition at a lower taxonomical scale included symbionts like *Wolbachia*, *Arsenophonus*, and Candidatus *Midichloria mitochondrii*.

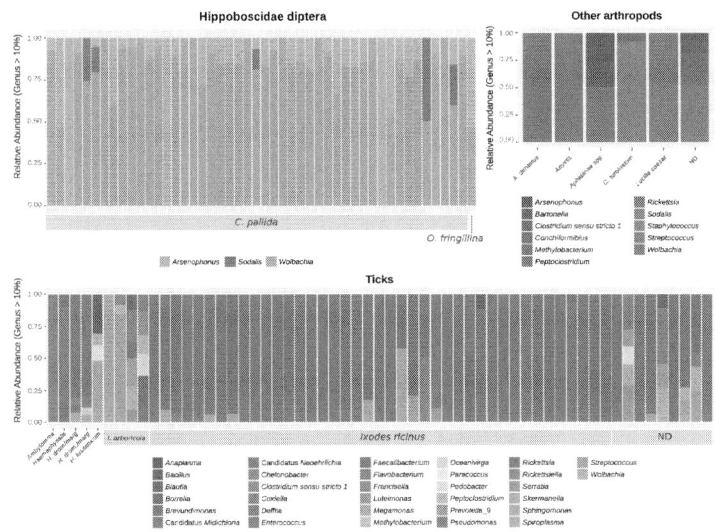

Figure 1. Genus composition accounting for 90% abundance for each sample. Samples are grouped by type of parasite (*Hippoboscidae* diptera, OA, and Ticks).

https://doi.org/10.1371/journal.pone.0202270.g001

Table 1. Most abundantgenera (%) in the whole data set.

Genus	**Relative abundance (%)**
Wolbachia	35.79
Candidatus_*Midichloria*	25.45
Rickettsia	12.57
Arsenophonus	8.19
Sodalis	1.60
Coxiella	1.22
Clostridium_sensu_stricto_1	0.96

https://doi.org/10.1371/journal.pone.0202270.t001

Table 2. Composition of class, order, family, and genus (expressed as relative abundance, %) with >1% relative abundance reported for each sample group.

	Class		Order		Family		Genus	
***Hippoboscidae* Diptera**	*Alphaproteobacteria*	78.28	*Rickettsiales*	78.27	*Rickettsiaceae*	78.26	*Wolbachia*	78.23
	Gammaproteobacteria	21.69	*Enterobacteriales*	21.69	*Enterobacteriaceae*	21.69	*Candidatus Phlomobacter*	17.95
							Sodalis	2.61
Ticks	*Alphaproteobacteria*	79.77	*Rickettsiales*	76.35	*Midichloriaceae*	51.33	*Candidatus Midichloria*	52.81
	Gammaproteobacteria	5.03	*Clostridiales*	4.77	*Rickettsiaceae*	25.29	*Rickettsia*	23.27
	Clostridia	4.77	*Legionellales*	3.30	*Coxiellaceae*	3.15	*Clostridium*	2.52
	Bacilli	2.18	*Entomoplasmatales*	1.83	*Clostridiaceae*	2.58	*Coxiella*	2.36
	Mollicutes	1.85	*Lactobacillales*	1.41	*Anaplasmataceae*	1.88	*Wolbachia*	2.01
			Sphingomonadales	1.38	*Sphingomonadaceae*	1.30	*Candidatus Neoehrlichia*	1.68
			Rhizobiales	1.04	*Enterococcaceae*	0.96	*Sphingomonas*	1.08
							Enterococcus	0.96
OA	*Alphaproteobacteria*	42.52	*Rickettsiales*	37.82	*Rickettsiaceae*	37.77	*Wolbachia*	29.46
	Gammaproteobacteria	24.98	*Enterobacteriales*	24.15	*Enterobacteriaceae*	24.15	*Arsenophonus*	16.38
	Bacilli	17.31	*Bacillales*	11.03	*Staphylococcaceae*	10.80	*Staphylococcus*	10.79
	Clostridia	9.48	*Clostridiales*	9.48	*Clostridiaceae*	4.55	*Rickettsia*	8.32
	Betaproteobacteria	1.79	*Lactobacillales*	5.61	*Streptococcaceae*	4.30	*Sodalis*	8.15
	Bacteroidia	1.30	*Rhizobiales*	3.10	*Lachnospiraceae*	2.03	*Clostridium*	4.35
			Neisseriales	1.48	*Methylobacteriaceae*	1.53	*Streptococcus*	4.29
			Bacteroidales	1.30	*Neisseriaceae*	1.48	*Methylobacterium*	1.53
					Veillonellaceae	1.35	*Megamonas*	1.31

https://doi.org/10.1371/journal.pone.0202270.t002

To better visualize the distribution of the symbiont genera in the *Hippoboscidae* diptera, a bar plot of the relative abundance within the family is reported in Fig. Due to the large number of genera detected, *Rickettsiales* from ticks are reported in a bar plot. To better represent the diversity of the symbionts within parasite species, the number of RSVs belonging to each genus are summarized in Table 3. According to the DADA2 developers, this algorithm in able to detect true biological sequence variants that might be considered different bacterial strains. The composition in RSVs of the major symbiont genera of *Hippoboscidae* diptera and ticks are reported.

Table 3. Number of ribosomal sequence variants (RSVs) for each of the main symbiont genera characterized in the entire data set.

Genus	No. of RSVs
Arsenophonus	7
Sodalis	4
Wolbachia	12
Candidatus_Midichloria	10
Rickettsia	7

https://doi.org/10.1371/journal.pone.0202270.t003

Bioinformatic and Statistical Analysis

The reports generated by the integrated pipeline of the Miseq Reporter were analyzed. Based on these results, the samples were considered positive or negative for the detected potentially pathogenic genera. Table 4 reports the number of samples in which the MSR found the corresponding genus (MSR Hits). The microbiota of the three groups of ectoparasites differed in microbial composition by both abundance and represented taxa. Box-plots illustrating alpha diversity are reported in Fig 2. The Shannon index showed no statistically significant difference between the groups (box-plot Fig 2). There was a statistically significant difference in the number of species for the three groups. The box-plot represents the low number of species in the *Hipposboscidae* diptera group as compared to the other two groups. The principal coordinates analysis based on the Bray-Curtis distance highlighted a difference in microbial composition in the microbiota of the *Hipposboscidae* diptera as compared to that of the ticks and the OA groups (PERMANOVA test implemented in the *adonis* function in *vegan*) (Fig 3).

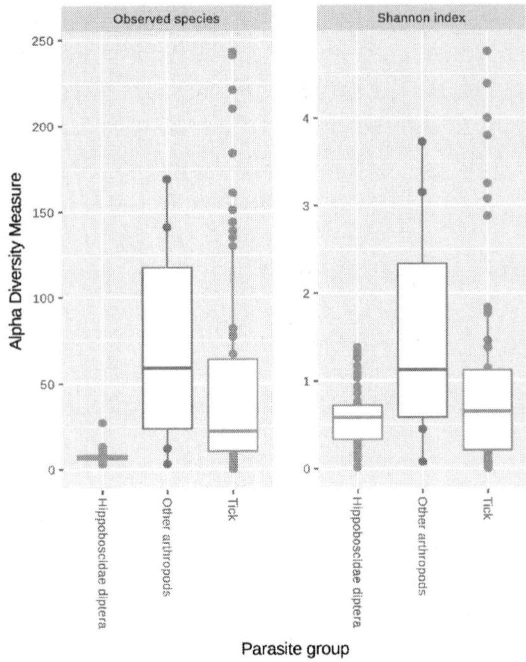

Figure 2. Box-plot of the main indexes for alpha diversity by parasite group. Indexes are observed species and Shannon index.

https://doi.org/10.1371/journal.pone.0202270.g002

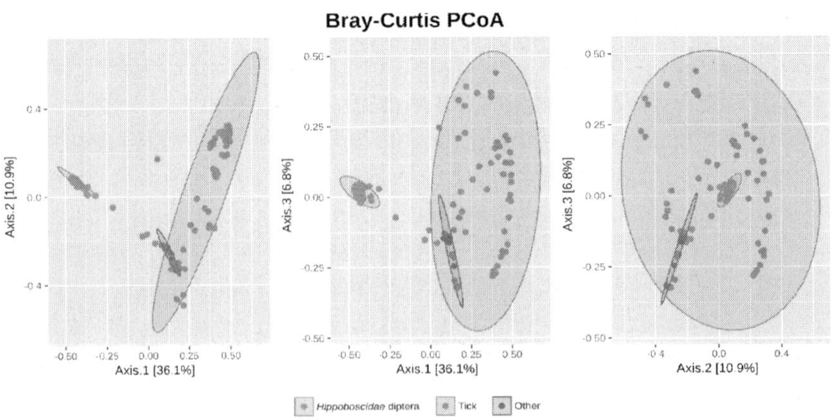

Figure 3. Principal coordinates analysis based on Bray-Curtis distances of the three separate groups. As reported by the axis label, the axis 1 shows 36.1% of variation in the samples.

https://doi.org/10.1371/journal.pone.0202270.g003

Table 4. Hits and PCR positivity to genera including known pathogen species. Hits (OTUs matching a given genus) were obtained by MiSeq Reporter (MSR) analysis. Prevalence was calculated on 116 samples. The 95% confidence interval (CI) of the prevalence is reported in brackets.

Genus	MSR Hits	MSR Prevalence	PCR confirmations	PCR prevalence
Rickettsia	110	94.8% (88.94–97.84)	47	40.5% (32.02–49.62)
Ehrlichia	98	84.5% (76.71–90.04)	-	-
Borrelia	17	14.7% (9.26–22.32)	10	8.6% (4.59–15.31)
Coxiella	6	5.2% (2.16–11.06)	0	0%
Francisella	4	3.4% (1.06–8.82)	2	1.7% (0.09–6.46)
Bartonella	4	3.4% (1.06–8.82)	0	0%
Anaplasma	3	2.6% (0.55–7.66)	2	1.7% (0.09–6.46)

https://doi.org/10.1371/journal.pone.0202270.t004

Molecular Diagnostics for Potential Pathogen Confirmation

As summarized in Table 4, *Rickettsia* and *Ehrlichia* were highly represented among the samples, with more than 80% of the samples having reads corresponding to these two genera. The diagnostic PCR for *Rickettsia* spp. confirmed only 47 positive samples (42% of hits, 40.5% of total samples), further characterized as *R. aeschlimannii*, *R. helvetica*, and *R. monacensis*. Some of the sequences were similar to unclassified endosymbionts, 6 of which were were close to *R. bellii*, a species found only in the Americas (USA, Brazil, Argentina, Costa Rica, Colombia, El Salvador, Peru) [36–42]. As described for *C. burnetii* [43], identification of *Rickettsia* endosymbionts by means of a PCR used for pathogen detection in routine work shows that these species may interfere with the correct diagnosis of pathogenic rickettsial species.

Borrelia spp. were confirmed in 10 samples, four of which were further confirmed by the PCR specific for *Borrelia burgdorferi* s.l. and classified as *B. valaisiana*. The two samples positive for *Francisella* spp. were *Francisella*-like endosymbionts, while the two *Anaplasma* positive samples belonged to *A. phagocytophilum*. *Ehrlichia* required a different approach, since the MSR identified *E. ovina* in 98 samples, a species poorly described in the literature. On the basis of this unexpectedly high prevalence, and in contrast to only three sequences registered in the NCBI database, we suspected a misclassification issue, so we randomly chose one sample and retrieved the reads assigned to this species. A Blast search against the NCBI 16S prokariotic rRNA database was performed and the results were then plotted in MEtaGenome ANalyzer. The output is reported the reads were classified as *Anaplasmataceae*, and at a lower taxonomic level as Candidatus *Midichloria mitocondrii*, *Anaplasma* spp., and *Wolbachia* spp. For this reason, the samples were not tested for *Ehrlichia* spp.

Among the PCR-confirmed samples, the following co-infections were observed: *Borrelia-Rickettsia* (n = 9), two of which occurred in individual samples, and *Anaplasma-Rickettsia* (n = 2). Considering only the ticks, the prevalence of confirmed *Rickettsia* spp. in the tick-only group was 60.32% (95% CI: 47.98–71.47), 15.87% (95% CI 8.86–26.81) for *Borrelia* spp., and 18.37% (95% CI 9.98–31.36) for *Ixodes*. *Rickettsia* spp. was present in numerous samples and detected in parasites collected from resident (n = 2), short-distance (n = 32), mid-distance (n = 2), and long-distance (n = 11) migratory birds. *Borrelia* spp. was detected only in ticks from short- and mid-distance migratory birds.

DISCUSSION

With this study we wanted to describe the microbiota of ectoparasites collected from migratory birds since they constitute a route of introduction for exotic vector-borne diseases. The parasites were divided into three groups based on taxonomical features and sample size: *Hippoboscidae* diptera, ticks, and other arthropods. The microbiota of the *Hippoboscidae* diptera was composed of a limited number of species, as expressed by the low value of the observed species. Considering the evenness (Shannon index), the diversity was comparable to the tick and the OA groups. The low number of species may be explained by the predominance of symbiont species among the most abundant genera observed, such as *Wolbachia*, *Arsenophonus*, and *Sodalis* (*Sodalis* endosymbionts were also detected in *Craterina melbae*) [44–47]. The difference in microbial composition by number and taxonomy of the RSVs in the three groups is supported by the significant difference in the alpha and beta diversity, suggesting that the bacterial communities are heavily influenced by the parasite they live with. Briefly, the most abundant symbionts were *Wolbachia*, *Rickettsia*, *Arsenophonus*, and C. *Midichloria*. They were closely associated with the type of arthopod: *Hippoboscidae* were mainly colonized by *Wolbachia* and *Arsenophonus* and ticks by *Rickettsia* and *C. Midichloria*.

Regarding the distribution of the main genera in the *Hippoboscidae*, the microbial population of all but three samples was almost totally composed of *Wolbachia*. The three *Wolbachia*-free samples were totally colonized by *Arsenophonus*, while *Sodalis* was present only together with other symbionts. Only in three samples *Wolbachia* was the unique genus, for a total of 6 individuals with a single symbiont. Although the relative abundance is based on the family, the majority of the *Hippoboscidae* was colonized by at least two symbiont genera.

Wolbachia strains seemed to be closely connected to host species; *C. pallida* was mainly colonized by one variant and *O. fringillina* by another. In contrast, *Arsenophonus* in *C. pallida* had a slight diversity (with RSVs similar by 99.30–99.53%), while in *O. fringillina* it was present only as one RSV (similar to other *Arsenophonus* symbionts of *Ornithomya* species). The high homogeneity of *Wolbachia* suggests that it may be an obligate symbiont vertically inherited by maternal lineage. Differently, the diversity of *Arsenophonus* within samples suggests that it may have been transmitted horizontally or by other ways. The high presence of *Wolbachia* define this genus as the predominant symbiont. While it might be an obligate symbiont within *C. pallida*, the presence of other genera suggests that they still play an important role in the survival of *Hippoboscidae*, but further data are needed.

To our knowledge, this report is the first identification of *Rickettsia bellii* and *R. monacensis* in the *Hippoboscidae C. pallida*. Strains of *R. bellii* have been reported only for the Americas; similar strains have been detected in Australia, Thailand, Réunion Island, and Japan [44–47]. Our report is the first identification of *R. bellii* in Italy and Europe. This finding raises the question as to whether *C. pallida* behaves as an accidental vector for rickettsiosis or, if not competent for transmission, whether it might play a role as a sentinel parasite for the spread of arthropod-borne pathogens.

In ticks, the endosymbiont Candidatus *Midichloria* accounted for half of the RSV abundance in the samples, followed by *Rickettsia* spp. When *Rickettsia* spp. was present, it had the highest prevalence in almost all samples; only in one sample, *Rickettsia* was present but the predominant genus was *Midichloria*. As reported elsewhere, *Wolbachia* symbionts in ticks are rare [48]. Unlike a recent study on ticks in France, our study noted no relevant presence of *Acinetobacter*, but we did observe co-infection with pathogens and symbionts in our samples [49]. In the samples with only one tick, we observed co-infection mainly between *Rickettsia* and *Midichloria*; also *Wolbachia*, *Rickettsiella*, *Neoehrlichia*, and *Spiroplasma* were present together with other symbionts. Candidatus *Midichloria* was present mainly in *Ixodes ricinus* ticks, where it was represented by a unique variant. Unfortunately, the sample size was too small to make further observations for other species like *I. arboricola* and *Hyalomma* spp.

Candidatus *Midichloria* was first described in 2006 as an endosymbiont of *Ixodes ricinus*, and was later also detected in other hard ticks (*Ixodidae*) in Italy [50–51]. The detection of circulating DNA and the presence of antibodies against an antigen against *M. mitochondrii* in humans and mammals suggest that it might represent a novel group of vector-borne agents [52, 53].

The role of endosymbionts in arthropods has been partially described and a strong correlation with pathogen replication and transmission has been shown in some cases. For example, infection of *Wolbachia*(+) and *Wolbachia*(-) *Culex quinquefasciatus* colonies with WNV revealed a greater proportion of *Wolbachia*(-) infected mosquitoes developing high virus titers in saliva, which is necessary for virus dissemination and transmission [54]. This observation led to the suggestion that the difference in susceptibility to WNV infection between *Cx. quinquefasciatus* and *Culex tarsalis* might be partially explained by the difference in *Wolbachia* infection between these two species, since *Cx. tarsalis* is not infected with *Wolbachia* [54].

By applying the metabarcoding approach, we were able to detect several pathogenic species and to confirm several of them by species-specific or genus-specific PCRs. As for *Rickettsia* and *Borrelia* genera, the prevalence in our data set is shared by similar studies in Italy and Europe [55–59]. In addition, our findings show that *Rickettsia* seems to be widespread among residential and migratory birds, while *Borrelia* was detected only in short- and mid-distance migratory birds, suggesting different patterns in its transmission.

The observation of bacterial genera in the metabarcoding results not confirmed by the species-specific or genus-specific PCR tests may be explained by the presence of *Rickettsia*-, *Coxiella*-, and *Francisella*-like symbionts. The primer pairs used for the diagnostic tests were retrieved from published studies on the detection of pathogenic species of these genera. Most likely, these tests fail to detect a symbiont species of the targeted genus. As reported for *Coxiella*, the genetic diversity of symbiont organisms is very high, and little is known about their spread in arthropods, which may explain the discordance between the results of 16S rRNA gene metabarcoding and diagnostic PCRs [60]. Alternatively, it has been shown that some molecular tests that are specific for *C. burnetii* also detect *Coxiella*-like bacteria, leading to overestimation of the pathogen species. Indeed, the molecular characterization of bacterial endosymbionts plays a pivotal role in the design of targeted molecular tests for the sole detection of pathogenic species.

Recent studies have shown that *C. burnetii* could have originated from a tick-associated ancestor, while the *Francisella*-like endosymbiont of the hard tick *Ambylomma* probably evolved from a pathogenic strain of *Francisella*, indicating that tick endosymbionts can evolve from mammalian pathogens [14,15]. Little is known about these recently uncovered symbionts, perhaps because the research was biased towards the pathogenic species. Such is the case of the *Coxiella* genus, which only has two species (i.e., *burnetii* and *cheraxi*). The majority of studies have described *C. burnetii* because most isolates were collected from humans or domestic ruminants during Q fever outbreaks. More information on novel *Coxiella*-like organisms in non-vertebrate species like ticks has been acquired via 16S rRNA gene metabarcoding [14].

Finally, we observed a critical point in the bioinformatics analysis of our data. The first point is the erroneous identification of *E. ovina* by the MSR, not confirmed by deeper analysis. This issue may concern the used database, since, as reported in the Illumina manual, the Metagenomics workflow uses an Illumina-curated version of the Greengenes database. The choice of database may also lead to different results in taxa assignment. In our analyses, the SILVA database allowed us to assign RSVs to *Wolbachia*, C. *Midichloria*, and *Arsenophonus*. Since these three genera represent the majority of the microbial community, being symbionts, correct taxonomic assignment is very relevant for these kind of studies. We suggest the use of the SILVA database for future projects investigating arthropod microbiota by 16S rRNA metabarcoding.

Our metabarcoding analysis showed that the microbiota living with (and within) arthropods is complex, closely related to the host species, and that its major component comprises endosymbiont-related species. This approach provides a global overview of the bacteria present in/on ectoparasites collected live from migratory birds. Because it employs a universal primer set for prokaryotic metabarcoding, this approach was also useful for identifying in one shot the genera that include pathogen species. Since the method does not often discriminate beyond the genus level, a second-level, genus- or species-specific investigation was required to confirm the presence of the pathogen species in some samples. Without an overview provided by the metabarcoding method, multiple tests for each pathogen in all the samples would have been needed.

ACKNOWLEDGMENTS

The authors wish to thank Carla Lo Vecchio for laboratory assistance with the PCR tests, Giovanni Savini (Istituto Zooprofilattico Sperimentale dell'Abruzzo e del Molise Giuseppe Caporale, Teramo), Riccardo Orusa (Istituto Zooprofilattico Sperimentale del Piemonte, Liguria e Valle d'Aosta, Aosta), Piergiovanni Piatti (Camera di Commercio, Torino), and Santo Caracappa (Istituto Zooprofilattico Sperimentale della Sicilia, Palermo) for their support in the study.

The authors are grateful to the two anonymous reviewers for their helpful suggestions in improving the manuscript.

REFERENCES

1. Saldaña MA, Hegde S, Hughes GL. Microbial control of arthropod-borne disease. Mem Inst Oswaldo Cruz. 2017;112: 81–93. 10.1590/0074-02760160373
2. Fauci AS, Morens DM. Zika virus in the Americas—yet another arbovirus threat. N Engl J Med. Massachusetts Medical Society; 2016;374: 601–604. 10.1056/NEJMp1600297
3. European Centre for Disease Prevention and Control (ECDC). Ixodes ricinus—Factsheet for experts [Internet]. 2014 [cited 13 Dec 2017]. Available: https://ecdc.europa.eu/en/disease-vectors/facts/tick-factsheets/ixodes-ricinus
4. Hoogstraal H, Kaiser MN, Traylor MA, Guindy E, Gaber S. Ticks (Ixodidae) on birds migrating from Europe and Asia to Africa 1959–61. Bull World Health Organ. World Health Organization; 1963;28: 235–262.
5. Lindeborg M, Barboutis C, Ehrenborg C, Fransson T, Jaenson TGT, Lindgren P-E, et al. Migratory birds, ticks, and Crimean-Congo hemorrhagic fever virus. Emerg Infect Dis. 2012;18: 2095–2097. 10.3201/eid1812.120718
6. Jameson LJ, Morgan PJ, Medlock JM, Watola G, Vaux AGC. Importation of *Hyalomma marginatum*, vector of Crimean-Congo haemorrhagic fever virus, into the United Kingdom by migratory birds. Ticks Tick Borne Dis. 2012;3: 95–99. 10.1016/j.ttbdis.2011.12.002
7. Duron O, Jourdain E, McCoy KD. Diversity and global distribution of the *Coxiella* intracellular bacterium in seabird ticks. Ticks Tick Borne Dis. Urban & Fischer; 2014;5: 557–563. 10.1016/j.ttbdis.2014.04.003
8. Duron O, Cremaschi J, McCoy KD. The high diversity and global distribution of the intracellular bacterium *Rickettsiella* in the polar seabird tick I*ixodes uriae*. Microb Ecol. 2016;71: 761–770. 10.1007/s00248-015-0702-8
9. Gofton A, Oskam C, Lo N, Beninati T, Wei H, McCarl V, et al. Inhibition of the endosymbiont "Candidatus *Midichloria mitochondrii*" during 16S rRNA gene profiling reveals potential pathogens in *Ixodes* ticks from Australia. Parasit Vectors. BioMed Central; 2015;8: 345 10.1186/3071-015-0958-3
10. Finney CAM, Kamhawi S, Wasmuth JD, Janelle J, Patrel D. Does the arthropod microbiota impact the establishment of vector-borne

diseases in mammalian hosts? Bliska JB, editor. PLOS Pathog. Public Library of Science; 2015;11: e1004646 10.1371/journal.ppat.1004646

11. Narasimhan S, Fikrig E. Tick microbiome: The force within. Trends Parasitol. Elsevier Ltd; 2015;31: 315–323. 10.1016/j.pt.2015.03.010
12. Novakova E, Woodhams DC, Rodríguez-Ruano SM, Brucker RM, Leff JW, Maharaj A, et al. Mosquito microbiome dynamics, a background for prevalence and seasonality of West Nile virus. Front Microbiol. Frontiers; 2017;8: 526 10.3389/fmicb.2017.00526
13. Mohanty I, Rath A, Mahapatra N, Hazra RK. *Wolbachia*: A biological control strategy against arboviral diseases. J Vector Borne Dis. 2016;53: 199–207.
14. Duron O, Noël V, McCoy KD, Bonazzi M, Sidi-Boumedine K, Morel O, et al. The recent evolution of a maternally-inherited endosymbiont of ticks led to the emergence of the Q fever pathogen, *Coxiella burnetii*. PLoS Pathog. 2015;11: 1–23. 10.1371/journal.ppat.1004892
15. Gerhart JG, Moses AS, Raghavan R. A *Francisella*-like endosymbiont in the Gulf Coast tick evolved from a mammalian pathogen. Sci Rep. Nature Publishing Group; 2016;6: 33670 10.1038/srep33670
16. Schroeder A, Mueller O, Stocker S, Salowsky R, Leiber M, Gassmann M, et al. The RIN: an RNA integrity number for assigning integrity values to RNA measurements. BMC Mol Biol. BioMed Central; 2006;7: 3 10.1186/1471-2199-7-3
17. Winnebeck EC, Millar CD, Warman GR. Why does insect RNA look degraded? J insect Sci. 2010;10: 159 10.1673/031.010.14119
18. Hebert PDN, Cywinska A, Ball SL, deWaard JR. Biological identifications through DNA barcodes. Proc R Soc B Biol Sci. 2003;270: 313–321. 10.1098/rspb.2002.2218
19. Callahan BJ, McMurdie PJ, Rosen MJ, Han AW, Johnson AJA, Holmes SP. DADA2: High-resolution sample inference from Illumina amplicon data. Nat Methods. NIH Public Access; 2016;13: 581–583. 10.1038/nmeth.3869
20. R Development Core Team. R: A language and environment for statistical computing. 2008.
21. McDonald D, Price MN, Goodrich J, Nawrocki EP, DeSantis TZ, Probst A, et al. An improved Greengenes taxonomy with explicit ranks for ecological and evolutionary analyses of bacteria and archaea. ISME J. BioMed Central; 2012;6: 610–618. 10.1038/ismej.2011.139

22. Quast C, Pruesse E, Yilmaz P, Gerken J, Schweer T, Yarza P, et al. The SILVA ribosomal RNA gene database project: improved data processing and web-based tools. Nucleic Acids Res. 2013;41: D590–6. 10.1093/nar/gk 219
23. Cole JR, Wang Q, Fish JA, Chai B, McGarrell DM, Sun Y, et al. Ribosomal Database Project: data and tools for high throughput rRNA analysis. Nucleic Acids Res. 2014;42: D633–42. 10.1093/nar/gkt1244
24. McMurdie PJ, Holmes S. Phyloseq: An R package for reproducible interactive analysis and graphics of microbiome census data. Watson M, editor. PLoS One. Public Library of Science; 2013;8: e61217 10.1371/journal.pone.0061217
25. Oksanen J, Blanchet FG, Friendly M, Kindt R, Legendre P, McGlinn D, et al. Vegan: community ecology package. 2017.
26. Wickham H. ggplot2: elegant graphics for data analysis Springer-Verlag; New York; 2009.
27. Love MI, Huber W, Anders S. Moderated estimation of fold change and dispersion for RNA-seq data with DESeq2. Genome Biol. 2014;15: 550 10.1186/ 3059-014-0550-8
28. Regnery RL, Spruill CL, Plikaytis BD. Genotypic identification of rickettsiae and estimation of intraspecies sequence divergence for portions of two rickettsial genes. J Bacteriol. 1991;173: 1576–1589.
29. Sprong H, Wielinga PR, Fonville M, Reusken C, Brandenburg AH, Borgsteede F, et al. *Ixodes ricinus* ticks are reservoir hosts for *Rickettsia helvetica* and potentially carry flea-borne *Rickettsia* species. Parasit Vectors. 2009;2: 41 10.1186/1756-3305-2-41
30. Stuen S, Nevland S, Moum T. Fatal cases of Tick-borne fever (TBF) in sheep caused by several 16S rRNA gene variants of *Anaplasma phagocytophilum*. Ann N Y Acad Sci. 2003;990: 433–4. 10.1111/j.1749-6632.2003.tb07407.x
31. Courtney JW, Kostelnik LM, Zeidner NS, Massung RF. Multiplex real-time PCR for detection of *Anaplasma phagocytophilum* and *Borrelia burgdorferi*. J Clin Microbiol. 2004;42: 3164–8. 10.1128/JCM.42.7.3164-3168.2004
32. Skotarczak B, Wodecka B, Cichocka A. Coexistence DNA of *Borrelia burgdorferi* sensu lato and *Babesia microti* in *Ixodes ricinus* ticks from north-western Poland. Ann Agric Environ Med. 2002;9: 25–8.
33. World Organisation for Animal Health, Cossio MLT, Giesen LF, Araya

G, Pérez-Cotapos MLS, VERGARA RL, et al. Manual of diagnostic tests and vaccines for terrestrial animals In: OIE; 2015. pp. 1–25. 10.1007/ 3398-014-0173-7.2

34. Berri M, Laroucau K, Rodolakis A. The detection of *Coxiella burnetii* from ovine genital swabs, milk and fecal samples by the use of a single touchdown polymerase chain reaction. Vet Microbiol. 2000;72: 285–293. 10.1016/S0378-1135(99)00178-9
35. Forsman M, Sandström G, Sjöstedt A. Analysis of 16S ribosomal DNA sequences of *Francisella* strains and utilization for determination of the phylogeny of the genus and for identification of strains by PCR. Int J Syst Bacteriol. Microbiology Society; 1994;44: 38–46. 10.1099/00207713-44-1-38
36. Tomassone L, Nuñez P, Ceballos LA, Gürtler RE, Kitron U, Farber M. Detection of "*Candidatus* Rickettsia sp. strain Argentina" and *Rickettsia bellii* in *Amblyomma* ticks (Acari: Ixodidae) from Northern Argentina. Exp Appl Acarol. 2010;52: 93–100. 10.1007/ 0493-010-9339-y
37. Barbieri ARM, Romero L, Labruna MB. *Rickettsia bellii* infecting *Amblyomma sabanerae* ticks in El Salvador. Pathog Glob Health. 2012;106: 188–189. 10.1179/2047773212Y.0000000022
38. Ogrzewalska M, Literak I, Cardenas-Callirgos JM, Capek M, Labruna MB. *Rickettsia bellii* in ticks *Amblyomma varium* Koch, 1844, from birds in Peru. Ticks Tick Borne Dis. 2012;3: 254–256. 10.1016/j.ttbdis.2012.05.003
39. Miranda J, Mattar S. Molecular detection of *Rickettsia bellii* and *Rickettsia* sp. strain colombianensi in ticks from Cordoba, Colombia. Ticks Tick Borne Dis. 2014;5: 208–212. 10.1016/j.ttbdis.2013.10.008
40. Troyo A, Moreira-Soto RD, Calderon-Arguedas ?lger, Mata-Somarribas C, Ortiz-Tello J, Barbieri ARM, et al. Detection of rickettsiae in fleas and ticks from areas of Costa Rica with history of spotted fever group rickettsioses. Ticks Tick Borne Dis. 2016;7: 1128–1134. 10.1016/j.ttbdis.2016.08.009
41. Hecht JA, Allerdice MEJ, Krawczak FS, Labruna MB, Paddock CD, Karpathy SE. Development of a *Rickettsia bellii*- specific taqman assay targeting the citrate synthase gene. J Med Entomol. Oxford University Press; 2016;53: 1492–1495. 10.1093/jme/tjw111
42. Blanco CM, Teixeira BR, da Silva AG, de Oliveira RC, Strecht L,

Ogrzewalska M, et al. Microorganisms in ticks (Acari: Ixodidae) collected on marsupials and rodents from Santa Catarina, Paran and Mato Grosso do Sul states, Brazil. Ticks Tick Borne Dis. 2017;8: 90–98. 10.1016/j.ttbdis.2016.10.003

43. Jourdain E, Duron O, Barry S, Gonzalez-Acuna D, Sidi-Boumedine K. Molecular methods routinely used to detect *Coxiella burnetii* in ticks cross-react with *Coxiella*-like bacteria. Infect Ecol Epidemiol. Co-Action Publishing; 2015;5: 29230 10.3402/iee.v5.29230
44. Vilcins IME, Old JM, Deane E. Molecular detection of *Rickettsia*, *Coxiella* and *Rickettsiella* DNA in three native Australian tick species. Exp Appl Acarol. 2009;49: 229–242. 10.1007/ 0493-009-9260-4
45. Sumrandee C, Hirunkanokpun S, Doornbos K, Kitthawee S, Baimai V, Grubhoffer L, et al. Molecular detection of *Rickettsia* species in *Amblyomma* ticks collected from snakes in Thailand. Ticks Tick Borne Dis. 2014;5: 632–640. 10.1016/j.ttbdis.2014.04.013
46. Dietrich M, Lebarbenchon C, Jaeger A, Le Rouzic C, Bastien M, Lagadec E, et al. *Rickettsia* spp. in seabird ticks from western Indian Ocean Islands, 2011–2012. Emerg Infect Dis. 2014;20: 838–842. 10.3201/eid2005.131088
47. Hayashi M, Watanabe M, Yukuhiro F, Nomura M, Kageyama D. A nightmare for males? A maternally transmitted male-killing bacterium and strong female bias in a green lacewing population. Bourtzis K, editor. PLoS One. R Foundation for Statistical Computing; 2016;11: e0155794 10.1371/journal.pone.0155794
48. Varela-Stokes AS, Park SH, Kim SA, Ricke SC. Microbial communities in North American ixodid ticks of veterinary and medical importance. Front Vet Sci. 2017;4: 179 10.3389/fvets.2017.00179
49. Moutailler S, Valiente Moro C, Vaumourin E, Michelet L, Tran FH, Devillers E, et al. Co-infection of ticks: the rule rather than the exception. PLoS Negl Trop Dis. 2016;10: 1–17. 10.1371/journal.pntd.0004539
50. Sassera D, Beninati T, Bandi C, Bouman EAP, Sacchi L, Fabbi M, et al. *Candidatus* Midichloria mitochondrii', an endosymbiont of the *Ixodes ricinus* with a unique intramitochondrial lifestyle. Int J Syst Evol Microbiol. 2006;56: 2535–2540. 10.1099/ijs.0.64386-0
51. Epis S, Sassera D, Beninati T, Lo N, Beati L, Piesman J, et al. *Midichloria mitochondrii* is widespread in hard ticks (Ixodidae)

and resides in the mitochondria of phylogenetically diverse species. Parasitology. 2008;135: 485–494. 10.1017/S0031182007004052

52. Mariconti M, Epis S, Gaibani P, Dalla Valle C, Sassera D, Tomao P, et al. Humans parasitized by the hard tick *Ixodes ricinus* are seropositive to *Midichloria mitochondrii*: is *Midichloria* a novel pathogen, or just a marker of tick bite? Pathog Glob Health. 2012;106: 391–6. 10.1179/2047773212Y.0000000050
53. Bazzocchi C, Mariconti M, Sassera D, Rinaldi L, Martin E, Cringoli G, et al. Molecular and serological evidence for the circulation of the tick symbiont *Midichloria* (Rickettsiales: *Midichloriaceae*) in different mammalian species. Parasites and Vectors. 2013;6: 1–7. 10.1186/1756-3305-6-1
54. Glaser RL, Meola MA. The native *Wolbachia* endosymbionts of *Drosophila melanogaster* and *Culex quinquefasciatus* increase host resistance to West Nile virus infection. PLoS One. 2010;5 10.1371/journal.pone.0011977
55. Hornok S, Kováts D, Csörgő T, Meli ML, Gönczi E, Hadnagy Z, et al. Birds as potential reservoirs of tick-borne pathogens: first evidence of bacteraemia with *Rickettsia helvetica*. Parasit Vectors. BioMed Central; 2014;7: 128 10.1186/1756-3305-7-128
56. Martello E, Selmi M, Ragagli C, Ambrogi C, Stella MC, Mannelli A, et al. *Rickettsia slovaca* in immature *Dermacentor marginatus* and tissues from *Apodemus* spp. in the northern Apennines, Italy. Ticks Tick Borne Dis. 2013;4: 518–21. 10.1016/j.ttbdis.2013.07.002
57. Amore G, Tomassone L, Grego E, Ragagli C, Bertolotti L, Nebbia P, et al. *Borrelia lusitaniae* in immature *Ixodes ricinus* (Acari: Ixodidae) feeding on common wall lizards in Tuscany, central Italy. J Med Entomol. 2007;44: 303–7.
58. Pintore MD, Ceballos L, Iulini B, Tomassone L, Pautasso A, Corbellini D, et al. Detection of invasive *Borrelia burgdorferi* strains in north-eastern Piedmont, Italy. Zoonoses Public Health. 2015;62: 365–74. 10.1111/zph.12156
59. Poupon M-A, Lommano E, Humair P-F, Douet V, Rais O, Schaad M, et al. Prevalence of *Borrelia burgdorferi* sensu lato in ticks collected from migratory birds in Switzerland. Appl Environ Microbiol. 2006;72: 976–9. 10.1128/AEM.72.1.976-979.2006

60. Machado-Ferreira E, Vizzoni VF, Balsemão-Pires E, Moerbeck L, Gazeta GS, Piesman J, et al. *Coxiella* symbionts are widespread into hard ticks. Parasitol Res. 2016;115: 4691–4699. 10.1007/s00436-016-5230-z

SECTION 2:
VECTORS

7

Disentangling Vector-Borne Transmission Networks: A Universal DNA Barcoding Method to Identify Vertebrate Hosts from Arthropod Bloodmeals

Miguel Alcaide[1], Ciro Rico[1], Santiago Ruiz[2], Ramón Soriguer[1], Joaquín Muñoz[1], Jordi Figuerola[1]

[1] Estación Biológica de Doñana (CSIC), Isla de la Cartuja, Sevilla, Spain,

[2] Diputación de Huelva, Área de Medio Ambiente, Huelva, Spain

ABSTRACT

Emerging infectious diseases represent a challenge for global economies and public health. About one fourth of the last pandemics have been originated by the spread of vector-borne pathogens. In this sense, the advent of modern molecular techniques has enhanced our capabilities to understand vector-host interactions and disease ecology. However, host identification protocols have poorly profited of international DNA barcoding initiatives and/or

Citation: Alcaide M, Rico C, Ruiz S, Soriguer R, Muñoz J, Figuerola J (2009) Disentangling Vector-Borne Transmission Networks: A Universal DNA Barcoding Method to Identify Vertebrate Hosts from Arthropod Bloodmeals. PLoS ONE 4(9): e7092. https://doi.org/10.1371/journal.pone.0007092

have focused exclusively on a limited array of vector species. Therefore, ascertaining the potential afforded by DNA barcoding tools in other vector-host systems of human and veterinary importance would represent a major advance in tracking pathogen life cycles and hosts. Here, we show the applicability of a novel and efficient molecular method for the identification of the vertebrate host's DNA contained in the midgut of blood-feeding arthropods. To this end, we designed a eukaryote-universal forward primer and a vertebrate-specific reverse primer to selectively amplify 758 base pairs (bp) of the vertebrate mitochondrial Cytochrome *c* Oxidase Subunit I (COI) gene. Our method was validated using both extensive sequence surveys from the public domain and Polymerase Chain Reaction (PCR) experiments carried out over specimens from different Classes of vertebrates (Mammalia, Aves, Reptilia and Amphibia) and invertebrate ectoparasites (Arachnida and Insecta). The analysis of mosquito, culicoid, phlebotomie, sucking bugs, and tick bloodmeals revealed up to 40 vertebrate hosts, including 23 avian, 16 mammalian and one reptilian species. Importantly, the inspection and analysis of direct sequencing electropherograms also assisted the resolving of mixed bloodmeals. We therefore provide a universal and high-throughput diagnostic tool for the study of the ecology of haematophagous invertebrates in relation to their vertebrate hosts. Such information is crucial to support the efficient management of initiatives aimed at reducing epidemiologic risks of arthropod vector-borne pathogens, a priority for public health.

INTRODUCTION

The control of emerging infectious diseases constitutes one of the most important concerns of global economies and human health. Recent studies have emphasized that the majority of the last pandemics have been originated by zooneses in the wild. The participation of vectors in the spread of zoonotic diseases was estimated to occur in about one fourth of pathogen outbreaks during the last century [1]. Importantly, human related activities are contributing to increase the impact of vector-borne diseases by increasing vector density (e.g. irrigation, urbanization, dam construction) or by introducing pathogens into areas in which they had been hitherto absent [2]. As a result, there has been the need to control the population of blood-feeding arthropods to reduce epidemiologic risks because they may harbor pathogens responsible for serious infectious diseases such as malaria, viral encephalitis, West Nile virus, Chagas disease, Lyme disease or African sleeping sickness. Although pathogen prevalence in vectors is usually low, a correct understanding of vector-hosts interactions is crucial to predict transmission patterns and for the development of efficient control policies.

The enhancement of species identification since the advent of molecular methods has revolutionized our view of such complex ecological networks at the same time that it has contributed to gain insights on the co-evolutionary relationships between vectors, pathogens and their hosts [reviewed in 3]. The emerging of international initiatives such as the Consortium for the Barcode of Life (www.barcoding.si.edu) is expected to greatly expand the benefits of molecular methods in this field. The DNA barcoding project aims to establish a 648-bp fragment of the Cytochrome *c* Oxidase Subunit I (COI) mitochondrial gene as universal method for the taxonomic classification of biodiversity [4], [5]. The coverage provided by barcoding databases will soon facilitate an accurate assignment of host-vector associations, and their use by ecologists and epidemiologists is believed to increase considerably during the next few years [6]. One of the most important advantages of the COI locus to identify vertebrate hosts is linked to its model of molecular evolution, which is believed to provide better resolution of deeper taxonomic affinities than other molecular markers [7], [8]. In this sense, precise host identifications might be crucial given that pathogen outcomes, exposure to vectors, demographic parameters and dispersal patterns may vary considerably even between closely related species [9]–[11]. Nevertheless, identification methods based on the partial sequencing of the vertebrate COI gene from arthropod bloodmeals have been underutilized in relation to other molecular approaches [3]. In fact, COI identification protocols conducted to date have focused exclusively on mosquito bloodmeals [3], [12]. There is therefore the need to ascertain and take advantage of the potential afforded by DNA barcoding tools in other vector-host systems of public health and veterinary importance.

In this study, we have designed a single pair of primers for the selective analysis of host COI sequences from arthropod bloodmeals. Our method was also aimed at accelerating host identification and reducing both laboratory efforts and costs.

RESULTS

Validation of Vertebrate-specific PCR Amplification

Primers M13BC-FW and BCV-RV1 were validated using high quality DNA extracts from avian (lesser kestrel *Falco naumanni*, Spanish Imperial eagle

Aquila adalberti), mammalian (water vole *Arvicola sapinus*), amphibian (natterjack toad *Bufo calamita*) and reptilian (Iberian wall lizard *Podarcis hispanica*) species. PCR experiments always yielded high concentration of amplicons of the expected size. The same PCR protocol failed to amplify the target fragment when applied to DNA extracts obtained from the abdomen of non-engorged arthropods (mosquitoes, ticks, culicoids and sandflies). Nonetheless, we successfully amplified and sequenced a fragment of the arthropod COI locus using the eukaryote-universal primers proposed for the 'DNA mini-barcoding' approach [13]. It is important to emphasize that eukaryote-universal primers preferentially amplified vector COI sequences when applied to mosquito bloodmeals. On the contrary, our vertebrate-specific primer set only replicated host COI genes.

Assignment of Unknown COI Sequences Replicated from Bloodmeals to Specific Vertebrate Hosts

The first PCR reaction with primers M13BC-FW and BCV-RV1 yielded suitable concentrations of PCR products for sequencing in 43 out of 100 mosquito bloodmeals. Non-suitable samples for sequencing displayed either low concentration of PCR products or a lack of bands of the expected size. Nonetheless, a nested PCR using M13 and BCV-RV2 primers generated suitable positives for sequencing in 97 out of 100 mosquito bloodmeals. The nested-PCR also proved satisfactory when applied to the bloodmeals of additional vector species (ticks, sandflies, culicoids and blood-sucking bugs).

The bioinformatics platform supported by the Barcode of Life Data Systems (BOLD) database permitted us to identify the origin of the bloodmeals contained in the midgut of several species of mosquitoes and other blood-fed ectoparasites (see Table 1). This is an online workbench that aids collection, management, analysis, and use of DNA barcodes by researches in different fields (www.barcodinglife.org). The inspection of sequencing electropherograms supported the amplification of the mitochondrial barcode locus from one single vertebrate host in the vast majority of cases. The similarity of our unknown vertebrate COI sequences with respect to those from museum voucher specimens was always >99% except for five cases. Various unknown vertebrate COI sequences could not be assigned at the species level. These COI sequences showed the highest sequence similarity (>90%) with species of the genus *Lepus*, *Lynx*, *Mus*, *Alectoris* and the Family *Herpestidae*. Given that the unique members of the genus *Lepus*, *Alectoris* and the Family *Herpestidae* inhabiting the south of Spain are not included within the BOLD System database yet, we deduced that these species could be the Iberian hare *Lepus granatensis*, the red-legged partridge *Alectoris rufa* and the Egyptian mongoose *Herpestes ichneumon*. Furthermore, we know that the two remaining COI sequences belonged to the Iberian Lynx *Lynx pardinus* and the Algerian mouse *Mus spretus* because ectoparasites were directly sampled on hosts. Overall, our molecular method allowed identifying 40 different vertebrate hosts, including 16 mammalian, 23 avian and one reptilian species (see Table 1). DNA extraction and PCR negative controls did not yield PCR bands and the analysis of our positive control always matched COI sequences of the mallard duck *Anas platyrhynchos*. Repeatability experiments performed over those bloodmeals reporting rare hosts (i.e. those found only in one bloodmeal) were also successful.

Table 1. Vertebrate hosts for different species of blood-feeding ectoparasites collected in South-western Spain.

Species	Mammalian host	Avian host
Anopheles algeriensis (Insecta: Culicidae)	Dama dama (1); Bos taurus (1)	
Anopheles atroparvus (Insecta: Culicidae)	Bos taurus (3); Oryctolagus cuniculus (1)	
Culex modestus (Insecta: Culicidae)		Anas platyrhynchos (2); Anser anser (1); Chen caerulescens (1); Branta canadensis (1); Egretta garzetta (1); Tadorna ferruginea (1); Anas strepera (1); Gallus gallus (1); Ardea cinerea (1); Anas acuta (1); Tadorna tadoma
Culex perexiguus (Insecta: Culicidae)	Rattus norvergicus (1); Canis familiaris (1)	Alectoris rufa (1); Streptopelia decaocto (1)
Culex pipiens (Insecta: Culicidae)	Homo sapiens (1); Herpestes ichneumon (1); Felis catus (1); Canis familiaris (2)	Passer domesticus (7); Turdus merula (3); Streptopelia decaocto (3); Galerida cristata (1); Sturnus vulgaris (2); Cairina moschata (1); Grus grus (1); Sylvia melanocephala (1); Alectoris rufa (1)
Culex theileri (Insecta: Culicidae)	Bos taurus (8); Cervus elaphus (4); Dama dama (2); Equus caballus (3); Homo sapiens (1); Lepus granatensis (2); Oryctola gus cuniculus (1); Sus scrofa (2)	Bubulcus ibis (1); Meleagris gallopavo (1)
Ochlerotatus caspius (Insecta: Culicidae)	Bos taurus (3); Canis familiaris (3); Capra hircus (1); Felis catus (1); Cervus elaphus (1); Equus caballus (2); Oryctolagus cuniculus (2); Ovis aries (2); Sus scrofa (1)	Gallus gallus (4); Podiceps nigricollis (1); Passer domesticus (2); Turdus merula (1); Sturnus vulgaris (1)
Culiseta longiareolata (Insecta: Culicidae)		Passer domesticus (2)
Phlebotomus sp. (Insecta: Phlebotominae)	Oryctolagus cuniculus (2)	
Culicoides sp. (Insecta: Ceratopogonidae)		Passer domesticus (2)
Dipetalogaster maximus (Insecta: Reduviidae)	Lynx pardinus (2)	
Rhipicephalus spp. (Arachnida: Ixodidae)	Mus spretus (1) Canis familiaris (2)	

We isolated either mammalian or avian DNA from arthropod bloodmeals, except for a *Culex modestus* mosquito that fed on a turtle (*Mauremys leprosa*, not included because of limitations in table format). Brackets indicate the number of bloodmeals reporting DNA from particular hosts in each vector species.
doi:10.1371/journal.pone.0007092.t001

https://doi.org/10.1371/journal.pone.0007092.t001

Identification of Mixed Bloodmeals

The alignment of ambiguous COI sequences (i.e. those displaying double peaks at different positions of sequencing electropherograms) with respect to our validated set of vertebrate COI sequences indicated four mixed bloodmeals. The inspection of sequencing electropherograms suggested the co-amplification of the COI locus from two different vertebrate hosts. A specimen of *Ochlerotatus caspius* and a specimen of *Culex pipiens* could have fed on a house sparrow *Passer domesticus* and a common blackbird

Turdus merula (see Fig. 1). A specimen of *Culex modestus* could have fed on a common goose *Anser anser* and a domestic snow goose *Chen caerulescens*, and a specimen of *Culex theileri* could have fed on a red deer *Cervus elaphus* and a cow *Bos taurus*. In some cases, particularly for the wild boar *Sus scrofa*, we suspected that double peaks could be related to the co-amplification of nuclear insertions of the mitochondrial genome or heteroplasmy. We submitted unambiguous stretches of COI sequences to the public databases to ensure that these DNA sequences could only be originated from the same species. Then, sequence identities larger than 99% were only reported with respect to the same, single vertebrate host.

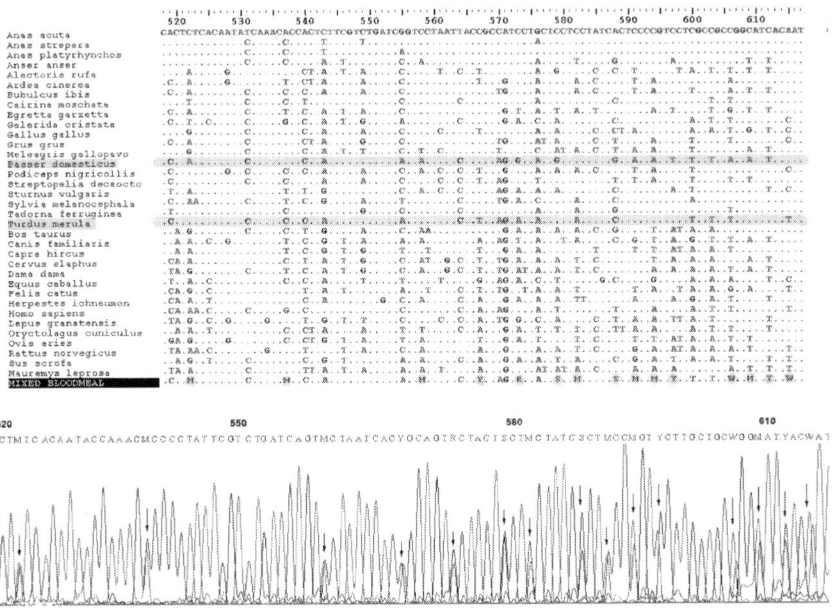

Figure 1. Identification of a mixed bloodmeal. The co-amplification of COI fragments from different hosts may generate several ambiguous nucleotide positions (in yellow and indicated by arrows). The ambiguous sequence can be subsequently compared with respect to a validated set of COI sequences from the hosts typically found in the study area. In this example, the sequencing electropherogram obtained could be the result of the simultaneous amplification of DNA from house sparrow *Passer domesticus* and common blackbird *Turdus merula*.

https://doi.org/10.1371/journal.pone.0007092.g001

DISCUSSION

In this study, we describe a straightforward and universal method for the PCR specific amplification and analysis of the vertebrate barcode COI locus from the midgut of blood-feeding arthropods. The suitability of a single pair of primers capable to cope with all candidate hosts across a broad spectrum of vectors constitutes the major improvement put forward by our method. Although single pairs of primers were formerly available for the amplification of the vertebrate COI [12] and Cytochrome *b* (Cyt *b*) genes [14], empirical support for their suitability in other vector species rather than mosquitoes and even the possible co-amplification of invertebrate DNA, respectively, could be considered as major limitations. In fact, the vast majority of studies have employed cocktails of primers to deal with different vertebrate hosts [reviewed in 3]. The inconvenience of using avian or mammalian specific primers for those vectors traditionally considered ornithophilic or mammalphilic is also highlighted in the present study. As an example, we found that some mosquito species such as *Culex pipiens*, *Cx. theileri* and *Ochlerotatus caspius* can feed occasionally on other vertebrates groups (see Table 1). Shedding light on this kind of relationships would provide valuable information to understand pathogen transmission between different vertebrate groups.

One of the most interesting methodological advantages of our assay relies on the suitability of the inexpensive HotSHOT protocol for a rapid and feasible DNA extraction from bloodmeals. Success of our nested-PCR approach was >90% for the bloodmeals analyzed, a similar value to other studies based on the amplification of a shorter fragment (<400 bp) of the Cyt *b* gene [e.g. 15]. Assuming a Taq DNA Polymerase error rate about 1×10^{-4} base substitutions/bp/cycle, the total number of mutations introduced by our nested PCR protocol is around 8.24 base substitution per sequence (i.e. $1 \times 10^{-4} \times 758$ bp$\times 110$ cycles (75 PCR cycles +35 Big Dye cycles)). This estimate represents 1% of overall nucleotide divergence and will not exceed in any case the 2% threshold of within-species variation established by the BOLD initiative. PCR costs and sample manipulation can be considerably reduced if the number of engorged arthropods is large enough to make unnecessary the analysis of those specimens containing tiny bloodmeals or bloodmeals into an advanced digestion stage. Direct sequencing methods, which represent >80% of overall costs, can be also replaced by other less expensive and more straightforward molecular methods [reviewed in 3] once researchers have accomplished a first preliminary survey using our general approach. The application of our method throughout hotspots of biodiversity is however strongly encouraged given that alternative methods can difficultly provide similar coverage thresholds and sample manipulation convenience.

Importantly, we show that the analysis of sequencing electropherograms could be useful for the identification of mixed bloodmeals from even closely related species (see Fig. 1). The development of specific software to assist the identification of mixed sequencing electropherograms using the sequence data deposited in the public databases must be encouraged in this respect. PCR reactions can be nevertheless highly competitive in mixed bloodmeals and some DNA data could be missing due to partial or total degradation of one of the host's DNA and/or because of the amount of ingested blood from different host species greatly differs. To solve this potential methodological limitation, some authors have satisfactorily used multiplexed primers targeting to different vertebrate groups [16], [17], [but see the multiplexing drawbacks reviewed in 3]. In the very next future, new generation and cost-effective sequencing technologies such as pyrosequencing are expected to revolutionize bloodmeal analyses because it will permit researchers to simultaneously screen a wider array of DNA sequences from mixed samples at different proportions [18], [19].

Our method could have a limited value, however, for those species parasitizing known hosts most of the time (e.g. ticks, fleas, mites or lices). Engorged ectoparasites can be nonetheless collected off-hosts using traps or drag sampling [20], [21]. Ticks deserve a particular consideration because of their mechanism of blood digestion differs from the rest of blood-feeding arthropods. DNA digestion seems to occur more quickly, and PCR-based methods have proven to vary considerably between different laboratories. Thus, recent research has encouraged the use of tryptic-digestion mass spectrometry of blood proteins to identify various hosts with a single tick bloodmeal [22]. Nevertheless, increased costs and a very limited availability of protein databases for non-model species are two important drawbacks.

In conclusion, we provide a novel, relatively straightforward and cost-effective molecular method that permits researchers to get deeper into the investigation of vector-borne disease ecology and co-evolutionary relationships. The reinforcement of our knowledge about blood-feeding behaviour of haematophagous arthropods and transmission patterns in wild species should decisively contribute to an efficient evaluation and modelling of epidemiologic risks and a better understanding of ecological networks.

MATERIALS AND METHODS

Sampling of Engorged Ectoparasites and Study Area

Wildlife engorged mosquitoes, culicoids and sand flies (Class *Insecta*, Order Diptera, Families *Culicidae*, *Ceratopogonidae* and *Phlebotomiae*) were captured using CDC traps supplied with dry ice to attract ectoparasites through light and CO_2. Traps were placed in several locations of South-western Spain, including the Doñana National Park, and the specimens collected the next day. This fact is important given that some studies have shown that the probability of amplifying host DNA after 36 hours from feeding decreases considerably [23]. Engorged ticks (Class *Arachnida*, Order Acari) were directly sampled from known parasitized hosts (mice and dogs). Individual ectoparasites were identified at the species or genus level. In addition, pathogen-free sucking bugs *(Dipetalogaster maximus*, Class *Insecta*, Order Hemiptera, Subfamily *Triatominae*) used for non-invasive blood sampling [see 24] of captive Iberian lynxs *Lynx pardinus* (www.lynxexsitu.es), were also obtained. Ectoparasites were preserved at −80°C until DNA extraction. Overall, we sampled 100 engorged mosquitoes plus a few specimens from other families of blood-feeding ectoparasites (Table 1).

DNA Extraction

Individual ectoparasites were processed in Petri plates. Using sterile tips, we pressed deeply on the abdomen of engorged individuals to release bloodmeals. According to the HotSHOT protocol [25], bloodmeals were pipetted into 50 μl of lysis solution (25 mM NaOH, 0.2 mM EDTA) and latter incubated at 95°C during 30 min. In those cases where bloodmeals could not be easily extracted from the midgut of the ectoparasite, we cut the entire abdomen, which was introduced and crushed into the lysis solution. In addition, at least two negative DNA extraction control (i.e. absence of tissue) were performed during PCR experiments. After incubation, the solution was put on ice for five minutes and then we added 50 μl of neutralization solution (40 mM Tris-HCl). Bloodmeals or abdomens were simultaneously processed using 96-thermowell plates or 8-thermowell individual strips and stored at −20°C until PCR amplification.

Primer Design Strategy

We downloaded all vertebrate COI sequences (N=18,928) from the Classes *Mammalia*, *Aves*, *Amphibia* and *Reptilia* that were available in the public domain managed by the BOLD Systems database in January 2009 (www.barcodinglife.org). We also downloaded 6,784 arthropod COI sequences from taxonomic groups that included blood-feeding species. This survey included species from the Classes *Arachnida* (Order Acari) and *Insecta* (Order Diptera, Families *Culicidae*, *Simuliidae*, *Ceratopogonidae*, *Hippoboscidae*, *Tabanidae*and *Glossinidae*; Order Hemiptera, Subfamiliy*Triatominae* and Order Phthiraptera). COI sequences were aligned using the software BioEdit 7.0.9.0 [26]. Our multiple alignments revealed several conserved nucleotide positions at the 5′ end of the COI gene. We designed a universal-forward primer with an M13-tail added at the 5′ end (M13BC-FW 5′-TGT AAA ACG ACG GCC AGT-HAA YCA YAA RGA YAT YGG NAC-3′), similar to other primers previously used in other barcode approaches [e.g. 13]. Then, we searched for a reverse primer that allowed the specific amplification of vertebrate COI sequences while avoiding the co-amplification of invertebrate DNA (BCV-RV1 5′-GC<u>Y C</u>AN AYY ATN CYY R<u>T</u>R (T)(A)-3′). Three nucleotide positions always differing between vertebrate and invertebrate COI sequences are underlined. An additional difference in base composition, which position varies depending on particular arthropod Families, is indicated in brackets. A positional nucleotide numerical summary carried out using BioEdit revealed that the matching of our forward primer to the target sequence was >99% at each nucleotide position, after analyzing more than 6,000 COI sequences. The vertebrate-specific reverse primer matched the target sequence with values >99% at every nucleotide position, after comparing a minimum number of 5,814 vertebrate COI sequences from the four taxonomic Classes. The mismatching of invertebrate COI sequences with respect to our vertebrate-specific primer was also >99% in at least all the four nucleotide positions mentioned above. This estimate was calculated after comparing 3,273 arthropod COI sequences.

PCR Amplification and Sequencing of Vertebrate COI Sequences

The expected length (excluding primers) of the PCR product obtained with primers M13BCV-FW and BCV-RV1 is 758 bp. An optimized PCR protocol consisted of 4 min at 94°C, followed by 35 cycles of 40 s at 45°C, 1 min at 72°C and 40 s at 94°C, with a final extension step of 7 min at 72°C. PCR reactions were carried out using a PTC-100 Programmable Thermal Controller (MJ Research) in a final volume of 30 μl containing 1 unit of a commercial Taq Polymerase (Bioline), 1X manufacturer-supplied buffer (Bioline), 2.5 mM $MgCl_2$, 0.25 mM of each dNTP, 5% DMSO, 10 μg of

BSA (Bovine Serum Albumin - Amersham corp.), 0.16 µM of primers M13BCV-FW and BCV-RV1 and 1 µl of extracted DNA. Bloodmeals can be sometimes partially digested, and thus, the number of DNA molecules to be used as templates decreases. Gradual digestion of DNA in arthropod's midgut should be heterogeneous across the sample, and depends on the time elapsed since the ectoparasite has fed. Thus, in order to standardize the number of amplified DNA copies before sequencing, we designed a PCR re-amplification protocol. Further, re-amplification of apparently negative PCR reactions has proven to be successful in previous studies [17]. In this second PCR reaction, we used the M13 primer and a modified reverse primer from that of the first PCR (BCV-RV2=5′-GCY CAN AYY ATN CYY RTR TAN CC-3′). Terminal cytosine nucleotides match to two conserved guanines of the COI locus in eukaryotes. This second PCR reaction consisted of 3 min at 94°C followed by a touch down protocol decreasing the annealing temperature from 60°C to 45°C during 40 s (−1°C/cycle), with 1 min extension and 40 s denaturalization steps at 72°C and 94°C, respectively, followed by 24 cycles of 40 s at 45°C, 72°C and 94°C, and a final elongation step of 7 min at 72°C. PCR reactions were carried out in a final volume of 30 µl containing 1 unit of Taq Polymerase (Bioline), 1X manufacturer-supplied buffer, 1.7 mM $MgCl_2$, 0.25 mM of each dNTP, 5% DMSO, 5 µg of BSA, 0.16 µM of M13 and BCV-RV2 primers and 1 µl of the PCR products obtained during the first amplification step (M13BC-FW/BCV-RV1). We used negative controls of PCR amplification (i.e. absence of DNA template) to detect contaminations derived from the presence of bovine (because of the use of BSA) and human DNA. As positive control, we used the bloodmeal of a mosquito sampled while feeding on a mallard duck captured for ringing. A further analysis of repeatability was accomplished by applying the same protocol to a couple of bloodmeals reporting rare hosts (i.e. hosts only found in one bloodmeal). The species used for the repeatability analyses were the crested lark *Galerida cristata* and the Egyptian mongoose *Herpestes ichneumon*. The COI locus from both species was amplified and sequenced independently in three different occasions. An identical PCR protocol was applied to DNA extracts obtained from the abdomen of non-engorged mosquitoes (*Culex* sp., N=2), ticks (*Hyalomma* sp., N=2), culicoids (*Culicoides* sp., N=2) and sand flies (*Phlebotomus* sp., N=2) to ensure the unsuccessful amplification of invertebrate DNA. Vertebrate DNA was used during these PCR experiments to control for systematic PCR failure. To identify problems related to the quality of the invertebrate DNA extracts, we performed additional PCR experiments using the eukaryote-universal mini-barcoding primers proposed by Meusnier and co-workers [13]. PCR products were visualized in 1.5% agarose gels and cleaned-up using the commercial ExoSAP-IT reagent (GE Healthcare Life Sciences). Sequencing reactions were performed according to the BigDye 1.1 technology (Applied Biosystems) using 5 pmoles of the BCV-RV2 primer. This primer is located around 100 bp downstream from the 3′ end of the molecular target for DNA barcoding. Labelled DNA fragments were resolved using an ABI 3130xl automated sequencer (Applied Biosystems).

Identifying Unambiguous and Ambiguous Unknown COI Sequences

We used the identification engine implemented in the BOLD-IDS platform (http://www.barcodinglife.org/views/idrequest.php) to assign unknown COI sequences to particular species. The BOLD-IDS component provides a species identification tool that accepts DNA sequences from the barcode region and returns a taxonomic assignment to the species level when possible. We always chose to search within the reference barcode dataset (i.e. validated subset of the full database with a minimum sequence length of 500 bp and containing only those species represented by three or more individuals showing less than 2% sequence divergence) when possible.

Additionally, haematophagous arthropods can feed on several hosts. If hosts belong to different species, the resulting sequencing electropherogram will contain a variable number of ambiguities (i.e. double peaks). We used BioEdit to introduce IUPAC degenerate codes (i.e. those representing different sets of nucleotides) across sequencing electropherograms. Ambiguous COI sequences were aligned with respect to a reference data set containing unambiguous COI sequences from the vertebrate species identified in our study area so far. Then, we removed from the alignment those COI sequences differing from the ambiguous unknown sequence in at least 2% of nucleotide positions (i.e. the within-species threshold proposed by the DNA barcoding initiative) until only a few candidate species remained.

ACKNOWLEDGMENTS

We are very grateful to Astrid Vargas, from the Ex-situ Conservation Program of the Iberian Lynx (www.lynxexsitu.es), and to the Institute of Wildlife and Zoological Research (IZW, Berlin, Germany, www.izw-berlin.de) for kindly providing blood-fed specimens of *Dipetalogaster maximum*. We also thank JL Barroso for providing engorged ticks and Esmeralda Pérez for her work collecting and classifying mosquito. We thank to Dr. Ng Lee-Ching for comments and suggestions on an earlier version of the manuscript.

REFERENCES

1. Jones KE, Patel NG, Levy MA, Storeyard A, Balk D, et al. (2008) Global trends in emerging infectious diseases. Nature 451: 990–994.
2. Gratz NG (1999) Emerging and resurging vector-borne diseases. Annu Rev Entomol 44: 51–75.
3. Kent RJ (2009) Molecular methods for arthropod bloodmeal identification and applications to ecological and vector-borne disease studies. Mol Ecol Res 9: 4–18.
4. Ratnasingham S, Hebert PDN (2007) BOLD: The Barcode of Life Data System. Mol Ecol Notes 7: 355–364.
5. Waugh J (2007) DNA barcoding in animal species: progress, potential and pitfalls. BioEssays 29: 188–197.
6. Valentini A, Pompanon F, Taberlet P (2008) DNA barcoding for ecologists. Trends Ecol Evol 24: 110–117.
7. Hebert PDN, Cywinska A, Ball SL, deWaad JR (2003) Biological identifications through DNA barcodes. Proc R Soc B 270: 313–321.
8. Tavares ES, Baker JA (2008) Single mitochondrial gene barcodes reliably identify sister-species in diverse clades of birds. BMC EvolBiol 8: 81.
9. Wheeler SS, Barker CM, Fang Y, Arnijos MV, Carrol BD, et al. (2009) Differential impact of West Nile Virus on California Birds. The Condor 111: 1–20.
10. Pérez-Tris J, Bensch S (2005) Dispersal increases local transmission of avian malarial parasites. Ecology Letters 8: 838–845.
11. Foley JE, Clueit SB, Brown RN (2008) Differential exposure to an aplasmaphagocytophilum in rodent species in northern California. Vector Borne Zoonitc Dis 8: 49–55.
12. Townzen JS, Brower AVZ, Judd DD (2008) Identification of mosquito bloodmeals using mitochondrial cytochrome oxidase subunit I and cytochrome b gene sequences. Med Vet Entomol 22: 386–393.
13. Meusnier I, Singer GAC, Landry JF, Hickey DA, Hebert PD, et al. (2008) A universal DNA mini-barcode for biodiversity analyses. BMC Genomics 9: 214.
14. Kocher TD, Thomas WK, Meyer A, Edwards SV, Pääbo S, et al. (1989) Dynamics of mitochondrial DNA evolution in animals: amplification

and sequencing with conserved primers. Proc Natl Acad Sci USA 86: 6196–6200.

15. Oshaghi MA, Chavshin AR, Vatandoost H, Yaaghoobi F, Mohtarami F, et al. (2006) Effects of post-ingestion and physical conditions on PCR amplification of host blood meal DNA in mosquitoes. ExperimParasitol 112: 232–236.
16. Kent RJ, Norris DE (2005) Identification of mammalian blood meals of mosquitoes by a multiplexed polymerase chain reaction targeting *cytochrome b*. Am J Trop Med Hyg 73: 336–342.
17. Kent RJ, West AJ, Norris DE (2004) Molecular differentiation of colonized human malaria vectors by 28S rDNA polymorphisms. Am J Trop Med Hyg 71: 514–51.
18. King RA, Read DS, Traugott M, Symondson OC (2008) Molecular analysis of predation: a review of best practice for DNA-based approaches. Mol Ecol 17: 947–963.
19. Traugott M, Bell JR, Broad GR, Powel W, Van Been FJF, et al. (2008) Endoparasitism in cereal aphids: molecular analysis of a whole parasitoid community. Mol Ecol 17: 3928–3938.
20. Woods MR, Montenieri JA, Eisen RJ, Zeidner NS, Borchert JN, et al. (2009) Identification of Flea Blood Meals Using Multiplexed Real-Time Polymerase Chain Reaction Targeting Mitochondrial Gene Fragments. Am J Trop Med Hyg 80: 998–1003.
21. Falco RC, Fish D (1992) A comparison of methods for sampling the deer tick, Ixodesdammini, in a Lyme disease endemic area. Exp Appl Acarol 14: 165–173.
22. Wickramasekara S, Bunikis J, Wysocki V, Barbour AG (2008) Identification of Residual Blood Proteins in Ticks by Mass Spectrometry Proteomics. Emerg Infect Dis 14: 1273–1275.
23. Rasgon J (2008) Stable isotope analysis can potentially identify completely-digested bloodmeals in mosquitoes. PLoS ONE 3: 1–3.
24. Thomsen R, Voigt CC (2006) Non-invasive blood sampling from primates using laboratory-bred blood-sucking bugs (*Dipetalogaste rmaximus*; Reduviidae, Heteroptera). Primates 47: 397–400.
25. Truett GE, Walker JA, Truett AA, Mynatt RL, Heeger P, et al. (2000) Preparation of PCR-quality DNA with hot sodium hydroxide and tris (HotSHOT). Biotechniques 29: 52–54.

26. Hall TA (1999) BioEdit: a user-friendly biological sequence alignment editor and analysis program for Windows 95/98/NT. Nucleic Acids Symp Ser 41: 95–98.

8

Correction: Disentangling Vector-Borne Transmission Networks: A Universal DNA Barcoding Method to Identify Vertebrate Hosts from Arthropod Bloodmeals

Miguel Alcaide[1], Ciro Rico[1], Santiago Ruiz[2], Ramón Soriguer[1], Joaquín Muñoz[1], Jordi Figuerola[1]

[1] Estación Biológica de Doñana (CSIC), Isla de la Cartuja, Sevilla, Spain,

[2] Diputación de Huelva, Área de Medio Ambiente, Huelva, Spain

In the Materials and Methods section two optimal primer sequences were switched by mistake with sub-optimal primer sequences. The correct sentences should be: "In this second PCR reaction, we used an M13 Primer (GTA AAA CGA CGG CCA GTG) and a modified reverse primer from that of the first PCR (BCV-RV2=5' ACY ATN CCY ATR TAN CCR AAN GG-3'." The sentence "Terminal cytosine nucleotides match to two conserved guanines of the COI locus in eukaryotes" should be omitted.

Citation: Alcaide M, Rico C, Ruiz S, Soriguer R, Muñoz J, Figuerola J (2010) Correction: Disentangling Vector-Borne Transmission Networks: A Universal DNA Barcoding Method to Identify Vertebrate Hosts from Arthropod Bloodmeals. PLoS ONE 5(10): 10.1371/annotation/ce5bff53-4638-4931-aa68-904b74db4b20. https://doi.org/10.1371/annotation/ce5bff53-4638-4931-aa68-904b74db4b20

9

One Health: the Importance of Companion Animal Vector-Borne Diseases

Michael J Day

School of Veterinary Sciences, University of Bristol, Langford BS40 5DU, UK

ABSTRACT

The international prominence accorded the 'One Health' concept of co-ordinated activity of those involved in human and animal health is a modern incarnation of a long tradition of comparative medicine, with roots in the ancient civilizations and a golden era during the 19th century explosion of knowledge in the field of infectious disease research. Modern One Health tends to focus on zoonotic pathogens emerging from wildlife and production animal species, but one of the most significant One Health challenges is rabies for which there is a canine reservoir. This review considers the role

Citation: Day, M.J. One health: the importance of companion animal vector-borne diseases. Parasites Vectors 4, 49 (2011). https://doi.org/10.1186/1756-3305-4-49

of small companion animals in One Health and specifically addresses the major vector-borne infectious diseases that are shared by man, dogs and cats. The most significant of these are leishmaniosis, borreliosis, bartonellosis, ehrlichiosis, rickettsiosis and anaplasmosis. The challenges that lie ahead in this field of One Health are discussed, together with the role of the newly formed World Small Animal Veterinary Association One Health Committee.

ONE HEALTH: HISTORY AND SCOPE

The 'One Health' concept states simply that there should be a seamless interaction between veterinary and human medicine with clinicians, researchers, agencies and governments working together for the benefit of domestic and wild animal and human health and the global environment. Such interactions may take place at many levels - from management of zoonotic infectious disease outbreaks in the field, to joint research programmes to integrated policy making and funding decisions.

The concept of One Health is nothing new. Dunlop and Williams in their *History of Veterinary Medicine*[1] catalogue the early recognition of the significance of zoonotic infectious disease. The ancient Mesopotamians collated laws into codes and the Eshuna Code of 2300 BC makes the owner of a rabid dog responsible for the containment of that animal. The Babylonian king Adad-apla-iddina (1068-1047 BC) constructed a temple to the healing goddess Gula who was represented with or as a dog and was worshiped as a healer and protector from rabies. It is also believed that the role of the dog as an adjunct to healing was recognized through the ability of these animals to provide comfort to patients. The ancient Egyptians provide the first documentation of animal diseases in the Kahun papyrus (1900 BC) and in the context of the present review; it is believed that this civilization had awareness of vector-borne disease through their use of mosquito netting. There is a rich history of comparative medicine within the ancient Chinese dynasties in which there were organized specialities of physicians, surgeons, dieticians and veterinarians - clearly emphasizing the importance of animals to this society. The principles of yin-yang as practiced through acupuncture applied equally to human and animal patients. In ancient Greece, Aristotle (384-322 BC) continued the tradition of Hippocrates (460-367 BC) and promoted comparative medicine through his extensive studies of animal anatomy and the pathology of animal diseases. This tradition of comparative anatomy was continued in ancient Roman civilization with the most prominent contribution made by Galen (130-200 AD). The

Hippocratic-Galenic principles continued throughout Medieval Europe to the Renaissance where the comparative anatomical legacy of Leonardo da Vinci (1452-1519) laid the way for future developments. Later, John Hunter (1728-1793) became a true practitioner of comparative medicine and was tutor to Edward Jenner (1749-1823) whose development of vaccination was a classical example of the One Health principle. The formalization of One Health is often attributed to Claude Bourgelat, the founder of the first veterinary school in Lyon in 1761, precisely 250 years ago. In writing on the relationship between human and animal medicine, Bourgelat stated that 'either medicine will mutually enlighten and perfect the other when we discard a derisory, harmful prejudice'. The greats of the golden era of infectious disease research - Louis Pasteur (1822-1895), Robert Koch (1843-1910), Rudolf Virchow (1821-1902) and John M'Fadyean (1853-1941), were all proponents of One Health and many of their advances were based in studies of animal disease [1–3].

It was perhaps in the 20th century that the medical and veterinary professions became slightly distant, a fact that was recognized by Calvin Schwabe (1927-2006) who promoted a 're-unification' from the 1960s and is attributed with coining the term 'One Medicine' or One Health [3, 4]. It was however, not until the past five years, that the One Health concept has truly gathered international momentum. Collaborative ventures between the British Medical Association and the British Veterinary Association [5] and the American Medical Association and the American Veterinary Medical Association [6] paved the way for the seminal concept note produced following a meeting of the World Organization for Animal Health (OIE), the World Health Organization (WHO), the Food and Agriculture Organization (FAO), the United Nations Children's Fund (UNICEF) and the World Bank http://www.oie.int/fileadmin/Home/eng/Current_Scientific_Issues/docs/pdf/FINAL_CONCEPT_NOTE_Hanoi.pdf. A number of organizations have endorsed the One Health concept (e.g. the Federation of Veterinarians of Europe) and One Health groups have been established; including the One Health Commission http://www.onehealthcommission.org/, the One Health Initiative http://www.onehealthinitiative.com/index.php and the Comparative Clinical Science Foundation http://www.onemedicine.org.uk/. These initiatives have culminated recently in the First International One Health Conference, held in Melbourne, Australia.

COMPANION ANIMALS IN ONE HEALTH

From the beginnings summarized above, One Health has now expanded rapidly with endorsement by numerous medical and veterinary organizations. The major focus of these proponents has been the very high-impact interactions between human and production animal and wildlife health with global zoonotic disease pandemics and 'emerging' infectious diseases deemed to have arisen in these animal species. Many of the best examples of such infections are viral, where virus mutation or re-assortment permits extension of the target host range (e.g. SARS coronavirus, H5N1 and H1N1 influenza virus, Nipah virus, Hendra virus, human immunodeficiency virus (HIV), but the role of parasitic diseases in One Health has also been discussed [7, 8]. The impact of environmental change (e.g. climate change, deforestation and urbanization) and lifestyle change including the increase in global human and animal movement [9] has also become part of the One Health concept.

Perhaps less often considered has been the enormous potential role of companion animals, and particularly the domesticated dog and cat, in One Health. In developed nations, pet ownership has reached unprecedented levels and these animals play a significant role in family life. In 2006, there were 72 million pet dogs in the USA (in 37% of households) and 81 million pet cats (in 32% of households) (2007 US Pet Ownership and Demographics Sourcebook). Estimates for 2010 in the UK are for 8 million pet dogs (in 23% of households) and 8 million pet cats (in 19% of households) http://www.pfma.org.uk. Increasingly over past decades, the small companion animal spends a majority of its life within the indoor domestic environment in very close physical association with its owners. Although many of these animals will enjoy a very high standard of healthcare with increasing longevity, there are a number of zoonotic infectious diseases that may be transmitted directly or indirectly from these species. Transmission of such diseases is dependent upon the lifestyle of the pet and is influenced by factors such as vaccination and parasite control, exposure to other domestic animals or wildlife (including urban wildlife such as foxes or small rodents) or exposure to particular environments (e.g. exercise in wooded areas with questing tick populations). The keeping of domestic companion or working small animals is not solely linked to affluence - in most developing nations there is individual or communal village ownership of dogs and cats, and in most of these countries there are extraordinary numbers of feral dogs and cats that have intensive contact with the human urban environment. In contrast, the majority of these animals may never receive veterinary attention and remain significant reservoirs of zoonotic infection.

There is no better example of this than rabies infection, which despite its long historical documentation, remains as much an issue today as at any time in history. Recent WHO estimates suggest that at least 55,000 people die each year from rabies infection; that 95% of these deaths occur in the developing nations of Asia and Africa and that 99% of cases are transmitted by dogs [10]. The significance of rabies virus infection is promoted through initiatives such as World Rabies Day http://www.worldrabiesday.org/ and the Afya vaccination programme in Africa http://www.afya.org/ and in September 2011 the OIE, WHO and FAO will host a Global Conference on Rabies Control in Seoul, Republic of Korea.

Less well-publicized are the remaining infectious diseases shared by man and dogs and cats that are well-summarized in a new text edited by Rabinowitz and Conti [11]. The present review is written in association with the 6th World Forum on Canine Vector-Borne Diseases and those of zoonotic potential will be highlighted in the section following.

COMPANION ANIMAL VECTOR-BORNE DISEASES IN ONE HEALTH

The major vector-borne infectious diseases of dogs and cats that also infect man are summarized in Table 1. Of single greatest significance is zoonotic visceral leishmaniosis caused by *Leishmania infantum* (*L. chagasi*) for which the domestic dog is the major reservoir for human infection via sandfly transmission. The disease is endemic in many countries throughout southern Europe, the Americas, northern Africa and Asia and high proportions of dogs in these areas are exposed to and often infected by the pathogen. Recent WHO estimates suggest that worldwide up to 12 million people are infected, with 2 million new cases identified annually and a population of some 350 million people at risk of infection; the risk of infection is significantly enhanced in HIV-infected individuals [12]. Control of leishmaniosis represents one of the greatest challenges for One Health and provides a clear example for the necessity of human and veterinary medicine to work together to develop strategies for management and elimination of this disease. Such measures have already occurred in the endemic focus of disease occurring in Brazil where the Ministry of Health oversees a programme of serologically testing dogs and culling positive animals [13]. Two commercially available vaccines for dogs have become available in Brazil and should form an important part of the control strategy. Data have already been published that show that where vaccination is widely practiced, the prevalence of both canine and human infection decreases due to reduced transmission of the organism [14].

Table 1. Canine and feline vector-borne diseases that also infect man

Disease	Vector
Leishmaniosis	Sand fly
Borreliosis	Tick
Bartonellosis	Flea Tick
Ehrlichiosis	Tick
Rickettsiosis	Tick Flea
Anaplasmosis	Tick
Dirofilariosis	Mosquito
Yersiniosis	Flea
Tularaemia	Tick
Coxiellosis	Tick
Tick-borne encephalitis	Tick
Louping ill	Tick
West Nile virus encephalitis	Mosquito
Trypanosomiosis	*Triatoma* bugs

A major concern is the extension of traditional endemic areas for *Leishmania* infection, which is now reported as autochthonous disease in northern European countries and North America [15–18]. This is associated directly with the increased mobility of pet animals across borders [16]. Pet travel has been responsible for establishment of reservoirs of Leishmania-infected dogs in these non-endemic areas and has incidentally contributed to the widening geographical range of arthropod vectors (e.g. the establishment of *Rhipicephalus sanguineus* in northern Europe). The unprecedented scale of global companion animal travel is readily demonstrated by the success of the European pet passport scheme. Since its inception in 2000 and up to August 2010, the number of pet animals entering or re-entering the UK alone was 717, 965 http://www.defra.gov.uk/. The focus of attention for health certification for international pet travel has been rabies vaccination (with some countries also requiring evidence of seroconversion after vaccination); however, apart from certain national requirements for tick and tapeworm treatment there has been little consideration of the spread of other zoonotic infectious diseases. Some exceptions to this rule exist; for example importation of dogs into Australia requires testing for monocytic ehrlichiosis, brucellosis, leishmaniosis and leptospirosis, but without such controls there is clear evidence of importation of infection from endemic to non-endemic areas [15]. Global pet travel therefore creates the potential for rapid dissemination of zoonotic infection and represents another major challenge for One Health.

Borreliosis (Lyme disease) is a significant disease of people in endemic areas (particularly in Europe and North America) associated with avian and wildlife reservoirs and *Ixodes* spp. ticks. The major species infecting man and companion animals are *Borrelia burgdorferi* sensu stricto (in North America) and *B. garinii* and *B. afzelii* (in Europe) [19]. Infected dogs and cats pose minimal threat to man, but do provide a means by which infected ticks can be carried into the domestic environment. There is also a risk of human infection should ticks be crushed during removal from a pet animal and tick salivary gland material be exposed to wounds on the hands of an owner. The dog in particular might be employed as a ‹sentinel› for monitoring the risk of human disease in an endemic area [20].

Bartonellosis should be regarded as one of the major potential emerging infections of man. Most is known about 'cat scratch disease' caused primarily by *Bartonella henselae* acquired directly from an infected cat and mostly likely transmitted between cats by the cat flea *Ctenocephalides felis*[21–23]. This infection becomes particularly significant in immunocompromised individuals (e.g. HIV-infected individuals with acquired immunodeficiency syndrome [AIDS], those receiving cancer chemotherapy or immunosuppression for transplantation or immune-mediated disease). The scale of the potential problem can be estimated by considering the extraordinary numbers of pet household cats (see above) and their very close physical association with owners, coupled with knowledge that in developing countries the prevalence of seropositivity amongst cats is on average 27% and the presence of active bacteraemia around 10% [24]. The major canine pathogen of these taxa is the tick-transmitted *B. vinsonii* subsp. *berkhoffii*, but a range of other *Bartonella* species are also reported from cats and dogs (and potential arthropod vectors) associated with a spectrum of clinical diseases. Attention is now focussed on these organisms due to a series of investigations showing that they may be associated with a wide range of human syndromes that were often considered chronic idiopathic diseases [21, 25]. There remains much to learn about the transmission of *Bartonella* from companion animals to man, but veterinary professionals have an occupational exposure risk due to their frequency of exposure to animal bites, arthropods, arthropod faeces and animal bodily fluids [21]. A target for One Health programmes should be further exploration of the significance of these pathogens in animal and human medicine.

A number of ehrlichial and rickettsial infections are shared by man and companion animals [26]. The cause of canine monocytic ehrlichiosis (*Ehrlichia canis* transmitted by *Rhipicephalus sanguineus*) is not considered an organism with significant zoonotic potential (although human cases of infection are reported); however, human and canine infections with *E. chaffeensis* and *E. ewingii* (transmitted by *Amblyoma americanum*) are well-documented although it is not clear whether the dog acts as a reservoir for human infection. Of increasing significance in human, canine and feline populations is the *Ixodes*-transmitted cause of granulocytic 'ehrlichiosis' (*Anaplasma phagocytophilum*). The reservoir species for this infection are wild rodents and wild and domestic ruminants, but human and companion animal infections are emerging as a significant problem in Europe and North America. The spotted fever group of rickettsiosis are also significant problems for man and dogs. In the Americas, Rocky Mountain spotted fever caused by *Rickettsia rickettsii* (transmitted by *Dermacentor andersoni*, *D. variablis*, *R. sanguineus* and *Amblyomma cajennese*) is the most important of this group, while in Europe, Asia and Africa, Mediterranean spotted fever caused by *R. conorii* (transmitted by *R. sanguineus*) is of greatest concern. The role of the dog in these human infections is again in transporting infected ticks into the domestic environment and the risk to owners of removing infected ticks from their pets. The early diagnosis of canine rickettsiosis by the veterinarian is of importance as this may precede the recognition of the pathogen in in-contact people. The flea-transmitted rickettsioses (*R. typhi* and *R. felis*) also fall into this group [23]. The reservoir potential for domestic pets for *R. typhi* is not proven, but there is much recent interest in the emergence of *R. felis* which is found in dogs, cats and cat fleas (*Ct. felis*) [23, 27]. As *R. felis* may be transmitted transovarially and trans-stadially, the domestic pet may introduce infected fleas into the home environment, which then become the means for human infection.

Table 1 also lists a range of other vector-transmitted diseases shared by man and companion animals, a number of which are relatively uncommon causes of human infection, but are important in animal species (e.g. dirofilarasis, West Nile fever).

In summary, the overall challenges for the One Health agenda in addressing these vector-borne diseases are several: (1) promoting awareness and ability to recognize these diseases by the human and veterinary medical professions, (2) undertaking 'joined-up' research programmes that investigate the agents, their vectors and epidemiology, geographical distribution, clinical significance and pathogenesis, (3) developing robust diagnostic tests and surveillance systems for mapping these infectious agents and their vectors globally, (4) identifying diseases for which domestic pet animals are true reservoirs of human infection (e.g. leishmaniosis) and formulating public health strategies that effectively control the disease in the reservoir (e.g. stray dog control, ectoparasiticide treatment, vaccination) in addition to the human patient, (5) investigating the contact between pet animals and peridomestic wildlife species that may act as true reservoirs of these diseases and developing strategies to minimize such contact, (6) promoting awareness of these diseases in the pet-owning public and the importance of regular ectoparasite control programmes for their pets and their domestic environment, and (7) identifying the risks and challenges imposed by increasing global mobility of pet animals and developing strategies to minimize the associated movement of zoonotic infectious disease.

COMPANION ANIMAL ONE HEALTH: THE WSAVA ONE HEALTH COMMITTEE

Given the growing importance of the One Health concept the World Small Animal Veterinary Association (WSAVA) has recently established a One Health Committee, which has as its remit the firm positioning of small companion animals within the global One Health programme [28]. The committee is chaired by the author and includes academic experts on small animal zoonotic infection (including rabies, leishmaniosis and vector-borne diseases), first opinion veterinary practitioner representation and delegates from the OIE and Centers for Disease Control and Prevention (CDC) in the USA. The committee is funded by the WSAVA Foundation through sponsorship by a consortium of the pet food and small animal pharmaceutical industry and commenced a three year programme of work in January 2011.

Although a major focus of the work of the One Health Committee will be zoonotic infectious disease, including those reviewed in this paper, there are two further key areas in One Health in which small companion animals should play a major role. The first of these is in the field of comparative and translational research. The historical perspective given earlier indicates clearly how throughout history human medicine has benefitted from comparative veterinary studies. To an extent, we seem to have forgotten this legacy and over recent decades research efforts have been polarized and veterinary science been very much the 'poor relation' of its larger brother. It is unquestionable that studies of spontaneously arising disease in relatively outbred and long-lived animals that so closely share our domestic environment must provide information of benefit to human medicine. The availability of the canine and feline genomes [29, 30] and the development of microarray genomic screening tools [31, 32] provide us with unprecedented ability to explore the basis of canine and feline diseases that so closely mimic those that occur in man. Promoting comparative clinical research will be the second major focus of the WSAVA One Health Committee.

The final aspect of the work of this group will be in supporting the most visible and simple aspect of the interaction between man and companion animals. The ancient Mesopotamians recognized the psychological benefit of companion animal interaction for human healing and in recent decades the 'human - companion animal bond' has been widely explored. This touches not only on 'pets as therapy' in hospitals or care homes but the widespread psychological and social impact of pets in society [33].

Acknowledgements

Publication of the CVBD6 thematic series has been sponsored by Bayer Animal Health GmbH

REFERENCES

1. Dunlop RH, Williams DJ: Veterinary Medicine: an Illustrated History. 1996, St Louis, Mosby-Year Book, Inc
2. Day MJ: One Health and the legacy of John M'Fadyean. J Comp Pathol. 2008, 139: 151-153. 10.1016/j.jcpa.2008.07.007.
3. Monath TP, Kahn LH, Kaplan B: Introduction: one health perspective. ILAR J. 2010, 51: 193-198.
4. Cardiff RD, Ward JM, Barthold SW: 'One medicine - one pathology': are veterinary and human pathology prepared?. Lab Invest. 2008, 88: 18-26. 10.1038/labinvest.3700695.
5. Alder M, Easton G: One medicine!. Vet Rec. 2005, 157: 699-
6. King LJ, Anderson LR, Blackmore CG, Blackwell MJ, Lautner EA, Marcus LC, Meyer TE, Monath TP, Nave JE, Ohle J, Pappaioanou M, Sobota J, Stokes WS, Davis RM, Glasser JH, Mahr RK: Executive summary of the AVMA One Health Initiative Task Force report. J Am Vet Med Assoc. 2008, 233: 259-261. 10.2460/javma.233.2.259.
7. Kaplan B, Kahn LH, Monath TP, Woodall J: One Health' and parasitology. Parasites Vectors. 2009, 2: 36-10.1186/1756-3305-2-36.
8. Paul M, King L, Carlin EP: Zoonoses of people and their pets: a US perspective on significant pet-associated parasitic diseases. Trends Parasitol. 2010, 26: 153-154. 10.1016/j.pt.2010.01.008.
9. Brown C: Emerging diseases: the global express. Vet Pathol. 2010, 47: 9-14. 10.1177/0300985809354351.
10. Bourhy H, Dautry-Varsat A, Hotez PJ, Salomon J: Rabies, still neglected after 125 years of vaccination. PLoS Negl Trop Dis. 2010, 4: 11-
11. Rabinowitz PM, Conti LA: Human - Animal Medicine: Clinical Approaches to Zoonoses, Toxicants and other Shared Health Risks. 2010, Maryland Heights, Saunders Elsevier
12. WHO: Report of the 5th Consultative Meeting on Leishmania/HIV Co-infection. 2007, Addis Ababa, Ethiopia
13. Nunes CM, Pires MM, Marques da Silva K, Assis FD, Filho JG, Perri SHV: Relationship between dog culling and incidence of human visceral leishmaniasis in an endemic area. Vet Parasitol. 2010, 170: 131-133. 10.1016/j.vetpar.2010.01.044.
14. Palatnik-de-Sousa CB, Silva-Antunes I, de Aguiar Morgado A, Menz I, Palatnik M, Lavor C: Decrease of the incidence of human and canine

visceral leishmaniosis after dog vaccination with Leishmune® in Brazilian endemic areas. Vaccine. 2009, 27: 3505-3512. 10.1016/j.vaccine.2009.03.045.

15. Shaw SE, Binns SH, Birtles RJ, Day MJ, Smithson R, Kenny MJ: Molecular evidence of tick-transmitted infections in dogs and cats in the United Kingdom. Vet Rec. 2005, 157: 646-648.
16. Shaw SE, Langton DA, Hillman TJ: Canine leishmaniosis in the United Kingdom: a zoonotic disease waiting for a vector?. Vet Parasitol. 2009, 163: 281-285. 10.1016/j.vetpar.2009.03.025.
17. Gaskin AA, Schantz P, Jackson J, Birkenheuer A, Tomlinson L, Gramiccia M, Levy M, Steurer F, Kollmar E, Hegarty BC, Ahn A, Breitschwerdt EB: Visceral leishmaniasis in a New York foxhound kennel. J Vet Intern Med. 2002, 16: 34-44. 10.1111/j.1939-1676.2002.tb01604.x.
18. Boggiatto PM, Ramer-Tait AE, Metz K, Kramer EE, Gibson-Corley K, Mullin K, Hostetter JM, Gallup JM, Jones DE, Petersen CA: Immunologic indicators of clinical progression during canine *Leishmania infantum* infection. Clin Vaccine Immunol. 2010, 17: 267-273. 10.1128/CVI.00456-09.
19. Little SE, Heise SR, Blagburn BL, Callister SM, Mead PS: Lyme borreliosis in dogs and humans in the USA. Trends Parasitol. 2010, 26: 213-218. 10.1016/j.pt.2010.01.006.
20. Hamer SA, Tsao JI, Walker ED, Mansfield LS, Foster ES, Hickling GJ: Use of tick surveys and serosurveys to evaluate pet dogs as a sentinel species for emerging Lyme disease. Am J Vet Res. 2009, 70: 49-56. 10.2460/ajvr.70.1.49.
21. Breitschwerdt EB, Maggi RG, Chomel BB, Lappin MR: Bartonellosis: an emerging infectious disease of zoonotic importance to animals and human beings. J Vet Emerg Crit Care. 2010, 20: 8-30. 10.1111/j.1476-4431.2009.00496.x.
22. Guptill L: Bartonellosis. Vet Microbiol. 2010, 140: 347-359. 10.1016/j.vetmic.2009.11.011.
23. McElroy KM, Blagburn BL, Breitschwerdt EB, Mead PS, McQuiston JH: Flea-associated zoonotic diseases of cats in the USA: bartonellosis, flea-borne rickettsioses, and plague. Trends Parasitol. 2010, 26: 197-204. 10.1016/j.pt.2010.01.001.

24. Birtles R: Bartonellosis. Arthropod-Borne Infectious Diseases of the Dog and Cat. Edited by: Shaw SE, Day MJ. 2005, London, Manson Publishing Ltd, 110-119.
25. Chomel BB, Boulouis HJ, Maruyama S, Breitschwerdt EB: *Bartonella* spp. in pets and effect on human health. Emerg Inf Dis. 2006, 12: 389-394.
26. Nicholson WL, Allen KE, McQuiston JH, Breitschwerdt EB, Little SE: The increasing recognition of rickettsial pathogens in dogs and people. Trends Parasitol. 2010, 26: 205-212. 10.1016/j.pt.2010.01.007.
27. Kenny MJ, Birtles RJ, Day MJ, Shaw SE: *Rickettsia felis* in the United Kingdom. Emerg Infect Dis. 2003, 9: 1023-1024.
28. Day MJ: One health: the small companion animal dimension. Vet Rec. 2010, 167: 847-849. 10.1136/vr.c6492.
29. Lindblad-Toh K, Wade CM, Mikkelsen TS, Karlsson EK, Jaffe DB, Kamal M, Clamp M, Chang JL, Kulbokas EJ, Zody MC: Genome sequence, comparative analysis and haplotype structure of the domestic dog. Nature. 2005, 438: 803-819. 10.1038/nature04338.
30. Pontius JU, Mullikin JC, Smith DR: Initial sequence and comparative analysis of the cat genome. Genome Res. 2007, 17: 1675-1689. 10.1101/gr.6380007.
31. Shearin AL, Ostrander EA: Leading the way: canine models of genomics and disease. Dis Models and Mech. 2010, 3: 27-34. 10.1242/dmm.004358.
32. Wilbe M, Jokinen P, Truve K, Seppala EH, Karlsson EK, Biagi T, Hughes A, Bannasch D, Andersson G, Hansson-Hamlin H, Lohi H, Lindblad-Toh K: Genome-wide association mapping identifies multiple loci for a canine SLE-related disease complex. Nature Genet. 2010, 42: 250-254. 10.1038/ng.525.
33. Johnson RA: Psychosocial and therapeutic aspects of human-animal interaction. Human-Animal Medicine: Clinical Approaches to Zoonoses, Toxicants and Other Shared Health Risks. Edited by: Rabinowitz PM, Conti LA. 2010, Maryland Heights, Saunders Elsevier, 24-36.

10

Immune Responses of a Native and an Invasive Bird to Buggy Creek Virus (Togaviridae: Alphavirus) and Its Arthropod Vector, the Swallow Bug (Oeciacus vicarius)

Carol A. Fassbinder-Orth[1], Virginia A. Barak[1], Charles R. Brown[2]

[1] Biology Department, Creighton University, Omaha, Nebraska, United States of America,
[2] Department of Biological Sciences, University of Tulsa, Tulsa, Oklahoma, United States of America,
Blood Systems Research Institute, United States of America,

ABSTRACT

Invasive species often display different patterns of parasite burden and virulence compared to their native counterparts. These differences may be the result of variability in host-parasite co-evolutionary relationships, the occurrence of novel host-parasite encounters, or possibly innate differences in physiological responses to infection between invasive and native hosts.

Citation: Fassbinder-Orth CA, Barak VA, Brown CR (2013) Immune Responses of a Native and an Invasive Bird to Buggy Creek Virus (Togaviridae: Alphavirus) and Its Arthropod Vector, the Swallow Bug (Oeciacus vicarius). PLoS ONE 8(2): e58045. https://doi.org/10.1371/journal.pone.0058045

Here we examine the adaptive, humoral immune responses of a resistant, native bird and a susceptible, invasive bird to an arbovirus (Buggy Creek virus; *Togaviridae*: *Alphavirus*) and its ectoparasitic arthropod vector (the swallow bug; *Oeciacus vicarius*). Swallow bugs parasitize the native, colonially nesting cliff swallow (*Petrochelidon pyrrhonota*) and the introduced house sparrow (*Passer domesticus*) that occupies nests in cliff swallow colonies. We measured levels of BCRV-specific and swallow bug-specific IgY levels before nesting (prior to swallow bug exposure) and after nesting (after swallow bug exposure) in house sparrows and cliff swallows in western Nebraska. Levels of BCRV-specific IgY increased significantly following nesting in the house sparrow but not in the cliff swallow. Additionally, house sparrows displayed consistently higher levels of swallow bug-specific antibodies both before and after nesting compared to cliff swallows. The higher levels of BCRV and swallow bug specific antibodies detected in house sparrows may be reflective of significant differences in both antiviral and anti-ectoparasite immune responses that exist between these two avian species. To our knowledge, this is the first study to compare the macro- and microparasite-specific immune responses of an invasive and a native avian host exposed to the same parasites.

INTRODUCTION

Different host species often vary considerably in their ability to immunologically resist certain parasites [1], [2]. Recently it has been suggested that one determinant of a host's susceptibility to parasites or pathogens may be whether or not a host exhibits an invasive life history [3]. Characteristics associated with success as an invasive may be either positively or negatively related to immune function [4], [5]. Consequently, the transmission dynamics of host-parasite systems can be altered in several possible ways when an invasive host colonizes an area and comes into contact with native hosts and parasites. Invasive host species may introduce a new pathogen (with which the invasive has co-evolved), causing naïve, native species to become susceptible hosts for the novel pathogen (the Novel Weapon Hypothesis [6], [7]). In these cases, the invasive host may be at an advantage over the native hosts who have had no evolutionary history with the introduced parasite. On the other hand, when an invasive species invades a new environment, it may also encounter a novel parasite and thus be a more susceptible host than those native hosts that have co-evolved with the parasite [8], [9].

One example of a natural host-pathogen system in which a non-native invasive species has recently assumed a role in the transmission cycle [9] is that of the alphavirus, Buggy Creek virus (BCRV; *Togaviridae*: *Alphavirus*), that historically was amplified only among its vertebrate host, the cliff swallow (*Petrochelidon pyrrhonota*; [10]–[12]). This virus is restricted to a unique ecological setting in which it is transmitted solely by its vector, the swallow bug (Hemiptera: Cimicidae: *Oeciacus vicarius*), to the cliff swallow, the primary host for the ectoparasitic bug [13], [14]. Following the introduction of European house sparrows (*Passer domesticus*) to North America in the 1800's [15] and their subsequent invasion of cliff swallow colonies where they usurp the swallows' nests [16], swallow bugs have switched to sparrows as alternative hosts in some cases, and in the process the bugs have exposed house sparrows to BCRV, a novel pathogen [17], [18].

The invasion of house sparrows into swallow colonies provides a number of opportunities to study how this invasive species may differ from the native species (cliff swallow) in its immune responses to both the virus and the ectoparasitic bugs. Although birds are the principal reservoirs for several alphaviruses of medical importance (e.g. western equine encephalomyelitis, eastern equine encephalomyelitis, Sindbis virus [19]), relatively little is known about avian immune responses to any of the alphaviruses or the factors causing variability in host susceptibility to either BCRV or, more generally, other arthropod-borne viruses (arboviruses) such as West Nile virus. Furthermore, ectoparasites such as the hematophagous swallow bugs can exert strong selection pressures on their avian hosts by decreasing survival and future reproduction [20]–[23] and presumably also affecting evolution of the hosts' immune response. The effect of ectoparasitic arthropods on disease etiology has been studied in some mammals [24]–[26], but little is known for avian species. In particular, we know nothing about how immune responses to the same arthropod vector may vary between an invasive and a native avian species.

In this study we compare the adaptive, humoral immune responses of two bird species – the native cliff swallow and the non-native house sparrow – to the alphavirus BCRV and to the virus's ectoparasitic arthropod vector, the swallow bug. We recognize that the evolutionary histories of cliff swallows and house sparrows are different and this may account in part for observed differences between them; however, these species are both passerine birds, and passerines of different species can show similar immune responses to arthropod-borne viruses [27]. Two opposite predictions are possible. One is that due to their long co-evolutionary history with both swallow bugs and

BCRV (and their heavy exposure to swallow bugs [21], [28]), cliff swallows should exhibit less strong responses (greater tolerance [29], [30]) to both the ectoparasite and the virus than should the immunologically naïve invasive house sparrow. The alternative prediction is that because invasive species sometimes invest less in their immune systems than non-invasives [3], house sparrows should show more muted responses to the ectoparasite and the virus than would the cliff swallow. We use these results to gain insight into (*i*) the potential impact of the invasive house sparrow on this host-parasite-pathogen system and (*ii*) how birds respond immunologically to alphaviruses and arthropods more generally.

METHODS

Ethics Statement

All procedures involving animals were approved by the Creighton University Institutional Animal Care and Use Committee under protocol 0915. Blood sampling of cliff swallows was approved under the Federal bird banding permit 20948, and Nebraska Game and Parks Scientific Permit no. 254. House sparrows are a non-native species, and federal and state permits are not required for sample collection from this species.

Study Organisms

House sparrows were introduced into North America from Europe in the mid 1800's [31], and have been present in Nebraska since about 1900 [15]. They are a widely distributed, semi-colonial species that often form aggregations of 2 to 20 nests in close proximity. They are sedentary, remaining at or near breeding sites year-round. House sparrows are multi-brooded, with nesting in our study area beginning in late April and ending in late July, with peak egg laying periods in mid May, late June, and late July. Nestlings fledge after about 14–18 days, and new broods are started soon after earlier ones fail or fledge. Cliff swallows are highly colonial, migratory passerines that breed across much of western North America [32]. They build gourd-shaped mud nests on the sides of cliff faces, inside highway and railroad culverts, and underneath bridges. In our study area, swallow colony sizes range from 2 to 6000 active nests. The mud nests persist from year-to-year and are frequently repaired and reused by cliff swallows for multiple seasons. Swallows arrive in our study area in early to mid May and typically raise a single brood, with most nestlings fledging by mid July. Individual colonies are highly synchronous and are quickly vacated by swallows after the nestlings fledge. Nestlings are in the nest for about 26 days before fledging.

During the 1960's, the construction of the interstate highway system provided alternative substrates (culverts, bridges) for cliff swallow nesting, resulting in cliff swallows' moving into peri-domestic settings and coming into contact with house sparrows. In our study area, most house sparrows have been using cliff swallow colonies for only about 30–40 years [16]. The swallow bug is a hematophagous nest ectoparasite of the cliff swallow that is not known to occur outside swallow colonies, and thus it was historically restricted to cliff swallows as hosts. The bugs take blood meals from cliff swallows and more recently, house sparrows [10], [33]. Swallow bugs disperse between nests within a colony by crawling on the substrate and disperse between colony sites by clinging to the feet/legs of cliff swallows that move from one site to another [34]. Sparrows are not known to move bugs between sites. The density of bugs in cliff swallow colonies can be quite high, with as many as 2600 bugs per cliff swallow nest and 2400 per house sparrow nest. In our study area approximately 25% of bug pools are infected with BCRV [35], [36], a single-stranded, positive-sense RNA virus antigenically and phylogenetically placed within the WEEV complex of alphaviruses [10]–[12], [37], [38]. BCRV is unusual among alphaviruses because it is vectored primarily by the ectoparasitic swallow bug, rather than by mosquitoes.

Sample Collection

The primary study site is near the Cedar Point Biological Station (CPBS; 41°12.591′ N, 101°38.969′ W) in Keith County, Nebraska, USA, and has been previously described by Brown and Brown [28]. During the spring and summer of 2011, five cliff swallow colonies situated under highway culverts and bridges were monitored for cliff swallow and house sparrow activity to determine the effects of seasonal nesting activity on BCRV-specific and swallow bug specific antibody levels in both species (Table 1). House sparrows also use cliff swallow colonies as year-round roosting sites, and resident sparrows were sampled in late March 2011 (about 4 weeks in advance of the breeding season) to represent a pre-exposure period before swallow bugs had terminated their winter dormancy and began taking avian blood meals. Post-nesting blood samples were collected from house sparrows at colony sites in September following late nesting (that can extend into August [39]). Pre-nesting samples were collected from cliff swallows in mid-May, immediately after their return to the study area from their South American wintering grounds [28]. Post-nesting samples from cliff swallows were collected in mid-July, immediately preceding their migratory exodus from the study area (Table 1).

Table 1. Sampling Details.

Species	Collection Site	Collection Date	Breeding Period	*n*
House sparrow	CPBS[3]	26 March 2011	Pre-nesting	25
	CPBS	24 September 2011	Post-nesting	20
	Council Bluffs, IA (Agricultural facility)	2 September 2011	N/A	18
Cliff swallow	CPBS	26 May 2011	Pre-nesting	18
	CPBS	13 July 2011	Post-nesting	20

[3]Cedar Point Biological Station, Keith County, Nebraska.

https://doi.org/10.1371/journal.pone.0058045.t001

Samples were also collected from house sparrows at an agricultural facility near Council Bluffs, Pottawattamie County, Iowa, USA, with no known cliff swallow colonies within a 15-km radius of the facility. Given that swallow bugs are uniquely associated with cliff swallow colonies, and that house sparrows are a largely sedentary species [40], it is unlikely that the house sparrows in this control group had ever been exposed to swallow bugs or BCRV and thus the Iowa birds served as unexposed negative controls for both swallow bug and BCRV serological surveys.

Adult cliff swallows were captured by mist netting, and adult house sparrows were captured by mist netting and nighttime removal from nests. Following capture, birds were bled by jugular venipuncture, 100 μl of blood were collected, and sera samples were separated and stored for later use in enzyme-linked immunosorbent assays (ELISAs). Following venipuncture, pressure was applied to the collection site to prevent hematoma formation. The collection site was then inspected to ensure the cessation of any residual bleeding and birds released back to their natural surroundings.

Antigen Production

Swallow Bug Homogenate

Swallow bugs were collected from nests at a cliff swallow colony near CPBS in July 2011. The nests were broken apart, and bugs were removed using forceps. Bugs were divided into pools of 10 bugs each and homogenized with 30 μl phosphate buffered saline (PBS) per bug. Homogenates were centrifuged at 10,000×*g* for 1 minute, and the supernatant was collected and filtered with a 0.2 μm filter. The protein concentration of the homogenate was determined using a Nanodrop spectrophotometer. The extract was diluted to 1 mg/mL using sterile PBS and stored at −80°C.

Viral RNA extraction from the bug homogenates was performed using the QIAmp Viral RNA Mini Kit (Qiagen, Valencia, CA). Negative and positive controls were included in each extraction procedure. Reverse transcription polymerase chain reaction (RT-PCR) was performed using the OneStep RT-PCR Kit (Qiagen, Valencia, CA), following the manufacturer's protocol to identify BCRV positive bug pools. Primers and thermocycler conditions were those of Moore et al. [36]. Amplification products were electrophoresed on a 2% UltraPure Agarose gel (Invitrogen, Carlsbad, CA). Four homogenate pools were found to be negative for BCRV (data not shown) and were combined for use in the ELISA.

Buggy Creek Virus Antigen

BCRV was cultured from whole blood samples diluted in BA-1 diluent according to O'Brien and Brown [17]. Briefly, Vero cells were grown in 25 cm^2 flasks in complete growth medium (EMEM with 10% heat inactivated FBS and 1% antibiotic/antimycotic). The virus was passaged twice and 200 μl of the second passage was used to infect two additional flasks of Vero cells. Infected flasks were incubated until 50–75% cytopathic effect was observed (approximately 2–3 days), then flasks were frozen overnight at −80°C. The cells were thawed on ice, centrifuged at 1700×*g* (4°C) for 20 minutes, and the supernatant and cellular fraction were separated. The cellular fraction was re-suspended in 1.5 mL PBS and refrozen at −80°C overnight. Samples were then thawed and 1.5 mL of 0.2 M glycine (9.5 pH) were added to each tube. The cells were homogenized with a sterile homogenizer tip on an Omni Homogenizer and placed in a 37°C water bath for 4.5 hours, vortexing every hour. Levels of virus were quantified by performing a plaque assay according to Moore et al. [36]. Briefly, we added 100 μl of the viral stock supernatant to a confluent monolayer of Vero cells in 6 well plates, incubated the plates for 1 hr at 37°C, 5% CO_2, then overlaid the monolayer with an agar overlay for plaque visualization. Plaques were scored 3–4 days later, and the final concentration of BCRV stock solution was 7.8×10^5 PFU/mL.

The BCRV stock solution was inactivated for use in the ELISA. To do this, 3 mL of PBS with 0.5% Triton X were added, and the mixture was incubated at 4°C for 2 hours, with vortexing every half hour. It was then centrifuged at 10,000×*g* for 10 minutes at 4°C. The supernatant was frozen at −80°C for later use in the ELISA.

ELISAS

A direct ELISA was performed to determine total IgY in house sparrows and cliff swallows. A flat-bottomed 96-well plate (Nunc, Roskilde, Denmark) was coated with 100 µl of pooled house sparrow and cliff swallow sera (n=15 individuals/species/pool) per well, diluted to 1:100 in coating buffer (0.015 M Na_2CO_3, 0.035 M $NaHCO_3$, pH 9.6). The plate was incubated overnight at 37°C. The coating solution was removed, 200 µl of blocking buffer (PBS with 5% non-fat dry milk, 0.05% Tween) were added to each well, and incubated at room temperature for 30 minutes. The plate was washed four times with wash buffer (PBS with 0.05% Tween) using a BioTek ELx50 Automated Strip Washer (Biotek Instruments, Winooski, VT). Fifty µl of the detecting conjugate goat anti-bird IgY-HRP (Bethyl Laboratories, Inc., Montgomery, TX) were added at 1:1000 in blocking buffer, incubated at 37°C for 1 hour, and washed. One hundred µl of tetramethylbenzidine (TMB)-peroxidase substrate (Kirkegaard & Perry Laboratories, Inc., Gaithersburg, MD) were added to each well, incubated for 5 minutes, and the reaction was stopped with 100 µl 1 M H_2SO_4. Optical density (OD) values were read at A_{450} using a BioTek Synergy HT automated microplate reader (Biotek Instruments, Winooski, VT). Total levels of IgY were not significantly different between species (Figure 1), indicating sufficient recognition of IgY in both avian species by the conjugate antibody.

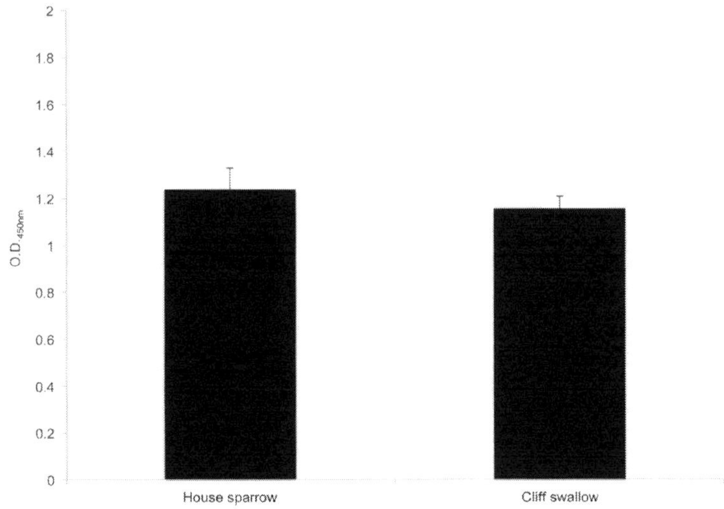

Figure 1. Total IgY ELISA Results. Values shown represent mean (+ SEM) optical density values (O.D.$_{450nm}$) for total IgY in cliff swallows and house sparrows. NS= non-significant difference (P>0.45).

https://doi.org/10.1371/journal.pone.0058045.g001

We performed indirect ELISAs that detected swallow bug and Buggy Creek virus (BCRV)-specific IgY. One hundred μl per well of 1:1000 swallow bug homogenate or 1:50 BCRV in coating buffer were added and incubated overnight at 4°C. After blocking with 200 μl of blocking buffer per well and incubating for 30 minutes, 100 μl of either cliff swallow or house sparrow sera were added at 1:100 in wash buffer and incubated at 37°C for 1 hour. Wells were washed, and 100 μl goat anti-bird IgY at 1:1000 in blocking buffer were added and incubated at 37°C for 1 hour. Wells were again washed, then 100 μl of TMB were added and allowed to incubate for 10 minutes at room temperature. The reaction was stopped with 100 μl of 1 M H_2SO_4 and plates were read at A_{450}.

Statistical Analyses

Values given are mean ± 1 SEM. Total IgY, and swallow bug and BCRV reactive antibodies were tested by ANOVA (GLM in SYSTAT [41]) according to experimental group (house sparrow control, house sparrow pre- and post- nesting, cliff swallow pre- and post-nesting). Subsequent ad hoc comparisons among experimental groups were made on the adjusted least squares from the ANOVA using the Tukey honestly significant difference test. The significance level was set at $P<0.05$.

RESULTS

Indirect ELISAs were successfully developed to measure swallow bug and BCRV-specific IgY levels in adult house sparrows and cliff swallows before and after nesting at selected cliff swallow colony sites in western Nebraska. Specific antibody levels in birds at the primary study area were compared to samples from a BCRV and the swallow bug- unexposed control group of house sparrows.

Seasonal variation in BCRV-specific antibodies was detected in this study (Figure 2). Pre-nesting levels of BCRV-specific antibodies in both house sparrows and cliff swallows were low and were not significantly different from the house sparrow control group ($P>0.62$). However, post-nesting BCRV-specific levels were significantly increased from pre-nesting levels in house sparrows ($P<0.0001$). In cliff swallows, post-nesting BCRV-specific antibody levels displayed a trend of significance when compared to pre-nesting levels ($P<0.079$), and were significantly higher than the house sparrow control group ($P<0.018$).

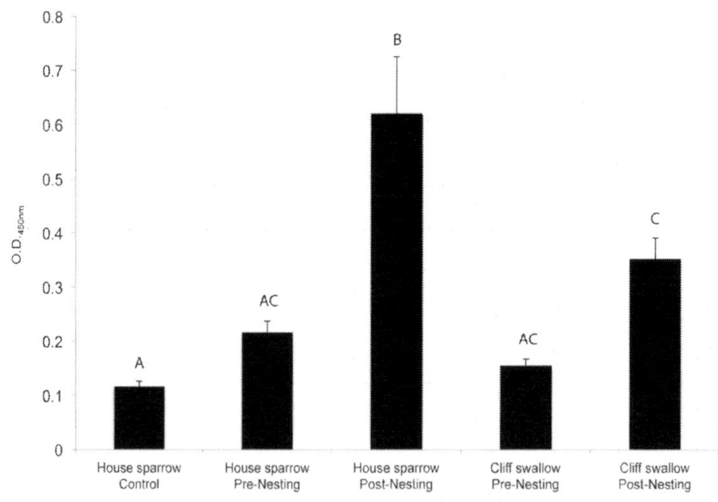

Figure 2. BCRV ELISA. Values shown represent mean (+ SEM) values for BCRV specific IgY levels in each group. House sparrow Control samples collected from house sparrows at Council Bluffs, Iowa (no known cliff swallow colonies in the area) in August, 2011. All other blood samples collected from

house sparrows and cliff swallows at swallow colony sites in western Nebraska in 2011. Letters above bars denote statistical differences among groups ($P<0.05$); bars not sharing the same letter are different.

https://doi.org/10.1371/journal.pone.0058045.g002

There were no significant effects of seasonality on swallow bug-specific antibodies in this study (Figure 3), as pre and post nesting swallow bug-specific IgY levels did not differ significantly within each species ($P>0.99$). Levels of swallow bug-specific IgY levels in cliff swallows were not significantly different from the house sparrow control group during both pre- and post-nesting periods ($P>0.99$), likely indicating that low levels of swallow bug-specific antibodies were present in adult cliff swallows in this study. However, significantly higher levels of swallow bug-specific IgY levels were found in the pre- and post-nesting house sparrow groups compared to all other groups ($P<0.002$).

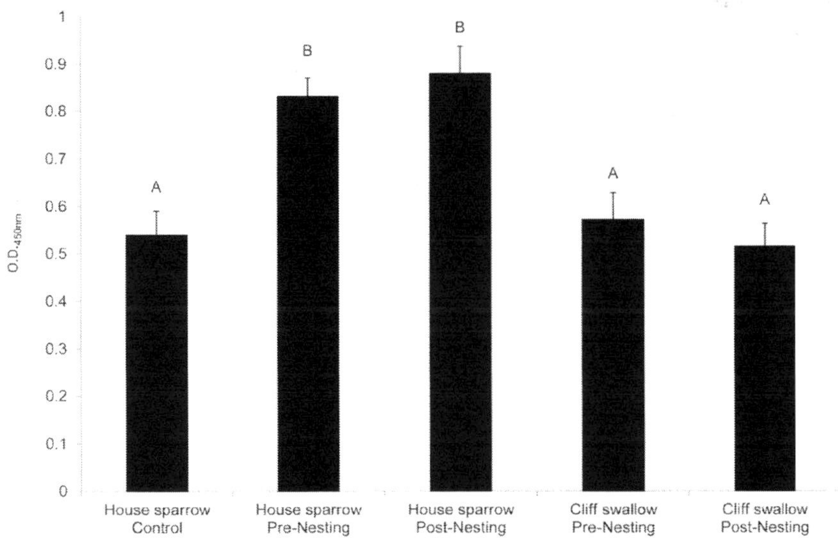

Figure 3. Swallow bug ELISA. Values shown represent mean (+ SEM) swallow bug-specific IgY levels in each group. Letters above bars denote statistical differences among groups ($P<0.05$); bars not sharing the same letter are different.

https://doi.org/10.1371/journal.pone.0058045.g003

DISCUSSION

Multiple surveys have been conducted that report on the prevalence of arboviruses or arbovirus-specific antibodies in diverse avian populations [42], [1]. However, few investigations have explored antibody levels to both arboviruses and their insect vectors in avian hosts, effects of seasonality on these antibody levels, or how invasive avian hosts may differ from native hosts in their antibody-mediated responses to viral or vector antigens. In this experiment, we detected significant differences in the antibody response patterns of house sparrows (the invasive host) compared to cliff swallows (the native host) to an alphavirus (BCRV) and its hemipteran vector (the swallow bug). BCRV-specific antibodies varied seasonally in both house sparrows and cliff swallows, but were significantly higher in house sparrows in the post-nesting season. Seasonal variation in the prevalence of both BCRV RNA and cytopathic virus has been previously documented in swallow bugs and house sparrows [33], [43]. In swallow bugs, BCRV prevalence is significantly lower during the winter months, with the majority of the virus detected being non-cytopathic during this time [13]. Termination of diapause and commencement of blood feeding is thought to initiate cytopathic viral growth in swallow bugs [43]. The seasonal changes in BCRV in bugs have been supported by data on virus prevalence in house sparrows: the highest levels of BCRV infection in nestling house sparrows (the primary amplifiers of BCRV) occur in mid-summer, which also coincides with the time of the largest swallow bug population sizes and the highest BCRV infection levels seen during the year [14]. Previous work therefore indicates that the increase in BCRV-specific antibodies following nesting in this experiment may be explained by the increased exposure of birds to cytopathic BCRV during the nesting period, brought about in part by higher biting rates as swallows bugs increase in abundance in mid-summer [34].

The finding that house sparrows had significantly higher levels of BCRV-specific antibodies compared to cliff swallows during the post nesting period could indicate that the house sparrow is exposed to higher levels of BCRV throughout the breeding season [9], or it may indicate that the humoral immune response of these species to this arbovirus is fundamentally different, with the cliff swallow displaying a less reactive response compared to the house sparrow. Recent research in mammals suggests that neutralizing antibodies may not be required for protection against some cytopathic viral infections [44]. Therefore, although the presence of BCRV-specific antibodies in house sparrows indicates humoral immune system recognition of the virus, it may not necessarily indicate the effectiveness of the antibody-mediated response of house sparrows to clear the BCRV infection.

The relatively high background levels of swallow bug-specific antibodies detected in this experiment may have been due to the possible detection of antibodies against other hematophagous nest or feather ectoparasites that have homologous salivary components to the swallow bug, as has been shown to occur with serological assays for other insect vectors [45]. However, the significantly higher level of swallow bug-specific antibodies that were detected in the western Nebraska house sparrows compared to the control house sparrows indicates a significantly different pattern of exposure to ectoparasites between these two groups of house sparrows. Given that house sparrows are frequently exposed to swallow bugs in the Nebraska study area [9], [13], it is likely that the increase in antibody levels are reflective of an increased humoral immune response to swallow bugs in these species.

Arthropod vector-specific immune responses are found in many vertebrate populations (e.g. humans, livestock, wild birds and mammals), and salivary protein-specific antibody responses have been used as a marker for vertebrate exposure to ticks, mosquitoes, sand flies, nest flies, black flies and triatomines [46]–[52]. We detected significant levels of swallow bug specific antibodies in house sparrows but not in cliff swallows, with no evidence for seasonal variation in antibody levels in house sparrows.

The high degree of host specificity that is found in many nest-based and feather parasites suggests that co-speciation of ectoparasites and avian hosts often occurs [53]. Given that swallow bugs are restricted exclusively to cliff swallow nests and parasite infestations are often high [21], [28], co-evolution of cliff swallow and swallow bug population biology is likely [14]. One way such co-evolution may be expressed is through the ectoparasite-specific immune response of cliff swallows' evolving to be more desensitized and tolerant in nature, compared to that of the house sparrow. Desensitization may be beneficial because robust ectoparasite-specific immune responses may negatively impact the outcome of concurrent or future arbovirus infections in some vertebrate hosts, making hosts either more susceptible to infection [54], [55] or resulting in decreased levels of parasite-specific antibodies [56]. To our knowledge, no previous studies have compared the ectoparasite-specific immune responses of a historically exposed native host to a more recently exposed invasive host.

The differences in antibody-mediated immune responses to the swallow bug and BCRV observed in this experiment may help to explain the differences in BCRV susceptibility that exist between cliff swallows and house sparrows. House sparrows are more likely to be infected, exhibit higher virus titers, and develop severe BCRV-induced pathology compared to cliff swallows [9], [18]. It is plausible that the house sparrow's pattern of immune recognition of swallow bug salivary components may alter the immune response in such a way as to enhance BCRV severity in these hosts. One way this may occur is by increased cellular recruitment at bite sites, dysregulated cytokine signaling, and altered humoral immune responses, as has been seen following mosquito, sand fly, and tick bites in other vertebrate hosts [57]–[59]. More research into the mechanistic details of the immune responses of house sparrows and cliff swallows to swallow bugs and BCRV is needed to identify all immunological factors that may be involved in determining disease resistance and susceptibility in these two species.

ACKNOWLEDGMENTS

We thank Neil Backer, Jordan Herman, Molly Hiatt, Nathan Miller, Ellecia Rainwater, and Tiffany Tran for field or laboratory assistance. The School of Biological Sciences at the University of Nebraska-Lincoln allowed us to use the Cedar Point Biological Station.

AUTHOR CONTRIBUTIONS

Conceived and designed the experiments: VAB CFO CRB. Performed the experiments: VAB CFO CRB. Analyzed the data: VAB CFO CRB. Contributed reagents/materials/analysis tools: VAB CFO CRB. Wrote the paper: VAB CFO CRB.

REFERENCES

1. Reisen WK, Wheeler SS, Garcia S, Fang Y (2010) Migratory birds and the dispersal of arboviruses in California. Am J Trop Med Hyg 83: 808–815.
2. Brown CR, O'Brien VA (2011) Are Wild Birds Important in the Transport of Arthropod-borne Viruses? Ornithol Monogr. 71: 1–64.
3. Lee KA, Klasing KC (2004) A role for immunology in invasion biology. Trends Ecol Evol 19: 523–529.
4. Lee KA, Martin LBII, Wikelski MC (2005) Responding to inflammatory challenges is less costly for a successful avian invader, the house sparrow (*Passer domesticus*), than its less-invasive congener. Oecologia 145: 244–251.
5. Lee KA, Martin LBII, Hasselquist D, Ricklefs RE, Wikelski MC (2006) Contrasting adaptive immune defenses and blood parasite prevalence in closely related *Passer* sparrows. Oecologia 150: 383–392.
6. Marzal A, Ricklefs RE, Valkiūnas G, Albayrak T, Arriero E, et al. (2011) Diversity, loss, and gain of malaria parasites in a globally invasive bird. PLOS One 6: e21905.
7. Callaway RM, Ridenour WM (2004) Novel weapons: invasive success and the evolution of increased competitive ability. Front Ecol Environ 2: 436–443.
8. Kelly DW, Paterson RA, Townsend CR, Poulin R, Tompkins DM (2009) Parasite spillback: a neglected concept in invasion ecology? Ecology 90: 2047–2056.
9. O'Brien VA, Moore AT, Young GR, Komar N, Reisen WK, et al. (2011) An enzootic vector-borne virus is amplified at epizootic levels by an invasive avian host. Proc R Soc B 278: 239–246.
10. Hopla CE, Francy DB, Calisher CH, Lazuick JS (1993) Relationship of cliff swallows, ectoparasites, and an alphavirus in west-central Oklahoma. J Med Entomol 30: 267–272.
11. Hayes RO, Francy DB, Lazuick JS, Smith GC, Gibbs EPJ (1977) Role of the cliff swallow bug (*Oeciacus vicarius*) in the natural cycle of a western equine encephalitis-related alphavirus. J Med Entomol 14: 257–262.
12. Calisher CH, Monath TP, Muth DJ, Lazuick JS, Trent DW, et al. (1980) Characterization of Fort Morgan virus, an alphavirus of the

western equine encephalitis virus complex in an unusual ecosystem. Am J Trop Med Hyg 29: 1428–1440.

13. Brown CR, Strickler SA, Moore AT, Knutie SA, Padhi A, et al. (2010) Winter ecology of Buggy Creek virus (*Togaviridae, Alphavirus*) in the central Great Plains. Vector Borne Zoonotic Dis 10: 355–363.
14. Brown CR, Padhi A, Moore AT, Brown MB, Foster JE, et al. (2009) Ecological divergence of two sympatric lineages of Buggy Creek virus, an arbovirus associated with birds. Ecology 90: 3168–3179.
15. Robbins CS (1973) Introduction, spread, and present abundance of the house sparrow in North America. Ornithol Monogr 14: 3–9.
16. Brown CR, Moore AT, O'Brien VA (2012) Prevalence of Buggy Creek virus (*Togaviridae*: *Alphavirus*) in insect vectors increases over time in the presence of an invasive avian host. Vector Borne Zoonotic Dis 12: 34–41.
17. O'Brien VA, Brown CR (2011) Group size and nest spacing affect Buggy Creek virus (*Togaviridae*: *Alphavirus*) infection in nestling house sparrows. PLOS One 6: e25521.
18. O'Brien VA, Meteyer CU, Ip HS, Long RR, Brown CR (2010) Pathology and virus detection in tissues of nestling house sparrows naturally infected with Buggy Creek virus (*Togaviridae*). J Wildl Dis 46: 23–32.
19. Calisher CH (1994) Medically important arboviruses of the United States and Canada. Clin Microbiol Rev 7: 89–116.
20. Hamilton WD, Zuk M (1982) Heritable true fitness and bright birds: a role for parasites? Science 218: 384–387.
21. Brown CR, Brown MB (1986) Ectoparasitism as a cost of coloniality in cliff swallows (*Hirundo pyrrhonota*). Ecology 67: 1206–1218.
22. Møller AP (1993) Ectoparasites increase the cost of reproduction in their hosts. J Anim Ecol 62: 309–322.
23. Fitze PS, Tschirren B, Richner H (2004) Life history and fitness consequences of ectoparasites. J Anim Ecol 73: 216–226.
24. Ramamoorthi N, Narasimhan S, Pal U, Bao F, Yang XF, et al. (2005) The Lyme disease agent exploits a tick protein to infect the mammalian host. Nature 436: 573–577.
25. Zeidner NS, Higgs S, Happ CM, Beaty BJ, Miller BR (1999) Mosquito feeding modulates Th1 and Th2 cytokines in flavivirus susceptible mice: an effect mimicked by injection of sialokinins, but not demonstrated in

flavivirus resistant mice. Parasite Immunol 21: 35–44.

26. Boppana VD, Thangamani S, Adler AJ, Wikel SK (2009) SAAG-4 is a novel mosquito salivary protein that programs host CD4+ T cells to express IL-4. Parasite Immunol 31: 287–295.
27. Reisen WK, Chiles RE, Martinez VM, Fang Y, Green EN (2003) Experimental infection of California birds with western equine encephalomyelitis and St. Louis encephalitis viruses. J Med Entomol. 40: 968–982.
28. Brown CR, Brown MB (1996) Coloniality in the cliff swallow: the effect of group size on social behavior. Chicago: University of Chicago Press. 580 p.
29. Medzhitov R, Schneider DS, Soares MP (2012) Disease tolerance as a defense strategy. Science 335: 936–941.
30. Restif O, Koella JC (2004) Concurrent evolution of resistance and tolerance to pathogens. Am Nat. 164: E90–E102.
31. Lowther PE, Cink CL (1992) House sparrow. In: Poole A, Frank BG, editors. The Birds of North America. Philadelphia: Academy of Natural Sciences.
32. Brown CR, Brown MB (1995) Cliff swallow (*Hirundo pyrrhonota*). In: Poole A, Gill F, editors. The birds of North America. Philadelphia: Academy of Natural Sciences.
33. O'Brien VA, Brown CR (2012) Seasonal variation and age-related correlates of Buggy Creek virus (*Togaviridae*) infection in nestling house sparrows. J Wildl Dis 48: 138–147.
34. Brown CR, Brown MR (2004) Empirical measurement of parasite transmission between groups in a colonial bird. Ecology 85: 1619–1626.
35. Brown CR, Komar N, Quick SB, Sethi RA, Panella NA, et al. (2001) Arbovirus infection increases with group size. Proc R Soc Lond B 268: 1833–1840.
36. Moore AT, Edwards EA, Brown MB, Komar N, Brown CR (2007) Ecological correlates of Buggy Creek virus infection in *Oeciacus vicarius*, southwestern Nebraska, 2004. J Med Entomol 44: 42–49.
37. Powers AM, Brault AC, Shirako Y, Strauss EG, Kang W, et al. (2001) Evolutionary relationships and systematics of the alphaviruses. J Virol 75: 10118–10131.
38. Pfeffer M, Foster JE, Edwards EA, Brown MB, Komar N, et al.

(2006) Phylogenetic analysis of Buggy Creek virus: evidence for multiple clades in the western Great Plains, United States of America. Appl Environ Microbiol 72: 6886–6893.

39. O'Brien VA, Meteyer CU, Reisen WK, Ip HS, Brown CR (2010) Prevalence and pathology of West Nile virus in naturally infected house sparrows, western Nebraska, 2008. Am J Trop Med Hyg 82: 937–944.
40. Anderson TR (2006) Biology of the ubiquitous house sparrow. New York: Oxford University Press. 560 p.
41. Wilkinson L (1999) SYSTAT. Chicago: SPSS Inc.
42. Dusek RJ, McLean RG, Kramer LD, Ubico SR, Dupuis AP Jr, et al. (2009) Prevalence of West Nile virus in migratory birds during spring and fall migration. Am J Trop Med Hyg 81: 1151–1158.
43. Brown CR, Moore AT, Young GR, Komar N (2010) Persistence of Buggy Creek virus (*Togaviridae*, *Alphavirus*) for two years in unfed swallow bugs (Hemiptera: Cimicidae: *Oeciacus vicarius*). J Med Entomol 47: 436–441.
44. Moseman EA, Iannacone M, Bosurgi L, Tonti E, Chevrier N, et al. (2012) B cell maintenance of subcapsular sinus macrophages protects against a fatal viral infection independent of adaptive immunity. Immunity 36: 415–426.
45. Rizzo C, Ronca R, Fiorentino G, Mangano VD, Sirima SB, et al. (2011) Wide cross-reactivity between Anopheles gambiae and Anopheles funestus SG6 salivary proteins supports exploitation of gSG6 as a marker of human exposure to major malaria vectors in tropical Africa. Malaria J 10: 206.
46. Sanders ML, Jaworski DC, Sanchez JL, DeFraites RF, Glass GE, et al. (1998) Antibody to a cDNA-derived calreticulin protein from *Amblyomma americanum* as a biomarker of tick exposure in humans. Am J Trop Med Hyg 59: 279–285.
47. Orlandi-Pradines E, Almeras L, Denis de Senneville L, Barbe S, Remoué F, et al. (2007) Antibody response against saliva antigens of *Anopheles gambiae* and *Aedes aegypti* in travelers in tropical Africa. Microbes Infect 9: 1454–1462.
48. Hostomska J, Rohousova I, Volfova V, Stanneck D, Mencke N, et al. (2008) Kinetics of canine antibody response to saliva of the sand fly *Lutzomyia longipalpis* . Vector Borne Zoonotic Dis 8: 443–450.

49. Schwarz A, Sternberg JM, Johnston V, Medrano-Mercado N, Anderson JM, et al. (2009) Antibody responses of domestic animals to salivary antigens of Triatomainfestans as biomarkers for low-level infestation of triatomines. Int J Parasitol 39: 1021–1029.
50. Huber SK, Owen JP, Koop JAH, King MO, Grant PR, et al. (2010) Ecoimmunity in Darwin's finches: invasive parasites trigger acquired immunity in the medium ground finch (*Geospiza fortis*). PLOS One 5: e8605.
51. Tsujimoto H, Gray EW, Champagne DE (2010) Black fly salivary gland extract inhibits proliferation and induces apoptosis in murine splenocytes. Parasite Immunol 32: 275–284.
52. King MO, Owen JP, Schwabl H (2011) Injecting the mite into ecological immunology: measuring the antibody response of house sparrows (*Passer domesticus*) challenged with hematophagous mites. The Auk 128: 340–345.
53. Proctor H, Owens I (2000) Mites and birds: diversity, parasitism and coevolution. Trends in Ecol Evol 19: 358–364.
54. Schneider BS, McGee CE, Jordan JM, Stevenson HL, Soong L, et al. (2007) Prior exposure to uninfected mosquitoes enhances mortality in naturally-transmitted West Nile virus infection. PLOS One 2: e1171.
55. Machain-Williams C, Mammen MP Jr, Zeidner NS, Beaty BJ, Prenni JE, et al. (2012) Association of human immune response to *Aedes aegypti* salivary proteins with dengue disease severity. Parasite Immunol 34: 15–22.
56. Peng Z, Beckett AN, Engler RJ, Hoffman DR, Ott NL, et al. (2004) Immune responses to mosquito saliva in 14 individuals with acute systemic allergic reactions to mosquito bites. J Allergy Clin Immunol 114: 1189–1194.
57. Belkaid Y, Kamhawi S, Modi G, Valenzuela J, Noben-Trauth N, et al. (1998) Development of a natural model of cutaneous leishmaniasis: powerful effects of vector saliva and saliva preexposure on the long-term outcome of Leishmania major infection in the mouse ear dermis. J Exp Med 188: 1941–1953.
58. Schneider BS, Soong L, Coffey LL, Stevenson HL, McGee CE, et al. (2010) *Aedes aegypti* saliva alters leukocyte recruitment and cytokine signaling by antigen-presenting cells during West Nile virus infection. PLOS One 5: e11704.

59. Wikel SK (1999) Tick modulation of host immunity: an important factor in pathogen transmission. Int J Parasitol 29: 851–859.

11

Bluetongue Disease Risk Assessment Based on Observed and Projected Culicoides obsoletus spp. Vector Densities

Katharina Brugger, Franz Rubel

Institute for Veterinary Public Health, University of Veterinary Medicine Vienna, Vienna, Austria

ABSTRACT

Bluetongue is an arboviral disease of ruminants causing significant economic losses. Our risk assessment is based on the epidemiological key parameter, the basic reproduction number. It is defined as the number of secondary cases caused by one primary case in a fully susceptible host population, in which values greater than one indicate the possibility, i.e., the risk, for a major disease outbreak. In the course of the Bluetongue virus serotype 8 (BTV-8) outbreak in Europe in 2006 we developed such a risk

Citation: Brugger K, Rubel F (2013) Bluetongue Disease Risk Assessment Based on Observed and Projected Culicoides obsoletus spp. Vector Densities. PLoS ONE 8(4): e60330. https://doi.org/10.1371/journal.pone.0060330

assessment for the University of Veterinary Medicine Vienna, Austria. Basic reproduction numbers were calculated using a well-known formula for vector-borne diseases considering the population densities of hosts (cattle and small ruminants) and vectors (biting midges of the *Culicoides obsoletus* spp.) as well as temperature dependent rates. The latter comprise the biting and mortality rate of midges as well as the reciprocal of the extrinsic incubation period. Most important, but generally unknown, is the spatio-temporal distribution of the vector density. Therefore, we established a continuously operating daily monitoring to quantify the seasonal cycle of the vector population by a statistical model. We used cross-correlation maps and Poisson regression to describe vector densities by environmental temperature and precipitation. Our results comprise time series of observed and simulated *Culicoides obsoletus* spp. counts as well as basic reproduction numbers for the period 2009–2011. For a spatio-temporal risk assessment we projected our results from the location of Vienna to the entire region of Austria. We compiled both daily maps of vector densities and the basic reproduction numbers, respectively. Basic reproduction numbers above one were generally found between June and August except in the mountainous regions of the Alps. The highest values coincide with the locations of confirmed BTV cases.

INTRODUCTION

Veterinary authorities are generally interested in a spatiotemporal risk assessment to establish efficient protection and control measures when facing diseases with high economic impact such as caused by the Bluetongue virus (BTV). Such risk assessments are frequently based on one common epidemiological parameter, the basic reproduction number R_0. It describes the number of secondary cases caused by a primary case in a completely susceptible population at the beginning of an epidemic. Thus, R_0 may be interpreted as a threshold parameter: a major disease outbreak may only occur for $R_0>1$, while for $R_0<1$ the disease will fade out [1].

The motivation for this study was the Bluetongue virus serotype 8 (BTV-8) epidemics in Europe, which emerged in 2006 at the border triangle of Belgium, Germany and the Netherlands [2,3]. Within a few years this arboviral disease rapidly spread across North-western and Central Europe [4,5] and caused substantial losses in agriculture amounting to millions Euro [6]. Beside movement restrictions and surveillance, the main veterinary measure was to vaccinate the susceptible populations beginning in 2008 [7].

In Austria, only a few cattle near the border to Germany were confirmed to be BTV-8 seropositive. Due to the German vaccination campaigns achieving a coverage of 78% [8] and preventive vaccination in Austria the eastwards spread of the Bluetongue disease could be stopped at the German-Austrian border (Figure 1).

In the course of the European BTV-8 outbreak several methods for the calculation of R_0 have been proposed. Some of these methods are based on data from surveillance and monitoring programmes. For example, [9] estimated R_0 between herds for the Netherlands. Furthermore, a theoretical study on R_0 using a deterministic epidemic model was introduced by [10] to investigate the seasonal spread of BTV over several years and to evaluate the effectiveness of vaccination strategies. Most approaches presented so far considered temperature dependent vector parameters as well as host and vector densities. While [11] presented a general concept on the calculation of R_0 for which they discussed the influence of temperature on the transmission cycle, [12] and [13] generated spatial R_0 maps for the Netherlands and Switzerland, respectively. Furthermore, the uncertainty and sensitivity of such temperature-dependent models have been analysed and temperature turned out to be the most dominant factor for determining the magnitude of R_0 [14]. [15] estimated the possible impact of past and future climates on the basic reproduction number R_0 on an European scale. For their study they used input data from climate model runs applied to different emission scenarios of the Intergovernmental Panel on Climate Change (IPCC). The authors have used a similar approach for modelling another mosquito-borne disease outbreak, the Usutu virus epidemics in Vienna [16]. Here we present a predictive risk assessment method based on the basic reproduction number R_0 after [12] applied to a potential Bluetongue outbreak in Austria. Results comprise regions at risk on a spatio-temporal scale representative for a resolution of 10 km and 1 day.

In general, BTV circulates in a natural transmission cycle between vectors (Culicoides biting midges) and hosts (ruminants: mainly cattle, sheep, and goats). For risk assessment using the concept of R0 the knowledge of both the vector and the host densities are of fundamental importance. While the population densities of farm animals are well documented in national veterinary databases or available worldwide on a regular grid [17], the vector density is generally not well known or rather unknown. Therefore, a central part of our study is concerned with the estimation of the vector population. We established a continuously operating daily Culicoides spp. monitoring at the University of Veterinary Medicine Vienna, Austria, to quantify the seasonal abundance for at least three years. This approach is in contrast to former - mainly weekly - monitoring programs, such as the large-scale entomological surveillance program throughout the European Union to investigate the occurrence and geographical distribution of Culicoides spp. ([18]; results were published e.g. by [19,20]).

METHODS

Vector Monitoring

For the Culicoides spp. monitoring a black light suction trap of Onderstepoort type, as described by [21], was placed next to an automatic weather station at the university campus (λ=16:400 E geographical longitude and Q~48:250 N geographical latitude). The location in Vienna is characterised by a temperate, fully humid climate with warm summers, corresponding to Cfb climate following the Ko¨ppen-Geiger climate classification [22]. This standardized trap is commonly used in monitoring and surveillance programs, because it is very efficient compared to other suction light traps [23]. Following the guideline of [24], the trap was hung at a height of 1.5 m above the ground. The distance to the stables and paddocks with cattle, sheep, and horses was less than 20 m. A collection bottle rotator (model 1512, John W. Hock Company, FL, USA) with eight beakers was used to segregate collections at 24 h intervals. For species evaluation the catches were separated first in Culicoides spp. and other insects (bycatches) under a stereomicroscope. Afterwards the Culicoides species complexes were determined by the characteristic pattern and coloration of the wings according to [24–26].

During the monitoring period 2009–2011 a total of 18,952 Culicoides spp. were captured, in which the predominantly trapped complex was C. obsoletus (85.7%). The C. obsoletus complex comprises C. obsoletus, C. scoticus, C. dewulfi, and C. chiopterus. In addition we identified midges of the C. pulicaris complex (9.8%), amongst others (4.5%). In this study we used only the most frequent vectors, i.e. those of the C. obsoletus complex.

Vector Modelling

The aim of our vector modelling is to statistically describe the midge catches by operationally available meteorological quantities. Assuming that the relationship between meteorological quantities and C. obsoletus abundance is location-independent (but not independent on the host density), the model will be used to estimate the spatial distribution of C. obsoletus. Additionally, the C. obsoletus abundance may be projected into the future by using meteorological output parameters from numerical weather prediction models. The development of the statistical model was carried out in two steps. In the first step, we performed a cross correlation map (CCM) analysis to investigate which of the meteorological quantities correlates best with the observed C. obsoletus counts. In the second step, we used these meteorological quantities in a Poisson regression model. This procedure was done first for Vienna to demonstrate the performance of the statistical model by comparing it's output with observations and subsequently applied for the entire region of Austria.

The CCM analysis is an extension of the ordinary crosscorrelation by introducing a second time lag. This second time lag is used to average or accumulate meteorological quantities over a period beginning at time lag 1 and ending at time lag 2. Correlation coefficients are then calculated for these averaged quantities, in which the CCMs depict the averaging period resulting in the highest correlation coefficient (Figure 2). CCMs were adopted for example by [27] to investigate the relationship between Ochlerotatus sollicitans abundance and environmental parameters. Further applications comprise studies on Aedes sollicitans [28] as well as Aedes sollicitans and Culex salinarius [29].

Using a more mathematical notation, the statistical analysis reads as follows: For n_i representing a time series of the observed population of midges and $\overline{X}_{i-j,i-k}$ a meteorological quantity with time index i, the CCM considers $cor(log_e(n_i+1),\overline{X}_{i-j,i-k})$ for each time lag j and k, at which $j \geq k$. Generally, the population data were normalized using a $\log_e$ (n_i +1)-transformation and $\overline{X}_{i-j,i-k}$ is averaged over the time interval ranging from i- j. The correlations for all possible time intervals are displayed in a CCM with the time lags j and k range along the two axes. Note, that for the special case of j= k (450 -line in a CCM), the correlation coefficients are equal to those of a cross-correlogram. We developed a source code for the calculation of CCMs using the R statistical computing environment [30].

Based on the correlations found by CCMs, a Poisson regression was applied to describe daily vector abundance by meteorological quantities. Again, the analysis was performed using the R statistical computing environment.

Risk Assessment

For the different tasks we used vector densities, estimated from both observed and simulated counts of midges, to assess the risk of a potential BTV-8 outbreak in cattle and small ruminants. Therefore, we applied the basic reproduction number R_0, which may be interpreted as a threshold value for the possibility of a major outbreak ($R_0>1$). An analytical solution for R_0 is generally derived from epidemic differential equation models. Here the formula used in recent studies by [12], [15] and [31] is applied, which may be traced back to the historical developments by Ronald Ross and George Macdonald [32].

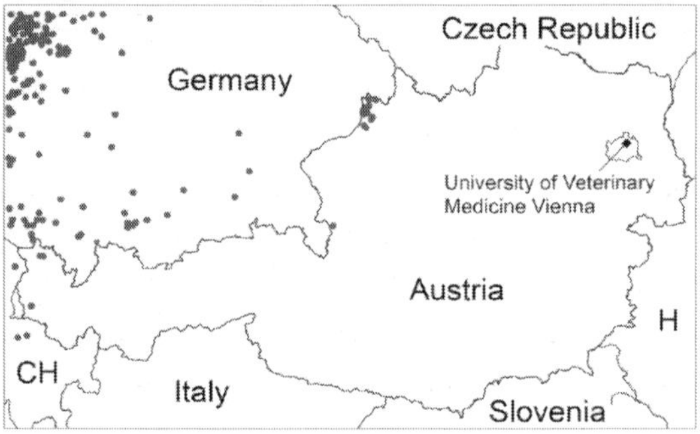

Figure 1. Reported locations of BTV-8 cases between 2006 and 2009 in Germany [8], Austria [43] and Switzerland [44].

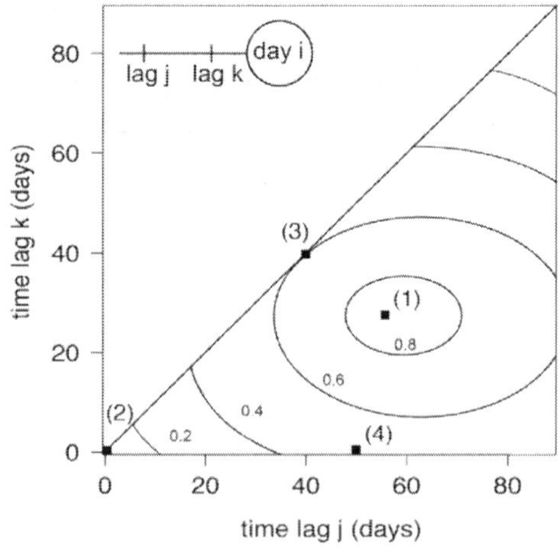

Figure 2. Schematic illustration of a cross correlation map (CCM). (1) Maximum correlation of r(60,30)=0:9 for a dependent vs. an independent quantity.

The latter averaged over the period of preceding days j = 60 to k = 30. (2) Commonly used correlation coefficient (without time lags, j = k = 0) of r(0,0)=0:1. (3) Crosscorrelation (with one time lag, j = k = 40) of r(40,40)=0:6, which is equal to a cross correlation with a single time lag. (4) Cross correlation of r(50,0), the independent quantity was averaged over the preceding 50 days.

$$R_0 = \sqrt{\frac{k(T)^2 p_M \gamma_M(T) N_M}{(N_C + N_S)^2 m_M(T)[\gamma_M(T) + m_M(T)]}\left(\frac{p_C N_C}{\alpha_C} + \frac{p_S N_S}{\alpha_S}\right)} \quad (1)$$

R_0 equations given by [12], [15] and [31] are formally equivalent but differ in the complexity of parameter formulation. The most comprehensive formulation by [31] provides a gamma distribution for the extrinsic incubation period, which reduces to equation (1) by setting the scale parameters for the gamma distributions to one. Model parameters are probabilities for an infection of cattle p_C, small ruminants p_S and midges p_M, the removal rate (recovery or death rate) for cattle α_C and small ruminants α_S, as well as some functions describing the temperature dependent behaviour of midges. The latter comprise the biting rate of midges k(T), the virus reproduction rate in midges $c_M(T)$, i.e. the reciprocal of the extrinsic incubation period, and the mortality rate of midges $m_M(T)$. All constant parameters as well as temperature dependent parameter functions were taken according to [12] and are summarized in Table 1. Note that the temperature dependent parameter functions are not well defined for low temperatures. Therefore, we calculated R_0 only for threshold values of T≥130 C. Note further that several other functions were proposed in the literature. A sensitivity study performed by [33] using alternative functions for midge mortality, extrinsic incubation period and biting rate results in similar R_0 values (see Text S1 for details). Additionally, the basic reproduction number depends on the host and vector densities. We distinguish between cattle N_C, small ruminants N_S and midges N_M.

RESULTS

Simulated Vector Population

In a first step we investigated the correlations between time series of midges of the Culicoides obsoletus complex and environmental quantities measured by the automatic weather station beside the light trap. The latter comprise amongst others air temperature, precipitation, wind speed and relative humidity. CCMs as discussed above were compiled. In accordance with the life cycle of midges we considered time frames of 120 days (4 months) preceding the catches. From all investigated quantities only temperature and precipitation were found to correlate quite well with the daily midge counts. Note that daily precipitation has a skewed frequency distribution and was normalized by logtransformation before CCMs were calculated (Figure 3).

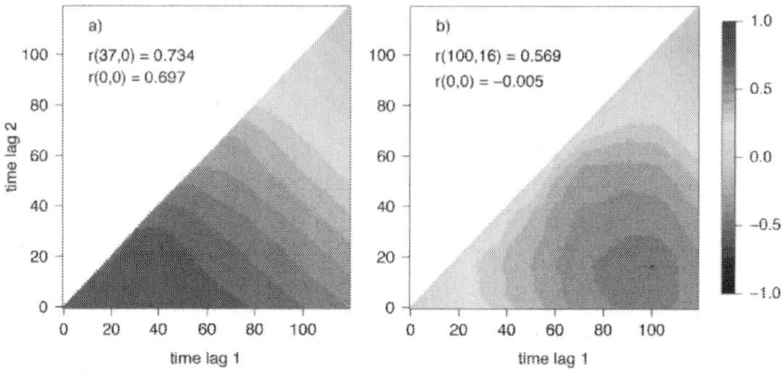

Figure 3. Cross correlation maps (CCMs) for daily time series in Vienna, Austria: Midges vs. temperature (a) and midges vs. precipitation (b). Note that midge counts and precipitation values were normalized by log-transformation. Period: Jan. 2009 to Dec. 2011.

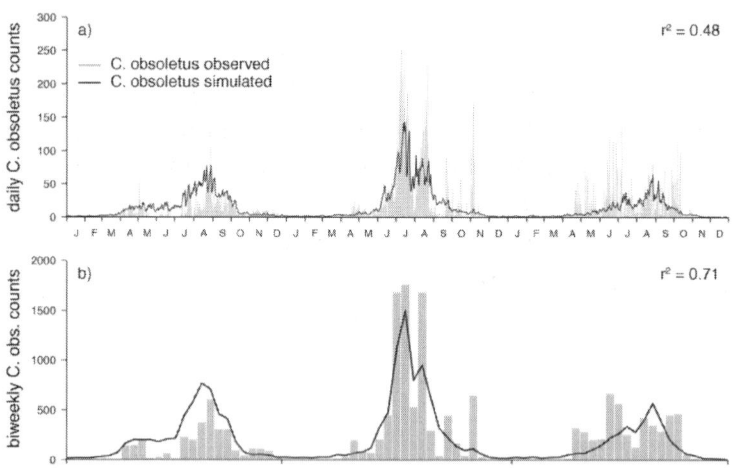

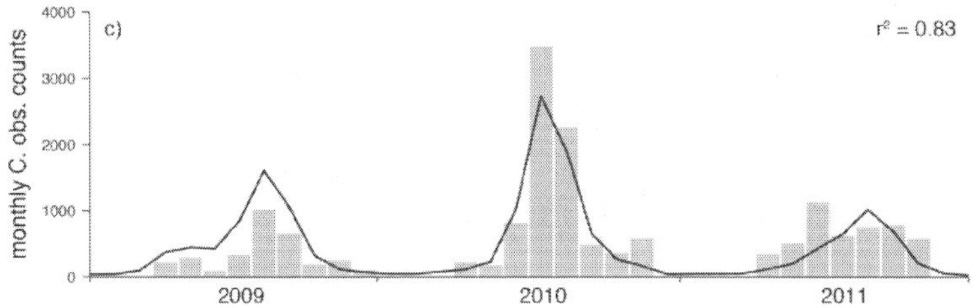

Figure 4. Daily (a), biweekly (b) and monthly (c) Culicoides obsoletus complex counts observed with the black-light trap (bars) vs. predicted by the Poisson regression model (lines). Period: January 2009 to December 2011; Vienna, Austria.

We found a maximum correlation for midges vs. temperature averaged over the preceding 37 days of r(37,0)=0:734 (Figure 3 left). This correlation is slightly higher than the corresponding dayto-day correlation of r(0,0) = 0.697. Between midges and precipitation we found no day-to-day correlation, but for the precipitation averaged over the period between the preceding days 100 to 16 a correlation of r(100,16) = 0.569 was calculated (Figure 3 right). Inserting these three positive correlations into the Poisson regression model leads to

$$log_e(n_i+1) = -0.856 + 0.076\, T_{i,i} + 0.048\, \overline{T}_{i-37,i} + 2.913\, \overline{P}_{i-100,i-16}, \quad (2)$$

with the daily temperature T, the mean temperature $\overline{T}$, and the mean logarithmic precipitation $\overline{P}$ (contribution of all coefficients significant with p<0:001). As a result, observed vs. predicted daily Culicoides counts for Vienna are depicted in Figure 4. According to that, temperature and precipitation explain about 48.1% of the daily variance of the C. obsoletus counts. We plotted also bi-weekly and monthly time series, although it is clear that the explained variances are higher for accumulated data, because they are usually appropriate for many epidemiological applications. For monthly data, for example, the explained variance increases up to 83.3%.

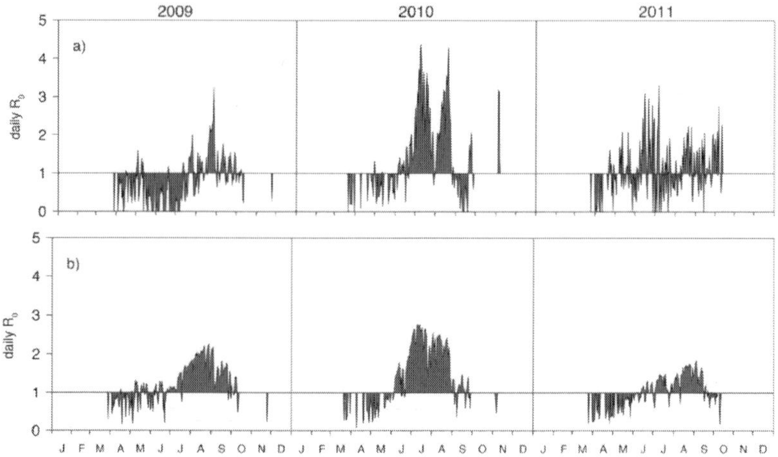

Figure 5. Daily basic reproduction number R_0 for Vienna based on observed (a) and simulated (b) Culicoides obsoletus time series. Red and green bars indicate $R_0>1$ and $R_0<1$, respectively.

Dynamic Bluetongue Risk Assessment for Vienna

We calculated the daily reproduction number R0 for Vienna by applying Eq. 1, in which vectors and hosts may be alternatively inserted as absolute numbers or densities. Assuming our university campus covers an area of 1 km^2, values given in absolute numbers or densities are equal. We used an average host density of cattle and small ruminants of $N_C=N_S=25$ animals/km^2. The true vector density N_M, however, is unknown. In every case it is much higher than the number of midges caught at the location of the light trap. In the absence of any other estimate we applied the reasonable assumption proposed by [12]. Accordingly, the trapped midges were assumed to reflect 1% of the local vector population, i.e. our catches were multiplied by a factor of 100 before entering Eq. 1 (N_M=n: 100). Additionally, the temperature dependent functions given in Table 1 are defined for temperatures w130 C, hence R_0 is calculated for only 46.3% of the days (mainly between April and October)

Figure 5 depicts the time series of the daily reproduction number R_0 for Vienna. While in Figure 5a the observed numbers of midges were used, Figure 5b depicts R_0 values based on the simulated numbers of midges. Taking the first as a gold standard we evaluated the influence of simulated midges NM on the calculation of R_0 by using the two verification measures sensitivity and specificity. We calculated a sensitivity of 0.81 (probability that days with $R_0>1$ were correctly realized using R_0 estimates based on simulated N_M) and a specificity of 0.53 (probability that days with $R_0<1$ were correctly realized).

Spatio-temporal Bluetongue Risk Assessment for the Entire Region of Austria

Assuming that R_0 calculations in Vienna based on simulated N_M values satisfy our requirements concerning the accuracy, we applied the R_0 estimation method to the entire region of Austria. Therefore we defined a grid with 10 km spacing, equidistant in geographical longitude $(\Delta\lambda = 0.15^\circ)$ and latitude $(\Delta\varphi = 0.09^\circ)$. Gridded data of temperature and precipitation were taken from the Austrian meso-scale numerical weather prediction model ALADIN [34]. Only data for the period 2010–2011 were used, because data for earlier years were not available. The advantage of ALADIN data over observations (in situ or satellite data) is that they are available on a higher temporal resolution, will be available in near future also on a much higher spatial resolution and gives us the opportunity to project R_0 values to the future. It should be mentioned that we found a few extraordinary high precipitation values within the ALADIN data leading to unrealistic high numbers of midges simulated by Eq. 2. To avoid such artefacts we applied a truncation for the accumulated logarithmic precipitation of $\overline{P} > 1.5$ mm/day. This cut-off coincides with annual precipitation maxima of about 1200 mm/year as observed in high Alpine regions [35].

Host and vector densities used for the R_0 calculation over Austria are depicted in Figure 6. The constant densities of cattle N_C (Figure 6a) and small ruminants N_S (Figure 6b) were taken from the Austrian veterinary database. The spatial distribution of the vector density N_M, a function of temperature, precipitation, N_C and N_S, is given in units of 103 midges/km^2. Again we assumed that the simulated midges reflect 1% of the real vector population. Further we adjusted the vector density for host densities following the findings of [36]. Thus, the vector density at a grid point with geographical coordinates λ and φ is calculated from the simulated vectors at this coordinate $n(\lambda,\varphi)$ multiplied by a factor of 100 as discussed above and normalized by the host density at the location of the University of Veterinary Medicine Vienna (with 50 ruminants/km^2).

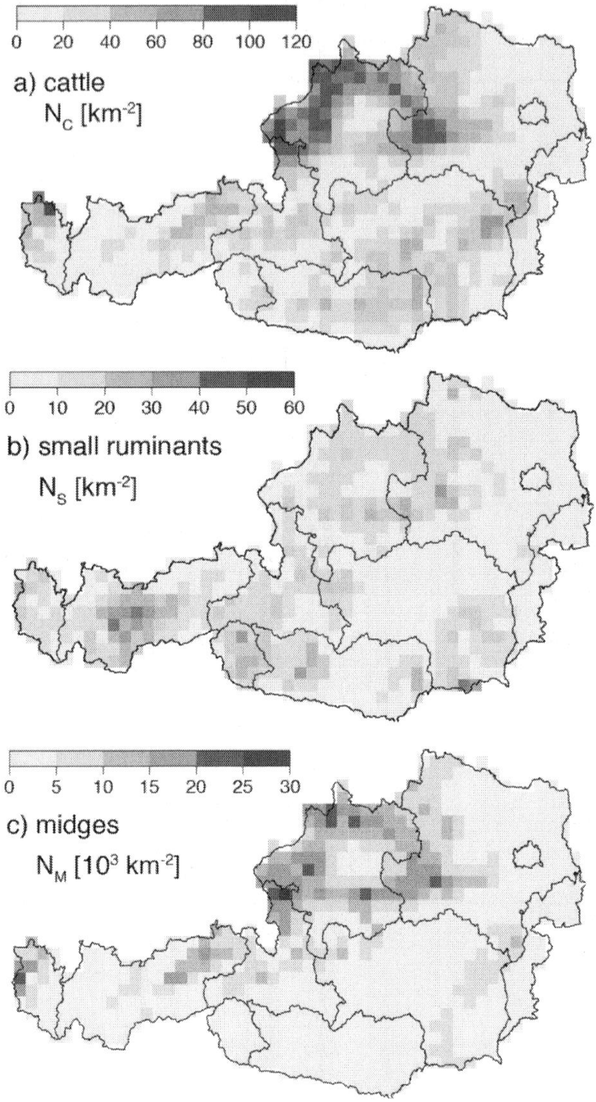

Figure 6. Constant host densities for cattle (a) and small ruminants (b), both given in animals/km^2, and simulated vector densities for 11 July 2010 in 10^3 midges/km^2 (c).

$$N_M(\lambda,\varphi)=n(\lambda,\varphi)\,100\left[\frac{N_C(\lambda,\varphi)+N_S(\lambda,\varphi)}{50}\right] \tag{3}$$

Note that the vector density varies temporally and spatially, because $n(\lambda,\varphi)$ is a function of temperature and precipitation. As an example, Figure

6c depicts the vector density for 11 July 2010. Using these daily vector densities, the host densities as well as the parameters and temperature dependent parameter functions we compiled daily risk maps for the period 2010–2011. Movies of simulated daily vector densities and Bluetongue risk maps are provided as supporting information or on our website http://epidemic-modeling.vetmeduni.ac.at/btvmodel. htm. Figure 7 depicts mean daily R_0 maps for June, July, August, and September 2010. While red areas indicate a potential risk for a major outbreak ($R_0>1$), green areas are associated with minor or no risk at all ($R_0<1$). For example, the R_0 values for July 2010 are within the range 0:1{4:6, in which the maximum indicates that from one infected animal on average 4.6 animals may be newly infected with BTV. Interestingly areas with a high vector density coincided with high host densities, even without the adjustment after [36]. Verification of the R_0 maps is currently not possible, because so far no major BTV outbreaks occurred in Austria. However, the regions with BTV-8 seropositive cases at the Austrian-German border (Figure 1) coincide with regions of high R0 values.

A visual comparison of the total numbers of days with $R_0>1$ between 2010 (Figure 8a) and 2011 (Figure 8b) shows that in 2010 many regions are at risk for longer, indicating a high inter-annual variation of the R_0 values.

DISCUSSION

We presented a risk assessment for a potential Bluetongue disease outbreak in Austria depicted by time series and maps of the basic reproduction number R_0. Input data were constant host densities and fluctuating vector densities simulated by a Poisson regression model using temperature and precipitation fields from the Austrian numerical weather prediction model ALADIN. Although the methodology presented here is generally accepted and was recently applied to BTV by [13], [12] and [15], there are some uncertainties in estimating R_0.

First of all, the applied R_0 formula is originated from an epidemic model, usually formulated by differential equations, by determining the largest eigenvalue of the so-called next-generation matrix. In the case of BTV it has never been verified by comparison with real outbreak data. Such a verification should demonstrate that the observed BTV-8 dynamics may be approximately reproduced by the underlying epidemic model with the parameter functions selected. Reasons for a missing verification may be due to the fact that it is extremely time-consuming, complex and expensive.

However, other epidemic models for arboviral diseases were successfully verified against outbreak data, e.g. for the well-documented Usutu virus outbreak in Vienna [37].

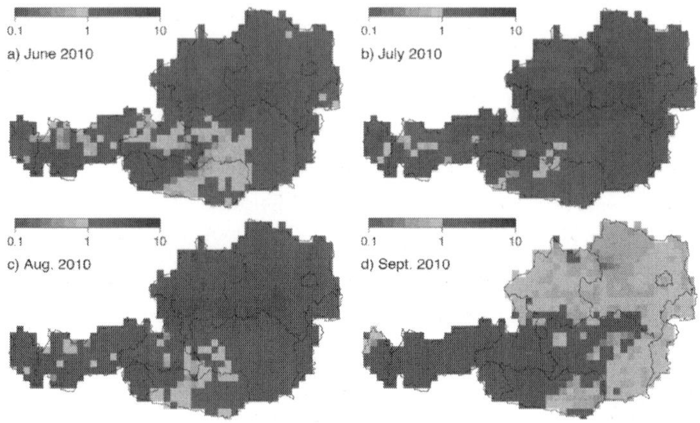

Figure 7. Mean daily basic reproduction number R0 for June to September 2010. The different maps represent the spatio-temporal distribution of the potential risk for a BTV outbreak in Austria. Green colours depict areas not under risk, red under moderate risk and dark red under high risk (note the logarithmic scaling).

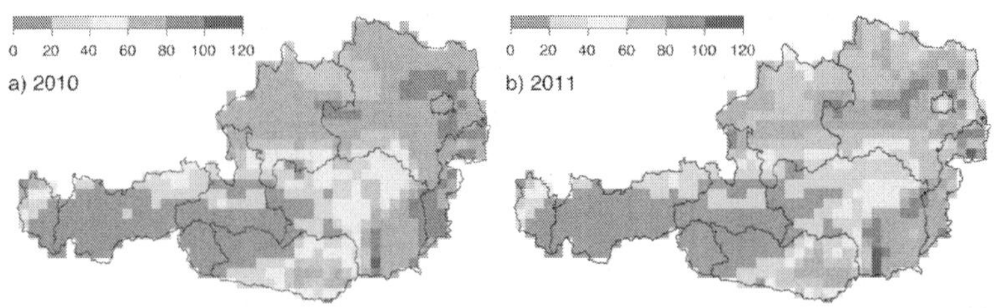

Figure 8. Numbers of days with $R_0>1$ for 2010 (a) and 2011 (b) demonstrating significant differences caused by inter-annual climate fluctuations. Regions with maxima of more than 100 days in the west and north-west of Austria coincide well with confirmed BTV cases depicted in Figure 1.

Further, it is difficult to estimate vector densities, even if numbers of trapped midges from a monitoring programme are available. We noticed that the official monitoring data sampled at 54 Austrian locations once a week according to the regulation of the European Union [18] provide an excellent overview on the species composition and on the seasonal activity of our native midge populations. Both were hitherto unknown or at least remain undocumented. A quantitative interpretation of the monitoring data, however, is hardly applicable. This is documented by the following small statistical experiment, where we compiled weekly time series based on Monday catches (after [18]) and compared them with weekly time series based on all daily catches. An explained variance of $r^2 = 0.59$ indicates that the numbers of midges sampled according to [18] were afflicted with significant undersampling errors. Further, the official sampling locations were neither representative for their vicinity nor comparable among each other. Also weather stations at short distances to the sampling locations were missing. Therefore, we established our own monitoring with continuously daily observations of midges and environmental parameters from 2009 to present. It allows us to calculate plausible R_0 values, although they should be interpreted with care. Due to the generally unknown magnitude of the vector density (we assumed numbers of midges per square kilometre to be two orders of magnitude higher than sampled at a point location), our R_0 values should be interpreted as relative rather than absolute measures.

Another uncertainty in the vector monitoring lies in the nature of the trapping method. [38] indicated that black light traps are adequate for measuring relative abundance, but not species composition. Further, we don't take into account that midges themselves can be transported by wind over considerable distances, although several models have been developed to investigate this wind-borne spread of BTV [39], [40]. An alternative study on *C. obsoletus* modelling applied a negative binomial regression model to simulate the seasonal cycle of daily catches at several locations in England [41]. It distinguishes between the influence of seasonality and meteorological quantities (temperature, precipitation and wind), but does not consider accumulated quantities as defined by CCMs. Further differences comprise the fraction of trapped midges of the *C. obsoletus* complex, which is 86% in Vienna and 48% in England. Unfortunately, due to differences in visualization and verification an objective comparison between [41] and the results presented here is not possible. Both approaches underestimate the extremely high variance of the observations, which is a general problem in vector modelling.

Similar to domestic ruminants most species of wild ruminants, e.g. red deer, are susceptible to BTV infection. We neglected the influence of these potential host species due to their still unclear epidemiological role in BTV transmission [42].

REFERENCES

1. Diekmann O, Heesterbeek J (2000) Mathematical epidemiology of infectious diseases. Chichester: Wiley.
2. Saegerman C, Berkvens D, Mellor PS (2008) Bluetongue epidemiology in the European Union. Emerg Infect Dis 14: 539–544.
3. Carpenter S, Wilson A, Mellor PS (2009) Culicoides and the emergence of bluetongue virus in northern Europe. Trends Microbiol 17: 172–178.
4. Mehlhorn H, Walldorf V, Klimpel S, Jahn B, Jaeger F, et al. (2007) First occurrence of Culicoides obsoletus-transmitted Bluetongue virus epidemic in Central Europe. Parasitol Res 101: 219–228.
5. Conraths FJ, Gethmann JM, Staubach C, Mettenleiter TC, Beer M, et al. (2009) Epidemiology of bluetongue virus serotype 8, Germany. Emerg Infect Dis 15: 433–435.
6. Wilson A, Mellor P (2008) Bluetongue in Europe: vectors, epidemiology and climate change. Parasitol Res 103 Suppl 1S69–S77.
7. Caporale V, Giovannini A (2010) Bluetongue control strategy, including recourse to vaccine: a critical review. Rev Sci Tech 29: 573–591.
8. Gethmann J, Hoffmann B, Probst C, Beer M, Conraths F, et al. (2010) A survey on three years with bluetongue disease serotype 8 in Germany (in German). Tierärztl Umschau 64: 4–12.
9. de Koeijer AA, Boender GJ, Nodelijk G, Staubach C, Meroc E, et al. (2011) Quantitative analysis of transmission parameters for Bluetongue virus serotype 8 in Western Europe in 2006. Vet Res 42: 53.
10. Charron MVP, Seegers H, Langlais M, Ezanno P (2011) Seasonal spread and control of Bluetongue in cattle. J Theor Biol 291: 1–9.
11. de Koeijer AA, Elbers A (2007) Modelling of vector-borne diseases and transmission of bluetongue virus in North-West Europe. Wageningen: Wageningen Academic Publishers.99-112.
12. Hartemink NA, Purse BV, Meiswinkel R, Brown HE, de Koeijer A, et al. (2009) Mapping the basic reproduction number (R_0) for vector-borne diseases: a case study on bluetongue virus. Epidemics 1: 153–161.
13. Racloz V, Venter G, Griot C, Stärk KDC (2008) Estimating the temporal and spatial risk of bluetongue related to the incursion of infected vectors into Switzerland. BMC Vet Res 4: 42.
14. Gubbins S, Carpenter S, Baylis M, Wood JLN, Mellor PS

(2008) Assessing the risk of Bluetongue to UK livestock: uncertainty and sensitivity analyses of a temperature-dependent model for the basic reproduction number. J R Soc Interface 5: 363–371.

15. Guis H, Caminade C, Calvete C, Morse AP, Tran A, et al. (2012) Modelling the effects of past and future climate on the risk of Bluetongue emergence in Europe. J R Soc Interface 9: 339–350.
16. Brugger K, Rubel F (2009) Simulation of climate-change scenarios to explain Usutu-virus dynamics in Austria. Prev Vet Med 88: 24–31.
17. Robinson TP, Franceschini G, Wint W (2007) The Food and Agriculture Organization's gridded livestock of the world. Vet Ital 43: 745–751.
18. EU (2007) Commission Regulation (EC) No 1266/2007.Official Journal of the European Union.
19. Anderle F, Sehnal P, Schneemann Y, Schindler M, Wöss G, et al. (2008) *Culicoides* surveillance in Austria (Diptera: Ceratopogonidae) - a snap-shot. Beiträge zur Entomofaunistik 9: 67–79.
20. Clausen PH, Stephan A, Bartsch S, Jandowsky A, Hoffmann-Köhler P, et al. (2009) Seasonal dynamics of biting midges (Diptera: Ceratopogonidae, *Culicoides spp.*) on dairy farms of Central Germany during the 2007/2008 epidemic of Bluetongue. Parasitol Res 105: 381–386.
21. Venter GJ, Meiswinkel R (1994) The virtual absence of *Culicoides imicola* (Diptera: Ceratopogonidae) in a light-trap survey of the colder, high-lying area of the eastern Orange Free State, South Africa, and implications for the transmission of arboviruses. Onderstepoort. J Vet Res 61: 327–340.
22. Kottek M, Grieser J, Beck C, Rudolf B, Rubel F (2006) World map of the Köppen-Geiger climate classification updated. Meteorol Z 15: 259–263.
23. Venter GJ, Labuschagne K, Hermanides KG, Boikanyo SNB, Majatladi DM, et al. (2009) Comparison of the efficiency of five suction light traps under field conditions in South Africa for the collection of *Culicoides* species. Vet Parasitol 166: 299–307.
24. Goffredo M, Meiswinkel R (2004) Entomological surveillance of bluetongue in Italy: methods of capture, catch analysis and identification of *Culicoides* biting midges. Vet Ital 40: 260–265.
25. Delécolle JC (1985) Nouvelle contribution a l'etude systématique et iconographique des espèces du genre *Culicoides* (Diptera:

Ceratopogonidae) du Nord-Est de la France [Ph.D. thesis]. Universitè Louis Pasteur de Strasbourg. 283 pp.

26. Rawlings P (1996) A key, based on wing patterns of biting midges (Genus: *Culicoides Latreille* - Diptera: Ceratopogonidae) in the Iberian Peninsula, for use in epidemiological studies. Graellsia 52: 57–71.
27. Curriero FC, Shone SM, Glass GE (2005) Cross correlation maps: a tool for visualizing and modeling time lagged associations. Vector Borne Zoonotic Dis 5: 267–275.
28. Shone SM, Curriero FC, Lesser CR, Glass GE (2006) Characterizing population dynamics of Aedes sollicitans (Diptera: Culicidae) using meteorological data. J Med Entomol 43: 393–402.
29. Walsh A, Glass G, Lesser C, Curriero F (2008) Predicting seasonal abundance of mosquitoes based on off-season meteorological conditions. Environ Ecol Stat 15: 279–291.
30. R Development Core Team (2012) R: A Language and Environment for Statistical Computing, version 2.14.1. R Foundation for Statistical Computing (Vienna, Austria). Available: http://www.R-project.org/.
31. Gubbins S, Hartemink N, Wilson A, Moulin V, Vonk Noordegraaf C, et al. (2012) Scaling from challenge experiments to the field: Quantifying the impact of vaccination on the transmission of bluetongue virus serotype 8. Prev Vet Med 105: 297–308.
32. Smith DL, Battle KE, Hay SI, Barker CM, Scott TW, et al. (2012) Ross, Macdonald, and a theory for the dynamics and control of mosquito-transmitted pathogens. PLoS Pathog 8: e1002588.
33. Jungblut C, Schober S (2012) Risk assessment methods for bluetongue disease (in German). Rubel F, Brugger K, editors. Austrian Contributions to Veterinary Epidemiology, volume 7 . Austria: Vetmeduni Vienna. 91 pp.
34. Wang Y, Bellus M, Wittmann C, Steinheimer M, Weidle F, et al. (2011) The Central European limited-area ensemble forecasting system: ALADIN-LAEF. Q J R Meteorol Soc 137: 483–502.
35. Frei C, Schär C (1998) A precipitation climatology of the Alps from high-resolution rain-gauge observations. Int J Climatol 18: 873900.
36. Garcia-Saenz A, McCarter P, Baylis M (2011) The influence of host number on the attraction of biting midges, Culicoides spp., to light traps. Med Vet Entomol 25: 113–115.
37. Rubel F, Brugger K, Hantel M, Chvala-Mannsberger S, Bakonyi

T, et al. (2008) Explaining Usutu virus dynamics in Austria: model development and calibration. Prev Vet Med 85: 166–186.

38. Carpenter S, Szmaragd C, Barber J, Labuschagne K, Gubbins S, et al. (2008) An assessment of Culicoides surveillance techniques in northern Europe: have we underestimated a potential bluetongue virus vector? J Appl Ecol 45: 1237–1245.
39. Ducheyne E, Lange M, Van der Stede Y, Meroc E, Durand B, et al. (2011) A stochastic predictive model for the natural spread of bluetongue. Prev Vet Med 99: 48–59.
40. Burgin LE, Gloster J, Sanders C, Mellor PS, Gubbins S, et al.. (2012) Investigating incursions of bluetongue virus using a model of long-distance Culicoides biting midge dispersal. Transbound Emerg Dis. doi:10.1111/j.1865-1682.2012.01345.x.
41. Sanders CJ, Shortall CR, Gubbins S, Burgin L, Gloster J, et al. (2011) Influence of season and meteorological parameters on flight activity of Culicoides biting midges. J Appl Ecol 48: 1355–1364.
42. Falconi C, López-Olvera JR, Gortázar C (2011) BTV infection in wild ruminants, with emphasis on red deer: a review. Vet Microbiol 151: 209–219.
43. Loitsch A, Sehnal P, Schneemann Y, Winkler M, Revilla-Fernández S, et al.. (2009). Durchführung der Bluetongue Überwachung in Österreich - BTV-Abschlussbericht 2008 & 2009 (in German). Austrian Agency for Health and Food Safety (AGES).
44. Schorer M, Schwermer H (2012) Blauzungenüberwachung in der Schweiz: Erklärung der Seuchen3 freiheit (in German). Federal Veterinary Office (FVO).
45. Mullens B, Gerry A, Lysyk T, Schmidtmann E (2004) Environment effects on vector competence and virogenesis of bluetongue virus in Culicoides: interpreting laboratory data in a field of context. Veter Ital 40: 160–166.
46. Carpenter S, Wilson A, Barber J, Veronesi E, Mellor P, et al. (2011) Temperature dependence of the extrinsic incubation period of orbiviruses in Culicoides biting midges. PLoS ONE 6: e27987.
47. Gerry A, Mullens B (2000) Seasonal abundance and survivorship of Culicoides sonorensis (Diptera: Ceratopogonidae) at a south California dairy, with reference to potential bluetongue virus transmission and persistence. J Med Entomol 37: 675–688.

12

Worldwide Niche and Future Potential Distribution of Culicoides imicola, a Major Vector of Bluetongue and African Horse Sickness Viruses

Sylvain Guichard[1,2], He´le`ne Guis[3,4], Annelise Tran[1,5], Claire Garros[3,4], Thomas Balenghien[3,4], Darren J. Kriticos[2,6]

[1]Cirad, UR AGIRs, F-34398, Montpellier, France,
[2]InSTePP, University of Minnesota, St. Paul, MN, United States of America,
[3]Cirad, UMR CMAEE, F-34398, Montpellier, France,
[4]INRA, UMR1309 CMAEE, F-34398, Montpellier, France,
[5]Cirad, UMR15 TETIS, F-34398, Montpellier, France,
[6]CSIRO Agriculture Flagship and Biosecurity Flagship, Canberra, ACT 2601, Australia

ABSTRACT

We modelled the ecoclimatic niche of *Culicoides imicola*, a major arthropod vector of midge-borne viral pathogens affecting ruminants and equids, at fine scale and on a global extent, so as to provide insight into current and

Citation: Guichard S, Guis H, Tran A, Garros C, Balenghien T, Kriticos DJ (2014) Worldwide Niche and Future Potential Distribution of Culicoides imicola, a Major Vector of Bluetongue and African Horse Sickness Viruses. PLoS ONE 9(11): e112491. https://doi.org/10.1371/journal.pone.0112491

future risks of disease epizootics, and increase current knowledge of the species' ecology. Based on the known distribution and ecology of *C. imicola*, the species' response to monthly climatic conditions was characterised using CLIMEX with 10′ spatial resolution climatic datasets. The species' climatic niche was projected worldwide and under future climatic scenarios. The validated model highlights the role of irrigation in supporting the occurrence of *C. imicola* in arid regions. In Europe, the modelled potential distribution of *C. imicola* extended further West than its reported distribution, raising questions regarding ongoing process of colonization and non-climatic habitat factors. The CLIMEX model highlighted similar ecological niches for *C. imicola* and the Australasian *C. brevitarsis* raising questions on biogeography and biosecurity. Under the climate change scenarios considered, its' modelled potential distribution could expand northward in the Northern hemisphere, whereas in Africa its range may contract in the future. The biosecurity risks from bluetongue and African horse sickness viruses need to be re-evaluated in regions where the vector's niche is suitable. Under a warmer climate, the risk of vector-borne epizootic pathogens such as bluetongue and African horse sickness viruses are likely to increase as the climate suitability for *C. imicola* shifts poleward, especially in Western Europe.

INTRODUCTION

Bluetongue (BT) and African horse sickness (AHS) viruses are vector-borne, causing diseases affecting ruminants and equids respectively. Each of these diseases causes significant economic losses and animal deaths [1]–[3].

Knowledge of the potential geographical range of these diseases can inform national and regional pest risk assessments, assisting biosecurity agencies to manage the invasion risks better by focusing scarce surveillance and prevention resources to protect endangered regions and assets [4].

Several species of the genus *Culicoides* (Diptera: Ceratopogonidae) transmit BT, AHS and other major animal viruses. *Culicoides imicola*, an afro-oriental species, is widespread and has been reported from South Africa to southern Europe, and from western Africa to southern China. This species has historically retained the interest because biological transmission of AHS and BT viruses by *Culicoides* was first demonstrated in this species, and to date, *C. imicola* remains the unique vector for both AHS and BT viruses in the afro-oriental region [5]–[7] whereas it has been suspected for other Afrotropical *Culicoides*, such as *C. bolitinos* among others [8]–[10]. It is also considered a key vector of both viruses in Africa, in the Middle East and the southern part of Europe due to the huge abundances it can attain [6].

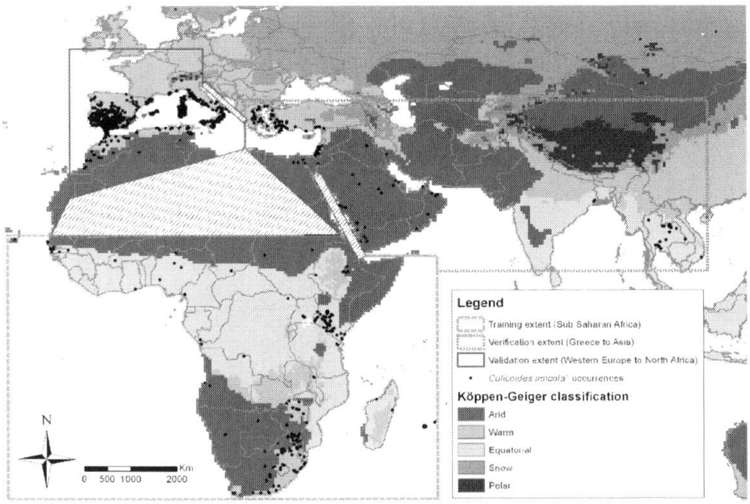

Figure 1. Distribution of *Culicoides imicola* **and study area:** *training*, *verification* **and** *validation* **zones for modelling (with respectively 187, 131, 561 occurrences at 10′ granularity) with Köppen-Geiger climate simplified classification [33].** Cross-hatched areas are zones of inhospitable habitat (Adriatic Sea, Sahara Desert and Red Sea) separating the 3 model-fitting zones.

https://doi.org/10.1371/journal.pone.0112491.g001

The distribution of *C. imicola*, especially in southern Europe, is an old and actual subject of interest as illustrated by AHS and BT history: AHS is endemic in areas of Sub-Saharan Africa, as highlighted by the recent outbreaks in Senegal in 2007 [1], and has sporadically caused devastating outbreaks outside Africa, *eg.* in 1959–61 in the Middle East and in Asia (where over 300 000 equids died), and in 1987–90 on the Iberian Peninsula [7]. Less than 10 years later, several serotypes of BT virus were transmitted intensively, leading to the death of hundreds of thousands sheep in the Mediterranean basin, in areas never previously affected by the virus [11]. Abundant populations of *C. imicola* were recorded in most of these areas, where it was thought to be absent [11]. Some authors have suggested that the distribution of *C. imicola* has expanded its range northwards in the last decades in Mediterranean Europe as climatic changes reduced the frequency and intensity of cold temperatures that limit overwintering potential [12], [13], whereas other studies challenged this hypothesis [14], [15]. Under future climate conditions, Acevedo and co-authors projected that the distribution of *C. imicola* is expected to remain constant in Spain while its abundance is expected to increase due to increasing precipitation and changes in seasonality [16].

Studies of the regional climatic niche of *C. imicola* had been developed from its distribution in Spain and Portugal [17], Morocco [18], Italy [19], the Mediterranean [20], [21] and South Africa [22] using correlative modelling methods. However, our study is the first to model the ecoclimatic niche of *C. imicola*, based on its global distribution.

It is generally well accepted that our climate is changing in response to anthropogenic emissions of CO_2 and other greenhouse gases. What is generally poorly appreciated is that our ability to forecast the future climatic conditions may be as limited by our ability to forecast factors such as patterns of socio-economic and technological development and human demography than by our ability to model the resulting atmospheric dynamics that drive global and regional climatic patterns [23]–[26]. These anthropogenic factors are difficult or impossible to predict at useful periods into the future, and they strongly influence the rate of emission of the greenhouse gases that are driving the observed and projected climatic changes. To address this problem the Intergovernmental Panel on Climate Change (IPCC) generated a series of social, economic and technological developmental storylines as a basis for the creation of future climate scenarios [27]. These storylines were intended to describe equally plausible future development paths that could be used to explore the sensitivity of climate-sensitive systems. Unfortunately, there has been a tendency to misunderstand or misuse the results of analyses based on these storylines, treating them as if they had predictive power. In this paper we apply a set of two scenarios to a biophysical model in order to highlight areas of potential future concerns for biosecurity risk managers. Note however, that in no way do we present these results as prediction of what will occur. Rather, they form a stress-test of the biosecurity risk patterns based on modelling using historical climatic data.

Ecoclimatic niche modelling has been used widely to explore the effects of climate change on the potential geographical range of many species of pest and conservation concern. CLIMEX [28] is a process-oriented niche model that can draw on both inferential and deductive modelling methods [29]. The model describes species' responses to climatic conditions and estimates species' potential distribution based upon climatic suitability. Due to its process-oriented formulation, CLIMEX is well suited for projecting species' potential distributions in novel climates such as inter-continental projections and future climatic scenarios [30]–[32].

As a primary step for evaluating the risks for these diseases at a particular location we investigated the ecoclimatic niche of *C. imicola*, a competent vector species for major animal viral diseases and projected its potential distribution under selected climate change scenarios.

METHODS

Geographical distribution of *Culicoides imicola*

We compiled a geo-database of 1 381 individual presence records of *C. imicola* from published sources (Appendix S1 in File S1). To address problems with duplication and auto-correlation in assessing the goodness of fit, the individual point locations were transformed into 879 unique cells with 10′ resolution. The 879 presence records of *C. imicola* were split into three subsets (Figure 1): 1) a *training* dataset to develop the model (Sub-Saharan Africa, 187 records), 2) a *verification* dataset, to adjust the model at unfitted locations (South-Eastern Europe, Middle East and Asia, 131 records) and 3) a *validation* dataset used to infer the strengths and limitations of the model (Western Europe and North Africa, 561 records). In selecting the distribution data sub-sets, we aimed to keep a maximum range of different climate categories in each zone, but also independence between zones with physical boundaries (Saharan desert, Red Sea and Adriatic Sea, cross-hatched areas in Figure 1). We used the Köppen-Geiger (KG) climate classification [33] and ensured that *training*, *verification* and *validation* zones shared similar climatic ranges, except for an equatorial climate, which was not represented in the validation zone. Africa was assigned to training because *C. imicola* was first described there (holotype from Kenya, Kieffer, 1913 cited in Borkent, [34]) and is considered as the historical distribution area [35]. The validation dataset encompassed western Europe where the range of *C. imicola*'s distribution is still under debate [15], [36].

Climate data

CLIMEX requires five climate variables: monthly values for minimum and maximum temperature, rainfall, and relative humidity at 9am and at 3pm. We used the CliMond 1975H_V1.1 fine spatial resolution (10′) global climate dataset [37]. CliMond includes monthly climate data averages centred on 1975, so-called historical climate dataset, and a total of eight future climate scenarios: *ie.* for 2030 and 2070, for two global climatic models, CSIRO-MK3.0 and Miroc-h, and two greenhouse gas emissions scenarios, A1B and A2 (4 future climate datasets for each year).

CLIMEX

CLIMEX is a dynamic bioclimatic niche model that integrates modelled weekly responses of a population to climate in order to create a series of climatic suitability indices that can be mapped or graphed [28], [29]. CLIMEX calculates an annual Growth Index (GI_A) to describe the potential for population growth as a function of weekly soil moisture and temperature during favourable conditions (Eq. 1).

$$GI_A = 100 \sum_{i=1}^{52} GI_{W_i}/52 \quad \textbf{(1)}$$

where GI_W is the weekly Growth Index, composed of the weekly Temperature Index multiplied by the weekly Moisture Index. The Temperature Index and the Moisture Index are specified with a functional form that accords with the Law of Tolerance [38]–[40] and are combined in a multiplicative manner that accords with the Law of the Minimum [41]. It uses up to eight stress indices (cold, wet, hot, dry, cold-wet, cold-dry, hot-wet and hot-dry; respectively *CS, WS, HS, DS, CWX, CDX, HWX, HDX*) to estimate the ability of the population to survive unfavourable conditions (Eqs. 2,3).

$$SI = \left(1 - \frac{CS}{100}\right)\left(1 - \frac{DS}{100}\right)\left(1 - \frac{HS}{100}\right)\left(1 - \frac{WS}{100}\right) \quad \textbf{(2)}$$

$$SX = \left(1 - \frac{CDX}{100}\right)\left(1 - \frac{CWX}{100}\right)\left(1 - \frac{HDX}{100}\right)\left(1 - \frac{HWX}{100}\right) \quad (3)$$

The Ecoclimatic Index (EI) summarises the balance between the opportunity for the species to grow during the favourable season(s) and the requirement to survive inclement season(s), and scales from 0 (unsuitable) to 100 (optimal) (Eq. 4).

$$EI = GI_A \times SI \times SX \quad \textbf{(4)}$$

CLIMEX models are best informed by fitting the stresses to the boundary of the known range of the species, using knowledge, when available, about its biology or ecology to inform thresholds and rate parameters. The temperature and soil moisture weekly growth indices (GI_w) were informed mostly by knowledge of the species biology and sometimes by using phenological observations and matched climates to infer growth limits and optima.

In our model, we tested the assumption that the presence of *C. imicola* in arid areas was due to irrigation maintaining suitable microclimate. The area under irrigation was identified using the data from Siebert *et al.* [42] describing presence of irrigation practices. In these areas we used CLIMEX irrigation scenario applying 3.5 mm per day as a "top-up", *ie.* during dry season, irrigation increases the rainfall equivalent to this limit [29].

Parameter-Fitting, Model Validation and Projection

Rainfall and temperature are known to drive activity, survival and seasonality of *Culicoides* [6], [43]. Initial values for the Growth and Stress parameters for the *C. imicola* CLIMEX model were informed by the following ecological traits. The species is present in warm and equatorial climatic conditions (Figure 1), but in regions where *C. imicola* is the main vector species for BT, it can persist over winter where daily maximum temperatures are greater than 12.5°C [44]–[46], and in Italy the probability of presence was close to zero where annual average of minimum daily temperatures fell below 9°C [47]. In laboratory studies Paweska *et al.* [9] noted that the minimum temperature for replication of bluetongue virus serotype 1 (BTV-1) in *C. imicola* was between 10°C and 15°C. In the Mediterranean region, the probability of *C. imicola* presence was greatest when minimum daily temperatures were greater than 16°C and increased with increasing relative humidity from 40% to its maximum at 60% [47]. In Morocco and Iberia [18], and in South Africa [22], minimum Normalized Difference Vegetation Index (NDVI), a variable correlated with soil moisture, is positively correlated with *C. imicola* abundance.

The natural rainfall model was fitted initially to *C. imicola*'s distribution only in the training zone until all occurrences receiving adequate rainfall to support a minimal amount of grass growth were modelled as being at least barely climatically suitable (minimising the climatically suitable area using biologically reasonable parameter values). The rationale for considering grass growth is that it is necessary to support the vertebrate hosts of *C. imicola*.

Cold Stress

We fitted a cold stress function to constrain survival to the coldest areas where *C. imicola* had been reported in traps.

Model parameters and intermediate steps in model development, successively using *training*, *verification* and *validation* datasets (Figure 1), are detailed in Appendix S2 in File S1.

The "immature model", fitted to the native range distribution data, was tested against the *verification* area (Figure 1), comparing the region where EI>0 with occurrence records in South Eastern Europe, the Middle East and Asia. Where necessary, we adjusted the model during the verification step which resulted in a second model named the "completed model". This step allowed us to make adjustments for any enemy release effects [48].

Each model was validated by calculating its sensitivity score: the proportion of records falling in locations modelled as being climatically suitable (EI>0). In addition, the degree of departure from random prediction was tested with the exact one-tailed binomial statistic [32] which measures the statistical significance of sensitivity scores given the modelled prevalence, *ie.* the proportion of the model universe that is climatically suitable. A useful model should have a high sensitivity score reflecting that recorded locations are modelled as being climatically suitable, and a significant binomial statistic score ensuring the model projection does not cover an excessive proportion of the study area. The completed model was then run using the global dataset under historical climate and future climate scenarios.

RESULTS

The geographical range of *C. imicola* appears primarily limited by cold stress and dry stress, and to a lesser extent wet stress. Its modelled potential distribution accords well with its known distribution in Africa, the Middle East and southern Europe (Figures 2, and SI2, SI3 and SI4 in File S1). The modelled climate suitability in France and continental Asia encompasses the reported distribution, but also extends well beyond the reported distribution. The model indicates large areas in the Americas and South-East Asia as being climatically suitable.

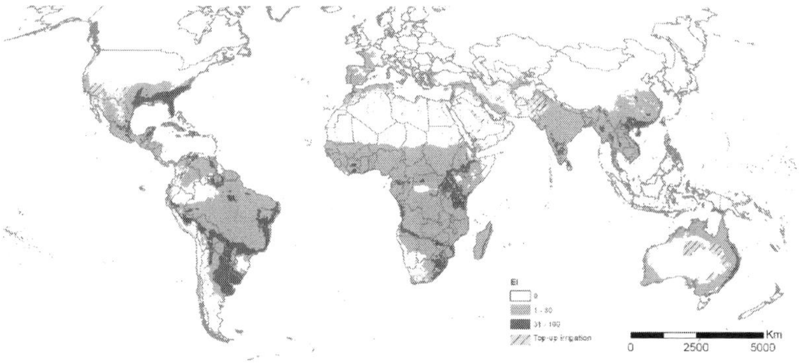

Figure 2. Global climate suitability for *Culicoides imicola* **under historical conditions as represented by the CLIMEX Ecoclimatic Index (EI)** for the completed model. Cross-hatching indicates suitable areas when irrigation is integrated in the model (Appendix S2 in File S1). Highly suitable areas in dark.

https://doi.org/10.1371/journal.pone.0112491.g002

The probable role of irrigation in extending the potential range of *C. imicola* into xeric regions is indicated by how well the composite risk map improved the model sensitivity score without affecting the model specificity significantly. For the three datasets, the completed model explained the presence of *C. imicola* in 833 grid cells at 10′ resolution and the addition of irrigation component explained an additional 38 records in arid areas. Only 8 occurrence records were not fitted in the composite irrigation model.

Under the natural rainfall future climate scenario, *C. imicola*'s projected potential distribution for climate scenario A2, generated by CSIRO-MK3.0 model, projected in 2070 (Figure 3) expands into areas that are presently too cold for persistence, and contracts from areas that are set to become more xeric. Similar changes in ecoclimatic suitability patterns were observed with other climate projections (*ie.* 2030 projections, A1B climatic scenario or Miroc-h climate model). In the scenarios considered, most of the potential range expansion due to a warming climate occurs in the Northern Hemisphere. In the Southern Hemisphere, where terrestrial habitat is limited to lower latitudes, most of the potential range expansion under the climate change scenario is into higher altitude areas. A similar response pattern was observed in equatorial regions (Figure 3).

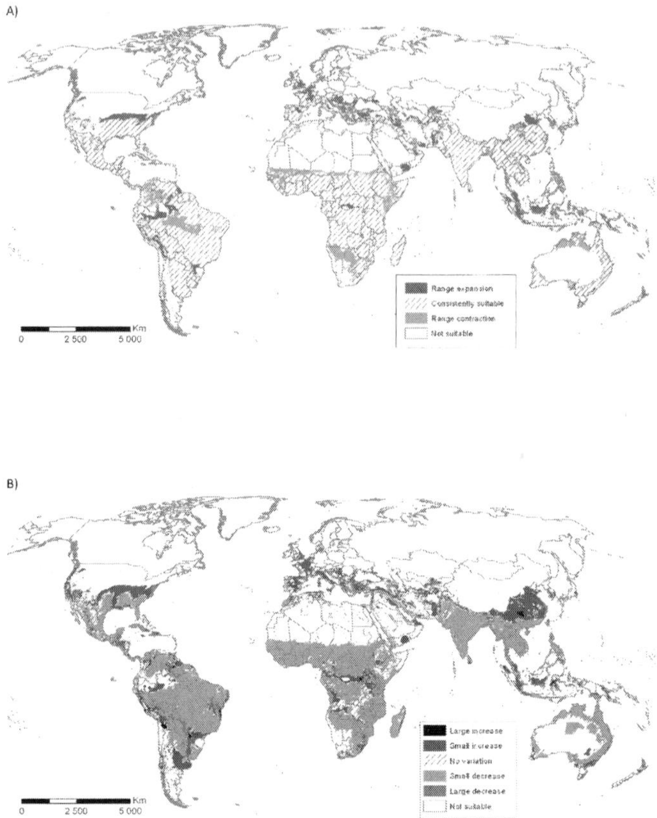

Figure 3. Forecast variation of ecoclimatic suitability between historical climate (current EI) and 2070 projected climate (future EI) from A2 climate scenarios using CSIRO-MK3.0 climate models. a) qualitative variation highlighted the expansion (current EI=0, future EI>0) and the contraction (current EI>0, future EI=0) of suitable areas for 2070 projection. b) quantitative variation illustrated climatic suitability increases or decreases. Small and large EI variations represented respectively differences in the range 3–30 and 31–100.

https://doi.org/10.1371/journal.pone.0112491.g003

DISCUSSION

There appears to be substantial opportunity for range expansion of *C. imicola* into the Americas and Australasia should it be translocated to these regions; the climate appears moderate to highly suitable over large areas, and domestic ruminant and equid hosts abound. Should *C. imicola* spread southwards into the Indonesian archipelago or Papua New Guinea, it is entirely feasible that it could spread to Australia [49], potentially increasing the risk of transmission of BT and AHS in Australia, and possibly the northern parts of New Zealand with a milder climate. In Australia, *C. imicola* would be adding to an extant community of competent vectors so the extent of the added risk is unknown.

The use of the irrigation information from Siebert *et al.* [42] allowed us to capture the effect of crop irrigation in supporting populations of *C. imicola* in xeric regions, without creating a risk map that indicated that vast areas of desert were climatically suited. The few location records in Figure 2 that fall in apparently climatically unsuitable locations are most likely the result of populations that are able to persist by virtue of small livestock watering points not captured by the irrigation dataset of Siebert *et al.* [42], perhaps signalling a scaling limitation regarding the use of this data. Agencies with biosecurity responsibilities in xeric regions may therefore wish to consider the irrigation climatic risk scenario independently of the irrigation locations indicated in Siebert *et al.* [42]. This risk map could indicate climate suitability for *C. imicola* in areas where soil moisture availability is not well modelled using natural rainfall and crop irrigation is not captured in the dataset of Siebert *et al.* [42].

The map of ecoclimatic suitability (Figure 2) approximates the fundamental niche of *C. imicola* (potential distribution negating limiting factors). As is commonly observed, the modelled potential distribution is wider than the observed distribution and the difference has several assumed origins:

- Biotic factors, not included in our model, and especially presence and abundance of hosts which are important to maintain a blood feeding species such as *C. imicola*. Considering abundances of wild and domestic hosts is particularly important when modelling *C. imicola*'s abundance [16].
- Incomplete sampling: apart from Europe, surveys have been implemented in a limited number of countries such as, for example, South Africa and Kenya, whereas other countries were under sampled (see point densities by countries). Further capture and identification of *Culicoides* species, especially in areas foreseen as highly suitable such as India and China, could

improve our understanding of the ecoclimatic requirements of the species.

- An ongoing (dynamic) process of colonization. Indeed, there have been no reports of *C. imicola* captured in the South West of France despite a dense and regularly surveyed trap network [36]; despite this area being modelled as suitable, and lies close to regions where some populations have been reported (South East of France, Venail *et al.* [36]; North West of Spain, Goldarazena *et al.* [50]). The colonization process is dependent on dispersal abilities, size of the source population, meteorological conditions and the presence of natural barriers. Venail *et al.* [36] discussed the confined distribution of *C. imicola* in South East France by physical altitudinal barriers.
- Geographically isolated landmasses such as America and Australia with territories where ecoclimatic conditions are suitable and therefore insect establishment may be possible (however note that other processes, not considered here, such as competition and parasitism for example, could prevent for species establishment). Identification of such territories is relevant for biosecurity purposes: even if intercontinental dispersal of the vector is unlikely, importation of hitchhiking insects along with trade has been reported previously for other insects (*eg*. Caton *et al.*, [51]) and is well known for Aedine mosquitoes. Besides, the introduction of exotic serotypes in Northern Europe in 2008 and the novel emergence of Schmallenberg virus in the same area highlight our misunderstanding of the possible entry routes.

We found that *C. imicola* shares temperature requirements for development and survival with one Australasian *Culicoides*: optimal temperatures for survival of larva of *C. brevitarsis* range between 20 and 35.5°C (unfortunately temperatures lower than 20°C were not tested) and temperatures greater or equal 38°C are lethal [52]. Murray and Nix [53] found that adult activity of *C. brevitarsis* started at 12°C. The modelled range for survival of *C. imicola* in our study is 12–36°C. No information is available on moisture requirements for *C. brevitarsis* but the known distribution of *C. brevitarsis* [54] is similar to the projected potential distribution of *C. imicola* in Australia (Figure 2). Furthermore, other evidence relates the two species: phylogenetic analyses of different *Culicoides* species show that *C. brevitarsis* sampled from Japan [55] are closely related to *C.*

imicola from South Africa [56], from France and Israel [57], [58]. Like *C. imicola*, *C. brevitarsis* is also competent for bluetongue virus [59]. Some authors have suggested a common *Culicoides* ancestral species for *C. brevitarsis* and *C. imicola* [35] but further study would be required to test this hypothesis. *Culicoides imicola* is the main vector of AHSV but we think that testing the competency of *C. brevitarsis* for AHSV would be informative for assessing biosecurity risks.

Under the future climate scenario, parts of Namibia, Botswana, Kenya, Somalia and Ethiopia could become unsuitable for *C. imicola* in the future, whereas the model projects a Northern expansion of suitable climate especially in Europe and China (Figure 3). However, presence of the vector is only one of the required conditions for vector-borne diseases to spread; appropriate hosts and presence of the pathogen are also required along with climatic conditions enabling vector activity, in particular vector-host contact, and pathogen replication. A complete risk assessment should also consider host densities and their relationship with vector densities [60], [61] and account for transmission routes from infected areas [49], [62].

Our study enables us to indirectly infer current and future risks of bluetongue and African horse sickness, diseases both transmitted by *C. imicola*, and thereby provides a framework for updating management and biosecurity strategies to target disease epizootics.

ACKNOWLEDGMENTS

We thank J.C. Dellécolle and T. Baldet for providing unpublished data and valuable comments. The funders had no role in study design, data collection and analysis, decision to publish, or preparation of the manuscript.

AUTHOR CONTRIBUTIONS

Conceived and designed the experiments: SG HG AT DJK. Performed the experiments: SG. Analyzed the data: SG HG DJK AT. Wrote the paper: SG HG DJK CG TB AT.

REFERENCES

1. Diouf ND, Etter E, Lo MM, Lo M, Akakpo AJ (2013) Outbreaks of African horse sickness in Senegal, and methods of control of the 2007 epidemic. Veterinary Record 172: 152.
2. MacLachlan NJ, Guthrie AJ (2010) Re-emergence of bluetongue, African horse sickness, and other Orbivirus diseases. Veterinary Research 41: 35.
3. Velthuis AGJ, Saatkamp HW, Mourits MCM, Koeijer AA de, Elbers ARW (2010) Financial consequences of the Dutch bluetongue serotype 8 epidemics of 2006 and 2007. Preventive Veterinary Medicine 93: 294–304.
4. Kriticos DJ, Venette R, Koch F, Rafoss T, Werf W Van der, et al. (2013) Invasive alien species in the food chain: Advancing risk assessment models to address climate change, economics and uncertainty. In: Kriticos DJ, Venette RC, editors. Neobiota. Pensoft Publishers, Vol. 18: 1–7.
5. Du Toit RM (1944) The transmission of bluetongue and horse sickness by *Culicoides* . Onderstepoort journal of veterinary science and Animal Industry 19: 7–16.
6. Mellor PS, Boorman J, Baylis M (2000) *Culicoides* biting midges: their role as arbovirus vectors. Annual review of entomology 45: 307–340.
7. Mellor PS, Hamblin C (2004) African horse sickness. Veterinary Research 35: 445–66.
8. Diarra M, Fall M, Fall A, Diop A, Seck M, et al. (2014) Seasonal dynamics of *Culicoides* (Diptera: Ceratopogonidae) biting midges, potential vectors of African horse sickness and bluetongue viruses in the Niayes area of Senegal. Parasites & Vectors 7: 147.
9. Paweska JT, Venter GJ, Mellor PS (2002) Vector competence of South African *Culicoides* species for bluetongue virus serotype 1 (BTV-1) with special reference to the effect of temperature on the rate of virus replication in *C. imicola* and *C. bolitinos* . Medical and Veterinary Entomology 16: 10–21.
10. Venter GJ, Wright IM, Van Der Linde TC, Paweska JT (2009) The oral susceptibility of South African field populations of *Culicoides* to African horse sickness virus. Medical and Veterinary Entomology 23: 367–378.

11. Mellor PS, Baylis M, Mertens PPC (2009) Bluetongue. Academic Press, London, UK.
12. Guis H, Caminade C, Calvete C, Morse AP, Tran A, et al. (2012) Modelling the effects of past and future climate on the risk of bluetongue emergence in Europe. Journal of The Royal Society Interface 9: 339–350.
13. Purse BV, Mellor PS, Rogers DJ, Samuel AR, Mertens PPC, et al. (2005) Climate change and the recent emergence of bluetongue in Europe. Nature Reviews Microbiology 3: 171–181.
14. Conte A, Gilbert M, Goffredo M (2009) Eight years of entomological surveillance in Italy show no evidence of *Culicoides imicola* geographical range expansion. Journal of Applied Ecology 46: 1332–1339.
15. Mardulyn P, Goffredo M, Conte A, Hendrickx G, Meiswinkel R, et al. (2013) Climate change and the spread of vector-borne diseases: using approximate Bayesian computation to compare invasion scenarios for the bluetongue virus vector *Culicoides imicola* in Italy. Molecular ecology 22: 2456–2466.
16. Acevedo P, Ruiz-Fons F, Estrada R, Márquez AL, Miranda MA, et al. (2010) A broad assessment of factors determining *Culicoides imicola* abundance: Modelling the present and forecasting its future in climate change scenarios. PLoS ONE 5: e14236.
17. Wittmann EJ, Mellor PS, Baylis M (2001) Using climate data to map the potential distribution of *Culicoides imicola* (Diptera: Ceratopogonidae) in Europe. Revue scientifique et technique (International Office of Epizootics) 20: 731–740.
18. Baylis M, Rawlings P (1998) Modelling the distribution and abundance of *Culicoides imicola* in Morocco and Iberia using climatic data and satellite imagery. Archives of Virology suppl. 14: 137–153.
19. Calistri P, Goffredo M, Caporale V, Meiswinkel R (2003) The distribution of *Culicoides imicola* in Italy: Application and evaluation of current Mediterranean models based on climate. Journal of Veterinary Medicine Series B 50: 132–138.
20. Purse BV, McCormick BJJ, Mellor PS, Baylis M, Boorman JPT, et al. (2007) Incriminating bluetongue virus vectors with climate envelope models. Journal of Applied Ecology 44: 1231–1242.
21. Tatem AJ, Baylis M, Mellor PS, Purse BV, Capela R, et al. (2003) Prediction of bluetongue vector distribution in Europe and north

Africa using satellite imagery. Veterinary Microbiology 97: 13–29.

22. Baylis M, Meiswinkel R, Venter GJ (1999) A preliminary attempt to use climate data and satellite imagery to model the abundance and distribution of *Culicoides imicola* (Diptera: Ceratopogonidae) in southern Africa. Journal of the South African Veterinary Association 70: 80–89.
23. Grübler A, Nakicenovic N (2001) Identifying dangers in an uncertain climate. Nature 412: 15.
24. IPCC (2013) Summary for Policymakers. Climate Change 2013: The Physical Science Basis. Contribution of Working Group I to the Fifth Assessment Report of the Intergovernmental Panel on Climate Change. Stocker F, Qin D, Plattner G-K, Tignor M, Allen SK, *et al.*, editors Cambridge University Press, Cambridge, United Kingdom and New York, NY, USA.
25. Moss RH, Edmonds JA, Hibbard KA, Manning MR, Rose SK, et al. (2010) The next generation of scenarios for climate change research and assessment. Nature 463: 747–756.
26. Schneider SH (2001) What is 'dangerous' climate change? Nature 411: 17–19.
27. IPCC (2000) Special report on emissions scenarios: a special report of working group III of the intergovernmental panel on climate change. Cambridge University Press, Cambridge, United Kingdom.
28. Sutherst RW, Maywald GF (1985) A computerised system for matching climates in ecology. Agriculture, Ecosystems & Environment 13: 281–299.
29. Sutherst RW, Maywald GF, Kriticos DJ (2007) CLIMEX version 3: user's guide. Hearne Scientific Software Pty Ltd, http://www.Hearne.com.au.
30. Kriticos DJ, Randall RP (2001) A comparison of systems to analyze potential weed distributions. In: Groves RH, Panetta FD, Virtue JG, editors. Weed risk assessment. pp. 61–79.
31. Sutherst RW, Bourne AS (2009) Modelling non-equilibrium distributions of invasive species: a tale of two modelling paradigms. Biological Invasions 11: 1231–1237.
32. Webber BL, Yates CJ, Le Maitre DC, Scott JK, Kriticos DJ, et al. (2011) Modelling horses for novel climate courses: insights from projecting potential distributions of native and alien Australian

acacias with correlative and mechanistic models. Diversity and Distributions 17: 978–1000.

33. Kottek M, Grieser J, Beck C, Rudolf B, Rubel F (2006) World Map of the Köppen-Geiger climate classification updated. Meteorologische Zeitschrift 15: 259–263.
34. Borkent A (2013) World Species of Biting Midges (Diptera: Ceratopogonidae). Available: http://www.inhs.illinois.edu/research/FLYTREE/Borkent.html.
35. Dyce A (2000) Biogeographic origins of species of the genus *Culicoides* (Diptera: Ceratopogonide) of the Australasian region. Arbovirus Research in Australia 8: 133–140.
36. Venail R, Balenghien T, Guis H, Tran A, Setier-Rio M-L, et al. (2012) Assessing diversity and abundance of vector populations at a national scale: example of *Culicoides* surveillance in France after bluetongue virus emergence. In: Mehlhorn H, editor. Arthropods as vectors of emerging diseases. Parasitology Research Monographs. Springer Berlin Heidelberg, Vol. 3: 77–102.
37. Kriticos DJ, Webber BL, Leriche A, Ota N, Macadam I, et al. (2012) CliMond: global high-resolution historical and future scenario climate surfaces for bioclimatic modelling. Methods in Ecology and Evolution 3: 53–64.
38. Shelford VE (1963) The ecology of North America. University of Illinois Press, Urbana, IL, USA.
39. Shelford VE (1918) A comparison of the responses of animals in gradients of environmental factors with particular reference to the method of reaction of representatives of the various groups from protozoa to mammals. Science 48: 225–230.
40. Shelford VE, Deere EO (1913) The reactions of certain animals to gradients of evaporating power of air. A study in experimental ecology. Biological Bulletin 25: 79–120.
41. Ploeg RR van der, Böhm W, Kirkham MB (1999) On the origin of the theory of mineral nutrition of plants and the Law of the Minimum. Soil Science Society of America Journal 63: 1055–1062.
42. Siebert S, Döll P, Hoogeveen J, Faures J-M, Frenken K, et al. (2005) Development and validation of the global map of irrigation areas. Hydrology and Earth System Sciences 9: 535–547.

43. Walker AR (1977) Seasonal fluctuations of *Culicoides* species (Diptera: Ceratopogonidae) in Kenya. Bulletin of Entomological Research 67: 217–233.
44. Mellor PS (1996) *Culicoides*: vectors, climate change and disease risk. Veterinary Bulletin 66: 301–306.
45. Sellers RF, Mellor PS (1993) Temperature and the persistance of virus in *Culicoides spp.* during adverse conditions. Revue scientifique et technique (International Office of Epizootics) 12: 733–755.
46. Venter GJ, Meiswinkel R (1994) The virtual absence of *Culicoides imicola* (Diptera, Ceratopogonidae) in a light trap survey of the Eastern Orange Free State, South Africa, and implications for the transmission of arboviruses. Onderstepoort Journal of Veterinary Research 61: 327–340.
47. Conte A, Giovannini A, Savini L, Goffredo M, Calistri P, et al. (2003) The effect of climate on the presence of *Culicoides imicola* in Italy. Journal of Veterinary Medicine, Series B 50: 139–147.
48. Keane RM, Crawley MJ (2002) Exotic plant invasions and the enemy release hypothesis. Trends in Ecology & Evolution 17: 164–170.
49. Eagles D, Walker PJ, Zalucki MP, Durr PA (2013) Modelling spatio-temporal patterns of long-distance *Culicoides* dispersal into northern Australia. Preventive veterinary medicine 110: 312–322.
50. Goldarazena A, Romón P, Aduriz G, Balenghien T, Baldet T, et al. (2008) First record of *Culicoides imicola*, the main vector of bluetongue virus in Europe, in the Basque Country (northern Spain). The Veterinary Record 162: 820–821.
51. Caton B, Dobbs T, Brodel C (2006) Arrivals of hitchhiking insect pests on international cargo aircraft at Miami International Airport. Biological Invasions 8: 765–785.
52. Allingham P (1991) Effect of temperature on late immature stages of *Culicoides brevitarsis* (Diptera: Ceratopogonidae). Journal of Medical Entomology 28: 878–881.
53. Murray M, Nix H (1987) Southern limits of distribution and abundance of the biting-midge *Culicoides brevitarsis* Kieffer (Diptera, Ceratopogonidae) in Southeastern Australia – an application of the growest model. Australian Journal of Zoology 35: 575–585.
54. St George TD, Baldock C, Bellis G, Bishop A, Cameron A, et al. (2001) The history of bluetongue, Akabane and ephemeral fever viruses and

their vectors in Australia 1975–1999. Animal Health Australia.

55. Matsumoto Y, Yanase T, Tsuda T, Noda H (2009) Characterization of Internal Transcribed Spacer (ITS1)-ITS2 region of ribosomal RNA gene from 25 species of *Culicoides* biting midges (Diptera: Ceratopogonidae) in Japan. Journal of Medical Entomology 46: 1099–1108.
56. Linton Y-M, Mordue AJ, Cruickshank RH, Meiswinkel R, Mellor PS, et al. (2002) Phylogenetic analysis of the mitochondrial cytochrome oxidase subunit I gene of five species of the *Culicoides imicola* species complex. Medical and Veterinary Entomology 16: 139–146.
57. Morag N, Saroya Y, Braverman Y, Klement E, Gottlieb Y (2012) Molecular identification, phylogenetic status, and geographic distribution of *Culicoides oxystoma* (Diptera: Ceratopogonidae) in Israel. PLoS ONE 7: e33610.
58. Perrin A, Cêtre-Sossah C, Mathieu B, Baldet T, Delécolle J-C, et al. (2006) Phylogenetic analysis of *Culicoides* species from France based on nuclear ITS1-rDNA sequences. Medical and Veterinary Entomology 20: 219–228.
59. Kirkland PD (2004) Bluetongue viruses, vectors and surveillance in Australia – the current situation and unique features. Veterinaria Italiana 40: 47–50.
60. Garcia-Saenz A, McCarter P, Baylis M (2011) The influence of host number on the attraction of biting midges, *Culicoides spp.*, to light traps. Medical and Veterinary Entomology 25: 113–115.
61. Lo Iacono G, Robin CA, Newton JR, Gubbins S, Wood JLN (2013) Where are the horses? With the sheep or cows? Uncertain host location, vector-feeding preferences and the risk of African horse sickness transmission in Great Britain. Journal of The Royal Society Interface 10: 20130194.
62. Sedda L, Brown HE, Purse BV, Burgin L, Gloster J, et al. (2012) A new algorithm quantifies the roles of wind and midge flight activity in the bluetongue epizootic in northwest Europe. Proceedings of the Royal Society B: Biological Sciences 279: 2354–2362.

13

Correction: Worldwide Niche and Future Potential Distribution of Culicoides imicola, a Major Vector of Bluetongue and African Horse Sickness Viruses

Sylvain Guichard[1,2], He´le`ne Guis[3,4], Annelise Tran[1,5], Claire Garros[3,4], Thomas Balenghien[3,4], Darren J. Kriticos[2,6]

[1]Cirad, UR AGIRs, F-34398, Montpellier, France,
[2]InSTePP, University of Minnesota, St. Paul, MN, United States of America,
[3]Cirad, UMR CMAEE, F-34398, Montpellier, France,
[4]INRA, UMR1309 CMAEE, F-34398, Montpellier, France,
[5]Cirad, UMR15 TETIS, F-34398, Montpellier, France,
[6]CSIRO Agriculture Flagship and Biosecurity Flagship, Canberra, ACT 2601, Australia

There are errors in the Funding section. Please see the complete, correct Funding statement here: This work was jointly funded by the Project "Climate Change and Impact Research: the Mediterranean Environment (CIRCE)," European Commission's 6th Framework Programme and also

Citation: The PLOS ONE Staff (2015) Correction: Worldwide Niche and Future Potential Distribution of Culicoides imicola, a Major Vector of Bluetongue and African Horse Sickness Viruses. PLoS ONE 10(3): e0119323. https://doi.org/10.1371/journal.pone.0119323

by HarvestChoice project with support of the Commonwealth Scientific and Industrial Research Organization (CSIRO), the University of Minnesota and the Bill and Melinda Gates Foundation. This study was partially funded by EU grant FP7–261504 EDENext and is catalogued by the EDENext Steering Committee as EDENext199 (http://www.edenext.eu). The contents of this publication are the sole responsibility of the authors and don't necessarily reflect the views of the European Commission. H. Guis, T. Balenghien and C. Garros were funded by the Ministère de l'agriculture, de l'agroalimentaire et de la forêt. The funders had no role in study design, data collection and analysis, decision to publish, or preparation of the manuscript.

There is an error in the first sentence of the Acknowledgements section. The correct sentence is: We thank J.C. Delécolle and T. Baldet for providing unpublished data and valuable comments [1].

REFERENCE

1. Guichard S, Guis H, Tran A, Garros C, Balenghien T, Kriticos DJ. (2014) Worldwide Niche and Future Potential Distribution of *Culicoides imicola*, a Major Vector of Bluetongue and African Horse Sickness Viruses. PLoS ONE 9(11): e112491. pmid:25391148

SECTION 3: MOSQUITOES

14

Environmental Drivers of Culicoides Phenology: How Important Is Species-Specific Variation When Determining Disease Policy?

Kate R. Searle[1], James Barber[2], Francesca Stubbins[2], Karien Labuschagne[3], Simon Carpenter[2], Adam Butler[4], Eric Denison[2], Christopher Sanders[2], Philip S. Mellor[2], Anthony Wilson[2], Noel Nelson[5], Simon Gubbins[2], Bethan V. Purse[1]

[1] Centre for Ecology and Hydrology, Bush Estate, Penicuik, Midlothian, United Kingdom,

[2] The Pirbright Institute, Ash Road, Pirbright, Woking, United Kingdom,

[3] PVVD, ARC-Onderstepoort Veterinary Institute, Private Bag X05, Onderstepoort, South Africa,

[4] Biomathematics and Statistics Scotland, James Clerk Maxwell Building, The King's Buildings, Edinburgh, United Kingdom,

[5] The Met Office, Based at the Pirbright Institute, Pirbright, Woking, Surrey, United Kingdom

Citation: Searle KR, Barber J, Stubbins F, Labuschagne K, Carpenter S, Butler A, et al. (2014) Environmental Drivers of Culicoides Phenology: How Important Is Species-Specific Variation When Determining Disease Policy? PLoS ONE 9(11): e111876. https://doi.org/10.1371/journal.pone.0111876

ABSTRACT

Since 2006, arboviruses transmitted by *Culicoides* biting midges (Diptera: Ceratopogonidae) have caused significant disruption to ruminant production in northern Europe. The most serious incursions involved strains of bluetongue virus (BTV), which cause bluetongue (BT) disease. To control spread of BTV, movement of susceptible livestock is restricted with economic and animal welfare impacts. The timing of BTV transmission in temperate regions is partly determined by the seasonal presence of adult *Culicoides* females. Legislative measures therefore allow for the relaxation of ruminant movement restrictions during winter, when nightly light-suction trap catches of *Culicoides* fall below a threshold (the 'seasonally vector free period': SVFP). We analysed five years of time-series surveillance data from light-suction trapping in the UK to investigate whether significant inter-specific and yearly variation in adult phenology exists, and whether the SVFP is predictable from environmental factors. Because female vector *Culicoides* are not easily morphologically separated, inter-specific comparisons in phenology were drawn from male populations. We demonstrate significant inter-specific differences in *Culicoides* adult phenology with the season of *Culicoides scoticus* approximately eight weeks shorter than *Culicoides obsoletus*. Species-specific differences in the length of the SVFP were related to host density and local variation in landscape habitat. When the *Avaritia Culicoides* females were modelled as a group (as utilised in the SFVP), we were unable to detect links between environmental drivers and phenological metrics. We conclude that the current treatment of *Avaritia Culicoides* as a single group inhibits understanding of environmentally-driven spatial variation in species phenology and hinders the development of models for predicting the SVFP from environmental factors. *Culicoides* surveillance methods should be adapted to focus on concentrated assessments of species-specific abundance during the start and end of seasonal activity in temperate regions to facilitate refinement of ruminant movement restrictions thereby reducing the impact of *Culicoides*-borne arboviruses.

INTRODUCTION

Northern Europe is currently experiencing an unprecedented series of incursions of arboviruses transmitted between ruminants by *Culicoides* (Diptera: Ceratopogonidae) [1]. Five separate strains of bluetongue virus (BTV) have been recorded in the region since 2006 [2]–[4], and a novel *Culicoides*-borne arbovirus, provisionally named Schmallenberg virus

(SBV), was discovered by metagenomic studies in Germany in 2011 [5], following unexplained clinical signs in dairy cattle. SBV has since spread rapidly across a large geographic area, and has been found to inflict foetal abnormalities in both cattle and sheep [6]. The route of entry of several of these arbovirus strains remains undefined [7]; hence the risk of further emergence of *Culicoides*-borne pathogens in this region cannot easily be estimated or mitigated.

In temperate ecosystems, the seasonal incidence and abundance of adult female *Culicoides* is a key parameter in determining the timing and intensity of arbovirus outbreaks and varies widely across geographical space [8]. It is thought that livestock-associated *Culicoides* in northern Europe overwinter in their final (fourth) larval instar and do not generally survive the winter as adults [9]. While live *Culicoides* adults have been recovered from animal housing in winter in northern Europe [10], [11], their numbers seem to be insufficient to drive BTV outbreaks since new confirmed clinical cases of BT are only very rarely recorded in winter (December to March).

The response to incursion of livestock arboviruses in Europe is dependent on strain pathogenicity and whether legislation exists that defines control measures. For BTV, emergence is notifiable to the World Organization for Animal Health (OIE). Movement and trade restrictions of susceptible stock in the surrounding area are imposed immediately to limit spread of BTV (defined under Council Directive 2000/75/EC of EU legislation). Although these measures reduce the speed and extent of spread of viruses, they also impose huge logistic and welfare costs on affected regions [12] that could be minimised with enhanced understanding of geographical and annual variation in adult vector seasons.

To date, the most damaging outbreak of bluetongue (BT) in northern Europe was inflicted by a serotype 8 strain of sub-Saharan origin [1], [13]. In response to this incursion, movement restrictions were imposed across the region but vaccination to combat spread was not available until spring 2008, some eighteen months after the initial incursion was identified. In 2007, to alleviate the impact of movement restrictions, a 'seasonal vector free period' (SVFP) scheme was defined by the European Union council enabling movement of susceptible livestock during winter under what was hypothesised to be an extremely low risk of BTV transmission (defined in Annexe V of Commission Regulation (EC) No. 1266/2007). The declaration enabled movement of these livestock from farms in the affected zones to winter sites or markets, substantially easing the economic difficulties of farmers in the affected region and enabling some 85,000 animal movements in the UK alone during the winter of 2007 [14].

Maintenance of the SVFP is reliant upon the operation of a network of light-suction traps designed to monitor adult *Culicoides* activity [15]. In northern Europe, a threshold of catching less than five parous (abdominally pigmented) female *Culicoides* per trap was set as a limit for declaring freedom of adult activity. In some countries (e.g., the UK), this restriction was additionally underpinned using data concerning the thermal limits of BTV replication [16]. Due in part to fact that the primary vectors of BTV-8 in northern Europe were not convincingly identified to species level [1], no attempt was made to account for potential variation in phenology in *Culicoides* species that could be involved in transmission.

Unlike BTV-8, SBV has not been declared a notifiable disease by a majority of affected countries partly because substantial geographical spread had already occurred before the pathogen was first detected. For mitigating SBV impacts, it is even more imperative to understand *Culicoides* phenology, because the extent of clinical birth malformations in lambs and calves is governed by whether the ewe or dam receives an infectious *Culicoides* bite during a particular period of pregnancy [17], [18], which is close to the last portion of the adult vector season. In the event that SBV persists in this region, or in the event of emergence of more pathogenic strains, this knowledge could be employed to make alterations in husbandry practises that would reduce the impact of the disease including changes in tupping schedules [19].

In this paper light-suction trap data collected over a five year period within the UK are analysed to quantify inter-specific differences in the phenology of *Culicoides*. Phenological metrics are related to remote environmental variables with the aim of understanding how variation in the SVFP may be produced under different climate, host and landscape conditions. The livestock-associated *Culicoides* fauna in northern Europe is dominated by species belonging to the subgenus *Avaritia* and four species of this group have been identified in the UK (*Culicoides obsoletus*, *Culicoides scoticus*, *Culicoides dewulfi* and *Culicoides chiopterus*). Due to poor levels of discrimination during vector competence studies it remains unclear which of these species were involved in transmission of BTV-8 and to what degree [1]. We examine the potential policy impact of treating these individual species as a single group and then identify key environmental drivers of the timing of each phenological metric for both the group and its constituent species. More specifically, we hypothesise that the four species differ significantly in their phenology and that these responses can be explained by varying responses to environmental variables characterising different habitat and climatic requirements.

In a broader sense, we also address the statistical complications that arise routinely when analysing phenological data. Phenological variables, which refer to the date on which particular events occur, are inherently circular (with Day=366, or Day=367 in leap years, being equivalent to Day=1). Standard statistical methods (such as linear regression, linear mixed models and GLMMs) fail to account for this circularity, and can therefore lead to biased results and potentially to the detection of spurious relationships. We present a hierarchical Bayesian method for dealing with the circular nature of phenological event data by using a natural extension of linear mixed modelling [20]. This method can be implemented using freely available software packages.

MATERIALS AND METHODS

Trapping Methods and Locations

Trapping was conducted from 2006–2010 using standard 8w ultraviolet Onderstepoort Veterinary Institute (OVI) light-suction traps, as previously recommended for *Culicoides* surveillance purposes [21]. OVI traps were hung at approximately 1.5 m as close to livestock as logistically possible to allow permanent positioning and in all cases were ≤25 m distance from ruminant hosts throughout the trapping period. A total of 29 trap locations on private land were used for a variable period of time ranging from one to four years across the UK (Figure 1; for further information regarding the precise location of trapping sites please contact KRS). Access to private land was granted by the landowners. Sampling was carried out by volunteers on one night from dusk until dawn each week with no attempt to synchronise trapping day across the network. Collections were made into approximately 250 ml of water with a drop of detergent to reduce surface tension and then later transferred to 70% ethanol and transported to the Pirbright Institute for identification. Fieldwork did not involve endangered or protected species.

Figure 1. Locations of UK trapping sites for *Culicoides* **surveillance (2006–2010).** Circle area is proportional to maximum catch ever recorded per site, where complete yearly data were available. Sites with crosses lacked complete yearly trapping profile.

https://doi.org/10.1371/journal.pone.0111876.g001

Following receipt, all non-*Culicoides* were removed from the samples based upon morphological features of wing pattern [22] and the absence of humeral pits diagnostic for *Culicoides*. The remaining portion of each sample was categorised into *Culicoides* species groups using wing patterns. Male members of the subgenus *Avaritia* were then identified to species level using genitalia morphology (hereafter referred to as '*Avaritia* males'), while females were treated as a single group (hereafter referred to as '*Avaritia* females'). Females of the subgenus *Avaritia* were characterised according to the appearance of the abdomen as pigmented, non-pigmented or gravid

[23]. Large catches (≥1000 individuals) were subsampled using volumetric methods.

Timing metrics

From the raw time-series data of weekly trap catches at each site in each year, the date of first appearance (start), last appearance (end), and the length of the inactive overwinter period (length overwinter) were derived for *Avaritia* females and for the individual species in the *Avaritia* male dataset. The start of the season for *Avaritia* females was defined as the first day of the year in which more than five pigmented females were caught (in accordance with the definition used by European Union Council to define the start and end of the SVFP in Europe). Site-by-year combinations were only included in the analysis if at least two trapping nights prior to this date had been recorded. The end of the season for *Avaritia* females was similarly determined as the last day of the year in which more than five pigmented females were caught (with the site-by-year combination being excluded from analysis if there were less than two subsequent trapping nights following this date). For the *Avaritia* males, the start and end of the season were defined as the first (start) and final (end) day of the season when more than five males were caught. The length of the overwinter period was determined as the difference in days between the start of the season in one year, and the end of the season the previous year. Relationships between these three timing metrics and a suite of environmental variables that have previously been found to be important in influencing the seasonal dynamics and abundance of these species were then investigated [24]–[27].

Environmental Covariates

We derived a series of meteorological variables for each site and year combination from the nearest UK Met Office weather station using raw data on total daily precipitation (mm), mean daily temperature (°C), and daily humidity (%) as measured at 15:00 hrs (Table 1). The mean distance to the nearest weather station was 3.9 km (range: 0.1 km−12.4 km). Photoperiod was calculated daily for each site using latitude and was expressed as the first day of the year at which nine hours of daylight were achieved in each year of observation. Percentage cover of two broad land-cover habitat classes – moorland and heathlands (*moors*) and broadleaf mixed woodland (*brdlf*) - in the surrounding 1 km around each trapping site were determined from the CEH Landcover Map 2007 [28]. The mean densities of cattle and sheep around each trapping site were determined for each year using the Edina AgriCensus data at a 2 km resolution (http://edina.ac.uk/agcensus/).

Table 1. Derived meteorological variables from UK Met Office weather station data for each of the trapping site by year combinations used in the timing analyses.

Seasonal metric	Meteorological variable	Definition	Notation
Start of season & lengthof overwinter	Mean wintertemperature **(°C)**	Mean daily temperatureover November 1st **to February 28th**	T_w
Start of season	Accumulated degreedays over winter	Accumulated degree daysgreater than 10°Cbetween Novemberand May to capturetemperature variationover the precedingwinter and current spring $\sum_{\text{Nov 1st}}^{\text{May 31st}}$ [mean temperature-10°C, 0]	DD_w
Start of season	Mean springtemperature **(°C)**	Mean daily temperatureover March 21st **to April 30th**	T_{spr}
Start of season	Total springprecipitation (mm)	Summed daily precipitationover March 1st **to May 31st**	P_{spr}
Start of season	Mean spring relativehumidity (%)	Mean daily (15:00 hrs)relative humidity over March 21stto April 30th.	RH_{spr}
End of season	Accumulated degreedays over summer	Accumulated degreedays greater than 10°Cbetween June and Septemberto capture temperaturevariation over the currentsummer and autumn $\sum_{\text{June 1st}}^{\text{Sept 30th}}$ [mean temperature-10 C, 0]	
End of season	Mean summertemperature **(°C)**	Mean daily temperature over1st **to September 30th**	T_{sum}
End of season	Total summerprecipitation (mm)	Summed daily precipitationover June 1st **to September 30th**	P_{sum}
End of season	Mean summerrelative humidity (%)	Mean daily (15:00 hrs)relative humidity overJune 1st **to September 30th**	RH_{sum}

https://doi.org/10.1371/journal.pone.0111876.t001

Statistical Models

We used statistical models to analyse relationships between seasonal timing metrics and environmental drivers. Three different response variables were considered: length of the overwinter period (in days), start of the season (Julian day) and end of the season (Julian day). Analyses were performed for the entire group of *C. obsoletus* complex parous females (*Avaritia* females) and, separately, for each of the individual species within this complex (*Avaritia* males). We constructed sets of environmental variables that were relevant for each seasonal timing metric. The five non-meteorological variables (photoperiod, percent cover of moorland (*moors*), percent cover of broadleaf woodland (*brdlf*), mean density of cattle, mean density of sheep) were assumed to be relevant to all three metrics, but distinct meteorological variables were identified for each seasonal timing metric (Table 1).

Models for length of the overwinter period (days) assumed that the response variable had a normal distribution, whilst models for the start and end of the season were based upon a Wrapped Normal distribution (WN). 'Start of season' and 'end of season' are circular variables. It would be inappropriate to assume that these variables have a normal distribution because this would imply that the difference between January 1st and December 31st (a difference in Julian days of 364) is twice as large as the difference between January 1st and July 1st (a difference in Julian days of 182). Distributions for circular data deal with this problem by treating the Julian date as a variable that lies on a circle rather than a line, so that the distance between December 31st and January 1st is equal to one rather than 364 [29]. Circular data also arise in other contexts – e.g. when modelling angles and directions. A range of distributions exist for modelling data of this form, but the Wrapped Normal (WN) is commonly used [29]. The WN distribution is symmetric and unimodel, and is obtained by wrapping a normal distribution on the real line around a circle [20]. The WN distribution contains two unknown parameters, which are directly analogous to the mean and variance of the normal distribution. The WN distribution is defined as a distribution for angles, which lie between zero and 2π, and variables that are defined in Julian days therefore need to be multiplied by $2\pi/365$ before the distribution is applied.

We assumed that the mean of the normal distribution (for length of overwinter period) or WN distribution (for start of season and end of season) had a linear relationship with environmental drivers. For *Avaritia* males the linear predictor also included a categorical variable to account for differences between species (as a fixed effect), and allowed for interactions between 'species' and the environmental drivers (species-environment interactions). Random effects were also included in models for both males and females in order to account for the structure of the data (and thereby avoid pseudo-replication). When performing analyses for the *Avaritia* females normally distributed random effects were included to account for variation between years (unstructured temporal heterogeneity) and sites (unstructured spatial heterogeneity). The model for *Avaritia* males included a single random effect to capture both site and year effects because the data were highly unbalanced with respect to site and year and so contained little information from which to separate out the effects of site, year and site-by-year interaction (leading to convergence problems in models for the phenological metrics with least data). To check that the use of a single random effect for males was reasonable and did not lead to pseudo-replication, we re-ran the final best-fitting model for the most data-rich phenological metric with additional random effects included (site, year, site-by-species, year-by-species) and found that the inferences regarding environmental relationships and species differences did not change.

Statistical Inference

All models were fitted via Bayesian inference using WinBUGS [30] and the package R2WinBUGS on the R platform [31]. Standard diffuse priors, Normal (0, 100000) for slope parameters and Uniform (0,100) for standard deviations, were assumed for all parameters other than the intercept of the WN model. That parameter is assumed to have a prior of the form Normal $(-\pi,\pi)$, because the use of a more diffuse prior may lead to problems with convergence (B. Reich, personal communication). The fitting of the WN model in WinBUGS is not trivial, and we follow the approach of Modlin *et al.* [20]: further details of this approach, along with the BUGS code.

Stepwise selection using Deviance Information Criterion [32] was used to select environmental variables (and, for males, species-by-environment interactions) in order to find the most parsimonious model for each seasonal timing metric. DIC is a generalisation of the Akaike Information Criterion (AIC), and is derived as the mean deviance adjusted for the estimated number of parameters in the model – DIC accounts for both model fit and complexity, and providing a relative measure of out-of-sample predictive performance [33]. DIC comparisons to 'null' models are presented for all best-fitting models, where the null model contained all fixed and random effects but no environmental variables.

Ideally, the regression models for the start, end and length of the overwinter period should include an estimate of the total abundance of each species at each site in each year. This is because more accurate estimates of the timing of emergence and disappearance are expected at sites with higher abundances of *Culicoides*. However, because of the unequal number and uneven timing of trapping across the sites, we were unable to create a meaningful estimate of overall site abundance for each year within this dataset. This problem is confounded by a degree of circularity between *Culicoides* phenology and abundance. We expect that trapping sites with higher abundance will produce more reliable estimates of the timing of phenological events such as the start and end of the season. However, it is also feasible that the phenology of the species may directly influence abundance, such that sites that are environmentally suitable for earlier spring emergence may also support greater population abundance. Therefore, disentangling phenology from abundance for these species is inherently difficult. To try and reduce this confounding we conducted a second analysis to compare the proportions of the total population of each of the four species that emerged in different seasonal periods for which trapping data were available using the *Avaritia* males dataset. This analysis allowed us to examine the influence of environmental drivers on the proportion of each species' population that was active at different periods of the year, bypassing the issue of not having a reliable estimate of overall abundance to include in the model. In so doing, we were able to determine that the same environmental variables affected both our measure of phenology, and proportional abundance, thereby providing a measure of the consistency in our inference regarding species-specific phenology (*'Multinomial analysis of Avaritia males phenology'*).

RESULTS

Using only those sites and years for which trapping captured the full seasonal profile of *Culicoides* abundance (n=69, 14 sites over four years), the maximum nightly catch of males per species was 123 (mean 44.1, s.d. 34.8) for *C. chiopterus*, 217 (mean 81.6, s.d. 68.3) for *C. dewulfi*, 460 (mean 86.4, s.d. 116.0) for *C. obsoletus,* and 120 (mean 44.6, s.d. 37.9) for *C. scoticus*. Of these sites, the *Avaritia* male data shows that *C. obsoletus* was caught most frequently (53% of the total maximum catch across all years), followed by *C. dewulfi* (24%), *C. scoticus* (13%) and *C. chiopterus* (10%).

Start of Season

Across all sites and years first appearance of *Culicoides* females occurred on average in early May (Table 2, Fig. 2) with the first record in late March (2007, site 3 'Chobham', Fig. 1) and the latest in late May (2009, site 25 'Winterslow', Fig. 1). Average appearance across all sites was remarkably consistent from year to year (early May), with the exception of 2007 in which the *Avaritia* females emerged on average two weeks earlier (Table 2), probably due to the warmer temperatures in winter and spring during that year relative to the other four years of observation (Fig. 3).

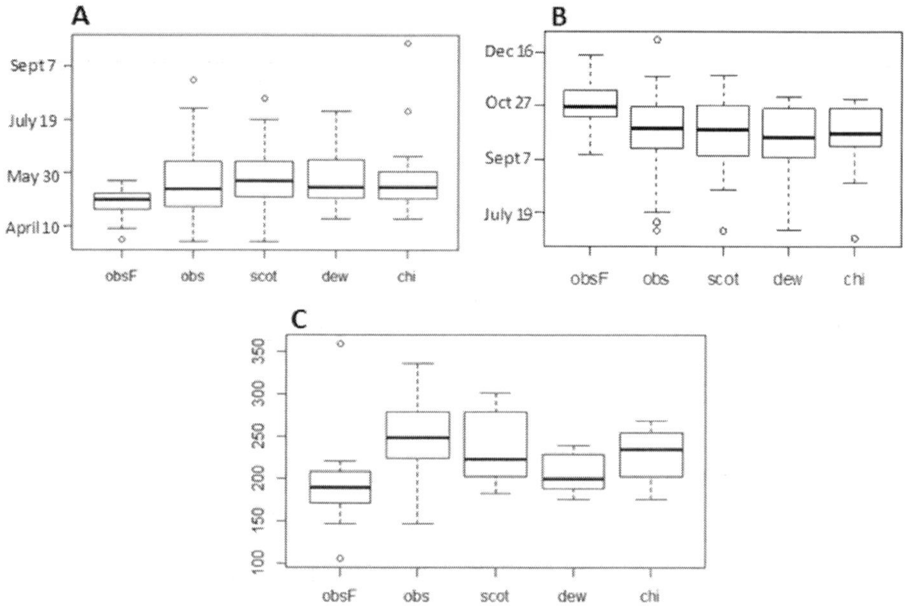

Figure 2. Raw data summarised for the start (A) and end (B) of seasonal activity (date), and length of overwinter period (C; days) derived from UK *Culicoides* **surveillance data during 2006 to 2010.**

Box plots show the median (central line), box denotes 25th and 75th percentiles, error bars represent 10th and 90th percentiles, and dots are points outside the 10th and 90th percentiles. Data are shown for the subgenus *Avaritia* (*Avaritia* females; obsF), *C. obsoletus Avaritia* males (obs), *C. scoticus Avaritia* males (scot), *C. dewulfi Avaritia* males (dew), and *C. chiopterus Avaritia* males (chi).

https://doi.org/10.1371/journal.pone.0111876.g002

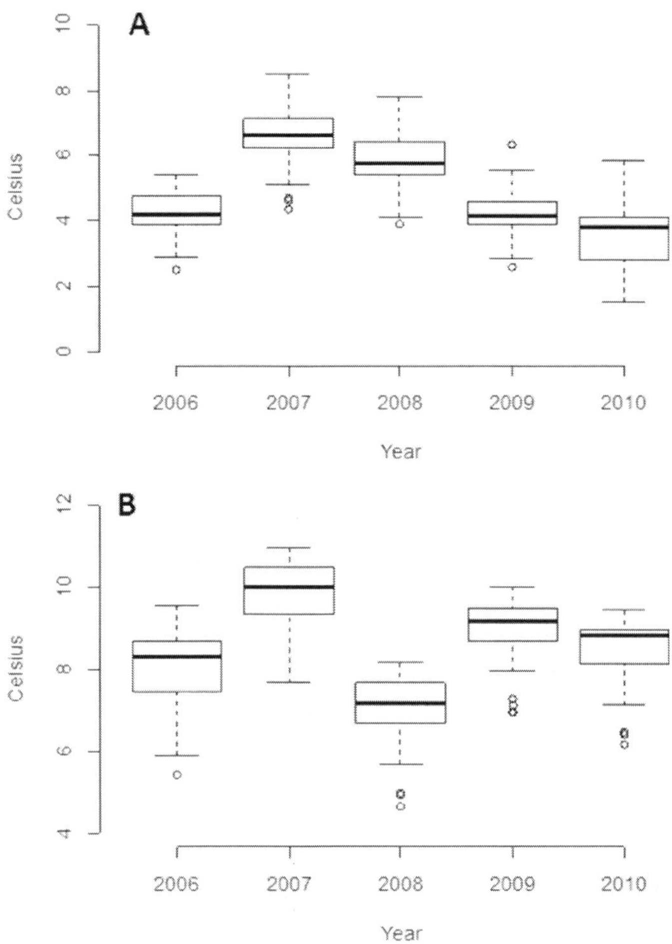

Figure 3. Mean winter (A) and spring (B) temperatures preceding *Culicoides* **activity season across all sites for each year of observation.**

Box plots show the median (central line), box denotes 25th and 75th percentiles, error bars represent 10th and 90th percentiles, and dots are points outside the 10th and 90th percentiles.

https://doi.org/10.1371/journal.pone.0111876.g003

Table 2. Summary of number of site-by-year combinations observed for each complex or species by timing metric used in the analysis.

								Mean by year				
START	observa-tions	sites	years	Min	Max	Range	Mean	2006	2007	2008	2009	2010
Subgenus *Avaritia*fe-males	49	20	5	88	143	55	123.7	126.3	115.6	127.3	126.1	125.7
C. obsoletus s.s	42	18	5	100	236	136	145.4	154.2	140.5	141.8	116.7	165.5
C. scoticus	19	10	4 (2009)	114	219	105	150.8	153	150.8	141	-	169
C. dewulfi	20	12	5	107	207	100	146.1	163.3	149	133.3	125	129
C. chiopterus	17	8	5	107	271	164	150.2	173	158	134.8	139	129
END												
Subgenus *Avaritia*fe-males	47	17	5	255	347	92	302	303.6	311.8	297.8	300.6	296
C. obsoletus s.s	47	18	5	183	362	179	270.7	275.3	256.6	259	288.8	272
C. scoticus	23	10	5	183	328	145	274.6	274.3	255.3	278.3	286.2	276.6
C. dewulfi	26	11	5	183	308	125	270.8	267.5	278.5	245.7	294.4	275.2
C. chiopterus	24	8	5	229	306	77	278.3	279.5	277.5	262.6	284.3	289.2
OVERWIN-TER												
Subgenus *Avaritia*fe-males	35	16	4	106	221	115	185.4	178.4	184.3	199.8	190.2	-
C. obsoletus s.s	26	13	4	147	336	189	247.7	246.3	250.9	235	248	-
C. scoticus	8	4	3 (2008)	182	301	119	236.5	219.3	241.3	-	255	-
C. dewulfi	8	3	3 (2008)	175	239	64	205.9	201.3	207.3	-	210.5	-
C. chiopterus	11	6	4	175	268	93	226.4	234	226.4	235	210.5	-

Years span from 2006–2010, missing years noted in parentheses. Summary of seasonal metrics (minimum, maximum and mean day of the year) across all sites, and mean for each year for *C. obsoletus* **group** (*Avaritia* **females),** *C. obsoletus* (*Avaritia* **males),** *C. scoticus* (*Avaritia* **males),** *C. dewulfi* (*Avaritia* **males), and** *C. chiopterus* (*Avaritia* **males). START: start of season (day of year), END: end of season (day of year), OVERWINTER: length of overwinter period (days).**

https://doi.org/10.1371/journal.pone.0111876.t002

Among the subgenus *Avaritia* males, all species emerged in late May, with *C. obsoletus* and *C. dewulfi* tending to emerge on average one week earlier than *C. chiopterus* and *C. scoticus* (Table 2, Fig. 2). The best-fitting model for this metric detected no significant species differences in timing of emergence, however, between *C. obsoletus* and the other three species (Table 3).

Table 3. Parameter estimates and 95% credible intervals for the fixed effects for the best-fitting model identified for each seasonal metric for individual subgenus *Avaritia Culicoides* **males.**

Variable	Estimate (95% credible interval)			
	Overall	*C. scoticus*	*C. dewulfi*	*C. chiopterus*
Start of season				
Intercept	149.5* (141.0, 158.7)			
Sp		0.22 (−12.82, 13.14)	1.91 (−10.84, 15.43)	−1.82 (−15.91, 12.54)
Cattle	0.32* (0.090, 0.56)			
Cattle:sp		−0.43* (−0.68, −0.18)	−0.43* (−0.68, −0.19)	−0.034 (−0.29, 0.23)
RH_{spr}	0.16 (−0.057, 0.38)			
RH_{spr}:sp		−0.068 (−0.30, 0.15)	−0.32* (−0.57, −0.059)	0.069 (−0.20, 0.33)
T_{spr}	1.94* (0.40, 3.53)			
P_{spr}	0.11 (−0.070, 0.29)			
End of season				
Intercept	273.7* (263.6, 283.9)			
Sp		−8.17 (−17.45, 1.56)	−6.31 (−16.14, 3.41)	−3.28 (−13.75, 7.72)
RH_{sum}	−2.76* (−5.23, −0.28)			
RH_{sum}:sp		1.43 (−0.084, 2.85)	1.62* (0.15, 3.33)	2.38* (0.00081, 4.85)
Sheep	0.024* (0.0058, 0.043)			
Photo	0.31 (−0.067, 0.68)			
Cattle	0.057 (−0.0041, 0.12)			
Length of overwinter period				
Intercept	262.5* (236.3, 286.8)			
Sp		60.09 (−18.53, 144.8)	−30.25 (−140.6, 88.21)	−12.27 (−67.10, 42.30)

Cattle	0.042 (−0.035, 0.12)			
Cattle:sp		−0.15* (−0.25, −0.046)	−0.11 (−0.25, 0.023)	−0.14* (−0.24, −0.029)
Sheep	−0.036* (−0.062, −0.0090)			
Sheep:sp		−0.0032 (−0.052, 0.041)	0.035 (−0.017, 0.085)	0.030 (−0.010, 0.071)
Moors	2.62 (−0.88, 6.23)			

Asterix denotes significance (95% credible interval does not bridge zero). Interactions with species are denoted '***:sp', and the parameter estimates associated with them refer to the differences in effect sizes relative to *C. obsoletus.* **The 'overall' column contains parameter estimates for main effects (which refer to all species if the corresponding interaction is not present in the model, and refer to** *C. obsoletus* **if the corresponding interaction is present).**

https://doi.org/10.1371/journal.pone.0111876.t003

The start of the season for *Avaritia* females was significantly negatively influenced by spring temperature, and close to being significantly positively influenced by photoperiod (more than 94% of posterior density mass greater than zero; Table 4). The best fitting model included both of these variables and received considerably more support in the data than the null model – containing no environmental variables (ΔDIC 4.5), although model fit to the data was poor (R^2 0.42, including both fixed and random effects).

Table 4. Parameter estimates and 95% credible intervals for the fixed effects within the best-fitting model identified for each seasonal metric for the subgenus *Avaritia* **female** *Culicoides*.

Seasonal metric	Variable	Estimate (95% credible interval)
Start of season	Intercept	123.9* (118.8, 130.9)
	T_{spr}	−3.27* (−6.06, −0.44)
	Photoperiod	1.05 (−0.13, 2.22)
End of season	Intercept	296.4* (294.3, 302.3)
	DD_s	0.031 (−0.0045, 0.064)
Length of	Intercept	184.3* (165.5, 201.9)

overwintering	Photoperiod	2.02 (−1.81, 6.46)
period	Moors	2.83 (−0.54, 6.26)
	Cattle	−0.037 (−0.11, 0.039)

Asterix denotes significance (95% credible interval does not bridge zero).

https://doi.org/10.1371/journal.pone.0111876.t004

The best-fitting model for the start of the season for the *Avaritia* males included a significant positive influence of cattle density on the start of seasonal activity for *C. obsoletus* with separate slopes for the other three species showing that start dates for *C. scoticus* and *C. dewulfi* were significantly less delayed by cattle density, and a non-significant difference for *C. chiopterus* (Table 3). A non-significant positive effect of spring humidity for *C. obsoletus* was also identified, with separate slopes for the other three species showing that start dates for *C. dewulfi* were significantly less affected by relative humidity than *C. obsoletus* (Table 3). In addition, a significant positive influence of spring temperature over all species was also determined and a non-significant positive influence of spring precipitation over all species occurred (best-fitting model: R^2=0.73; Table 3). An identical model without spring precipitation received essentially equal support in the data as the best-fitting model (ΔDIC 0.6), indicating this variable is perhaps less important than the other three included in the best model. The null model received almost no support in the data in comparison to the best-fitting model (ΔDIC 37.9).

End of the Season

Averaged over all sites and years, the last capture date for the *Avaritia* females occurred at the end of October (Table 2, Fig. 2). The earliest final capture occurred in mid-September (2006, site 4 'Compton', Fig. 1), while the latest was in mid-December (2006, site 3 'Chobham', Fig. 1). The four constituent species of the *Avaritia* males had mean dates of last capture within five days of each other, occurring in late September (Table 2, Fig. 2). *C. dewulfi* adults disappeared earliest, followed by *C. obsoletus*, *C. chiopterus* and *C. scoticus* (Table 2, Fig. 2). The best-fitting model for this metric detected no significant species differences between *C. obsoletus* and the other three species (Table 3). Overall, there was greater variability in the end dates for seasonal activity than for the start dates except for *C. chiopterus* (Table 2, Fig. 2).

No significant relationships were detected between environmental variables and the end of the season for *Avaritia* females (Table 4). Moreover, the null model received approximately similar support in the data as all other models, indicating that effects of climate and landcover on this phenological metric were not well captured when modelled as a complex (ΔDIC 1.3), and model fit was poor (R^2 best-fitting model 0.42).

The best-fitting model for the end of seasonal activity for the *Avaritia* males contained a significant negative influence of mean summer humidity for *C. obsoletus* and separate and significantly different slopes for *C. dewulfi* and *C. chiopterus* showing that *C. obsoletus* was inhibited more by summer humidity than both these dung-breeding species (Table 3). A significant positive effect of sheep density was also identified across all species, along with a non-significant positive effect of photoperiod across all species and a non-significant positive effect of cattle density across all species (R^2=0.85; Table 3). Adding mean summer temperature to this model resulted in the same level of support in the data (ΔDIC 0.1), as did dropping cattle density (ΔDIC 0.4) or dropping photoperiod (ΔDIC 0.6, Table S2). The best-fitting model received considerably more support in the data than the null model (ΔDIC 4.28). This suggests the predominant variables influencing the end of the season are summer relative humidity and sheep density; indeed, a model including only these two variables received very similar support in the data as the best fitting model (ΔDIC 0.7).

Length of Overwinter

The length of the overwinter period for the *Avaritia* females, when averaged over all sites and years, was approximately 185 days. The shortest overwinter period was 106 days (2006, site 3 'Chobham', Fig. 1), while the longest was 221 days (2006, site 4 'Compton', Fig. 1). The individual species making up the *Avaritia* males showed considerable variation in the length of the inactive overwinter period, with maximum overwinter lengths per species differing by up to 97 days (Table 2, Fig. 2). Averaged over all sites and years, *C. dewulfi* had the shortest overwinter period (206 days), followed by *C. chiopterus* (226 days), *C. scoticus* (237 days), and finally *C. obsoletus* (248 days) (Fig. 2). The best-fitting model for this metric strongly suggested that *C. scoticus* had a longer overwinter period (~60 days) than *C. obsoletus* (~93% of the posterior mass for this effect was greater than zero, Table 3), while both *C. chiopterus* and *C. dewulfi* showed no significant difference in overwinter length compared to *C. obsoletus* (Table 3).

The best-fitting model for the length of the inactive overwinter period for the *Avaritia* females included non-significant positive effects of photoperiod and percent cover of moorland and heathland, and a non-significant negative effect of cattle density (Table 4). However, model fit was relatively poor (R^2: 0.63), and the null model received essentially similar support in the data (ΔDIC 1.1).

The best-fitting model for the length of the inactive overwinter period for the *Avaritia* males included a non-significant positive effect of cattle density on *C. obsoletus* with separate slopes for the other three species showing that the length of the overwinter period for all three species was less affected by cattle density than *C. obsoletus,* with both *C. scoticus* and *C. chiopterus* significantly less affected (Table 3). In addition, a significant negative influence of sheep density was demonstrated for *C. obsoletus* with separate, but non-significant slopes for the other three species, and a close to significant positive influence of percent cover of moorland and heathland on all four species (~93% of posterior density mass greater than zero) (R^2=0.87). Several similar models received very close support in the data to the best-fitting model (ΔDIC<2 for next two best-fitting models, however, the null model received no support in the data compared to the best-fitting models (ΔDIC 15.8).

Variation across years and sites in the phenological models for *Avaritia* females were similar for the start of the season, although there was slightly more variation between years than sites for the end of the season, and slightly less variation between years than sites for the length of the overwinter period. For the *Avaritia* males, there was greatest variation across sites and years (combined site*year random effect) for the end of the season, followed by the length of the overwinter and the start of the season.

DISCUSSION

This study has demonstrated that the phenology of *Culicoides* in the UK is species-specific, but also exhibits considerable intra-specific variation between sites and years. The most significant finding in relation to disease control was the documentation of large variations in the length of the inactive overwinter period amongst the constituent species of the subgenus Avaritia. When assessed from catches of males in light-suction traps, *C. dewulfi* had on average the shortest overwinter period (206 days), followed by *C. chiopterus* (~2 weeks longer), *C. obsoletus* (~4 weeks longer) and *C. scoticus* (~8 weeks longer). Importantly, this variation in the overwinter period of males was related to underlying land-cover and host density variables. As such, when defined using data for males, the overwinter SVFP differed by up to eight weeks between the four species. Moreover, evidence from this nationwide surveillance dataset showed that the end of the flight season (autumn) is considerably more variable than the start (spring) across the species comprising the subgenus Avaritia. This is perhaps due to the synchronisation of overwintering larvae into the fourth instar stage resulting in relatively synchronous emergence as adults in the spring. This result implies that accurately predicting the start of the SVFP using current monitoring methods aggregated across species will be difficult. Importantly, when the subgenus *Avaritia* females were modelled as a group, we were unable to detect underlying links with environmental drivers for most of the phenological metrics, indicating that accounting for species-specific variation within this group is important for both understanding phenology and producing models that can be used to predict the SVFP from remote environmental variables.

Models describing the start and end of seasonal activity and the length of the overwinter period for subgenus *Avaritia* males were strongly supported at a species level. Species-specific drivers of phenology identified included increased cattle density (which led to a later start of season in *C. obsoletus*, but had less impact on *C. scoticus* or *C. dewulfi*) and percentage land-cover of moorland (with increasing cover tending to lead to longer overwinter across all species). Additionally, significantly shorter overwintering periods were documented at increased sheep density for *C. obsoletus*, and extended season end dates across all species. *Culicoides obsoletus* also exhibited significantly earlier season end dates at sites with higher mean summer humidity, but that both *C. dewulfi* and *C. chiopterus* were significantly less affected by this variable. *Culicoides dewulfi* was also significantly less influenced by relative humidity in the spring than *C. obsoletus* in relation to the timing of the start of seasonal activity. Warmer spring temperatures also resulted in significantly later start dates for all species using the *Avaritia* male data in direct contrast to the *Avaritia* females, which appeared earlier under warmer conditions.

The biological drivers of these relationships are challenging to interpret, although limited conclusions can be drawn given current knowledge of the ecology of each species. A key consideration is the contrasting types and availability of larval habitat used, with *C. obsoletus* and *C. scoticus* occupying a diverse range of development sites while *C. chiopterus* and *C. dewulfi* are restricted to cattle and horse dung [9], [34]. Somewhat counter-intuitively from an ecological point of view, season start dates for both *C. scoticus* and *C. dewulfi* were significantly less influenced by cattle density than *C. obsoletus*. This is in contrast to our expectations because cattle are known to be an important host for *C. obsoletus*, *C. dewulfi* and *C. chiopterus* [11], [35]. Cattle have been suggested as the most attractive hosts for Palaearctic *Culicoides* [36], [37], though, because hosts have not been enumerated in field studies, a robust host preference has not been conclusively demonstrated at the species level [11]. This observation could arise from interspecific differences in both intrinsic sensitivity of males to UV light and the availability and localisation of mating sites and resting areas. Numbers of *Culicoides* males occurring in UV or incandescent light-suction trap collections usually constitute only a small proportion of the total catch, as seen in both the current study and those conducted across Europe [38]–[40]. No attempt has been made, however, to quantify population responses at a species level in the presence of competing mating cues (e.g. pheromones) or to map resting populations at a farm level scale.

While speculative, similar factors could also drive the observations that greater sheep density resulted in both shorter overwinter periods and significantly later end dates across male populations of all species, although this could also be related to the greater provision of overwinter livestock accommodation of a suitable type to allow later survival of *Culicoides* as documented in northern Europe [10], [11]. Sheep are known to be important hosts for most of these species – *Culicoides scoticus*, *C. dewulfi* and *C. chiopterus* have been collected in abundance on the body of sheep [41], and all four species have been found to engorge on sheep in recent blood meal analysis studies in both farm and extensive pasture settings [35], [37], [42]–[45]. *Obsoletus* group catches in Wales were found to increase with the number of sheep on farms [46], a relationship also identified by a controlled study reporting a linear relationship between *C. obsoletus* trap catches and the number of sheep positioned beneath light traps [47]. It is likely that much finer resolution livestock density data and *C. chiopterus* population data from a passive trapping method would be needed to better understand the relationship and mechanisms whereby host density differentially affects the phenology of these species.

Biological factors underlying climatic drivers of populations similarly require further studies to confirm relationships. In this regard the differential response between dung-breeding species (*C. dewulfi* and *C. chiopterus*) and those utilising less localised and uniform larval development sites is of interest. The localisation of breeding habitats has the potential to lead to restrictions on suitable resting sites for males in order to maximise contact with gravid females. This close association with larval development sites may explain a reduced vulnerability to high humidity levels although adaptation to these conditions in dung breeding species may also play a role. Overall, our results suggest that the resolution and specificity of the freely available climate and ecological datasets we used were not sufficient to accurately decipher how temperature and precipitation interact to drive phenological events in these species, highlighting the need for further work at finer spatial scales.

Importantly, the evidence for considerable variation between species in the length of the inactive period overwinter revealed by analysing data for males in this study has significant implications for disease management via the SFVP in temperate zones. Transmission of vector-borne diseases is highly dependent on the host-vector ratio [48], [49], therefore if these four species are found to be differentially competent for arboviruses, the species-specific variation documented here in terms of their temporal phenology could impact strongly on the length and infection risk of the transmission season for arboviruses in the UK and elsewhere in northern Europe. A key caveat, is that our analysis of male trap catches is an acceptable representation of the phenology of female *Culicoides*. This is important because only female *Culicoides* blood-feed and are responsible for the transmission of viruses. Demonstrating correlation and synchrony in the phenology of males and females of the *Avaritia* species is difficult, and cannot be accomplished without using recently developed high-throughput qPCR assays for pooled samples (which are yet to be proven with collections arising from surveillance). To address this concern with our dataset, we analysed the extent of seasonal correlation in weekly trap catch abundance between males and parous females for two related Palaearctic *Culicoides* species (*C. pulicaris* and *C. impunctatus*) for which we have data spanning the same time period and sites as used in this study. This analysis demonstrated that both *C. pulicaris* and *C. impunctatus* exhibited good correlation between the seasonal abundance of male and parous female trap catches with 13 (*C. pulicaris*) and 10 (*C. impunctatus*) of the 15 sites examined for each species showing a correlation of greater than 0.5. In summary, we believe that these data demonstrate good correlation between seasonal trap catches of males and parous females for these two related *Culicoides* species. Moreover, there is no biological reason we are aware of that would suggest a different relationship between male and female seasonal activity for *C. pulicaris* or *C. impunctatus* in comparison to the four *C. obsoletus* complex species used in our main analysis. All of these species require tight synchronisation between male and female seasonal activity to ensure successful reproduction, and this is particularly true in temperate zones such as the UK where multiple generations of these species occur within a single year. Finally, although our analysis cannot define the length of the SFVP for these species because correlation in male and female seasonal abundance and phenology is by no means absolute, it does demonstrate a clear potential for species-specific variation in this important disease management tool. If these findings can be replicated using data on females the implications for disease management and spatio-temporal variation in risk are profound.

A key consideration for the current study and for future surveillance as part of the SVFP lies in the ability in the future to accurately differentiate female members of the subgenus *Avaritia* as these constitute the vast majority of light trap catches. The recent development of high-throughput real-time RT-PCR assays to differentiate species within pools of the subgenus *Avaritia* has great potential in offering a processing method that is sufficiently rapid to sustain surveillance trapping programmes [50]. While the current study has demonstrated that male populations of these species vary in their phenology, uptake of such techniques to examine such variation in females is likely to be determined by the ability to demarcate the role of specific species in the transmission of arboviruses and thereby provide a practical tool for estimating the risk of transmission. While this was not achieved for BTV-8 during the northern European incursion [1], results from the SBV outbreak strongly imply the presence of multiple vector species [51], [52]. This is likely to significantly complicate future attempts to model the risk of transmission according to season. A key observation, however, lay in the observation that the beginning of the adult *Culicoides* flight season was significantly more straightforward to predict than the end. This may enable at least partial prediction of SVFP's without recourse to costly and time consuming direct surveillance methods.

Recommendations for Management or Policy

As part of a surveillance system designed to allow ruminant movements during incursion of *Culicoides*-borne arboviruses, it was suggested that a SVFP could be maintained during which animal movement restrictions could be relaxed. Our study demonstrates that active surveillance of haematophagous female *Culicoides* vector populations cannot currently be replaced using remote models of abundance. This failure was most likely related to the diverse ecology of species conflated within this taxonomic grouping and was partially resolved by the use for species level modelling based on collections of male *Culicoides*.

The differences identified in this study of around eight weeks in the length of the overwinter period for the four species are particularly relevant to disease policy in the UK in relation to defining the SVFP. For instance, the Schmallenberg virus is known to have its greatest impact on mammalian hosts when infection occurs at a particular point in the gestation cycle of the host [5], [18], which coincides with the tail end of the *Culicoides* vector season in the UK. This coincidence in vector phenology and host susceptibility has been demonstrated to drive the extent and size of potential outbreaks of SBV in Scotland [19]. A key finding was that the timing of the end of the season may be more difficult to forecast, and should perhaps be treated with more caution by policy-makers than the beginning of the season, because it varies widely between species, years and locations in response to environmental heterogeneity. We recommend more intensive trapping across a range of climatic zones with species-level identification of *Culicoides* females wherever feasible to facilitate more accurate detection and understanding of the start of the SVFP in temperate zones.

ACKNOWLEDGMENTS

We thank the private landowners who allowed us access to their property to conduct the trapping. We particularly wish to thank Brian Reich for providing assistance in the development of the circular regression model in WinBUGS. We also thank two anonymous reviewers for their insights during the review process.

AUTHOR CONTRIBUTIONS

Conceived and designed the experiments: KRS SC AB PSM SG BVP. Performed the experiments: JB FS KL ED CS AW NN. Analyzed the data: KRS SC BVP. Contributed to the writing of the manuscript: KRS BVP SC PSM SG.

REFERENCES

1. Carpenter S, Wilson A, Mellor PS (2009) Culicoides and the emergence of bluetongue virus in northern Europe. Trends in Microbiology 17: 172–178.
2. Vandenbussche F, De Leeuw I, Vandemeulebroucke E, De Clercq K (2009) Emergence of Bluetongue Serotypes in Europe, Part 1: Description and Validation of Four Real-Time RT-PCR Assays for the Serotyping of Bluetongue Viruses BTV-1, BTV-6, BTV-8 and BTV-11. Transboundary and Emerging Diseases 56: 346–354.
3. van Rijn PA, Geurts Y, van der Spek AN, Veldman D, van Gennip RGP (2012) Bluetongue virus serotype 6 in Europe in 2008-Emergence and disappearance of an unexpected non-virulent BTV. Veterinary Microbiology 158: 23–32.
4. Hofmann MA, Renzullo S, Mader M, Chaignat V, Worwa G, et al. (2008) Genetic Characterization of Toggenburg Orbivirus, a New Bluetongue Virus, from Goats, Switzerland. Emerging Infectious Diseases 14: 1855–1861.
5. Hoffmann B, Scheuch M, Hoper D, Jungblut R, Holsteg M, et al. (2012) Novel Orthobunyavirus in Cattle, Europe, 2011. Emerging Infectious Diseases 18: 469–472.
6. Davies I, Vellema P, Roger P (2012) Schmallenberg virus - an emerging novel pathogen. In Practice, 34: 598–+.
7. Mintiens K, Meroc E, Mellor PS, Staubach C, Gerbier G, et al. (2008) Possible routes of introduction of bluetongue virus serotype 8 into the epicentre of the 2006 epidemic in North-Western Europe. Preventive Veterinary Medicine 87: 131–144.
8. Mellor PS, Boorman J, Baylis M (2000) Culicoides biting midges: Their role as arbovirus vectors. Annual Review of Entomology 45: 307–340.
9. Kettle DS, Lawson JWH (1952) The early stages of British biting midges *Culicoides* Latreille (Diptera: Ceratopogonidae) and allied genera. Bulletin of Entomological Research 43: 421–467.
10. Meiswinkel R, Goffredo M, Dlikstra EGM, van der Ven IJK, Baldet T, et al. (2008) Endophily in Culicoides associated with BTV-infected cattle in the province of Limburg, South-Eastern Netherlands, 2006. Preventive Veterinary Medicine 87: 182–195.

11. Viennet E, Garros C, Rakotoarivony I, Allene X, Gardes L, et al... (2012) Host-Seeking Activity of Bluetongue Virus Vectors: Endo/Exophagy and Circadian Rhythm of Culicoides in Western Europe. PLoS ONE: 7.
12. Schley D, Gubbins S, Paton DJ (2009) Quantifying the Risk of Localised Animal Movement Bans for Foot-and-Mouth Disease. PLoS ONE: 4.
13. Maan S, Maan NS, Ross-Smith N, Batten CA, Shaw AE, et al. (2008) Sequence analysis of bluetongue virus serotype 8 from the Netherlands 2006 and comparison to other European strains. Virology 377: 308–318.
14. Carpenter S, Wilson A, Mellor PS (2009a) Bluetongue virus and *Culicoides* in the UK: The impact of research on policy. Outlooks on Pest Management 20: 161–164.
15. EFSA, *Annual Report* 2008: European Food Safety Authority, European Union. 108.
16. Wilson A, Carpenter S, Gloster J, Mellor P (2007) Re-emergence of bluetongue in northern Europe in 2007. Veterinary Record 161: 487–489.
17. Gariglinany MM, Hoffmann B, Dive M, Sartelet A, Bayrou C, et al.. (2012) Schmallenberg virus in calf born at term with porencephaly, Belgium [letter]. Emerg Infect Dis [serial on the Internet]. 2012 June. http://dx.doi.org/10.3201/eid1806.120104.
18. Lievaart-Peterson K, Luttikholt SJM, Van den Brom R, Vellema P (2012) Schmallenberg virus infection in small ruminants – First review of the situation and prospects in Northern Europe. Small ruminant research 106: 71–76.
19. Bessell PR, Searle KR, Auty HK, Handel IG, Purse BV, et al... (2013) Epidemic potential of an emerging vector borne disease in a marginal environment: Schmallenberg in Scotland. Scientific Reports: 3.
20. Modlin D, Fuentes M, Reich B (2012) *Circular conditional autoregressive modeling of vector fields.* Environmetrics, 2012. 23(1): p. 46–53.
21. Mellor PS, Tabachnick W, Baldet T, Baylis M, Bellis G, et al. (2004) Conclusions of working groups. Group 2. Vectors. Proceedings of the 3rd OIE Bluetongue International Symposium, 26–29 October 2003, Taormina, Italy. Veterinaria Italia 40: 715–717.

22. Campbell JAP (1960) A taxonomic review of the British species of *Culicoides* Latreille (Diptera: Ceratopogonidae). Proceedings of the Royal Entomological Society of London: Series B. 67: 181–302.
23. Dyce AL (1969) The recognition of mulliparous and parous *Culicoides* (Diptera Ceratopogonidae) without dissection. Journal of the Australian Entomological Society 8: 11–15.
24. Blackwell A (1997) *Diel flight periodicity of the biting midge Culicoides impunctatus and the effects of meteorological conditions*. Medical and Veterinary Entomology, 1997. 11(4): p. 361–367.
25. Purse BV, Falconer D, Sullivan MJ, Carpenter S, Mellor PS, et al. (2012) Impacts of climate, host and landscape factors on Culicoides species in Scotland. Medical and Veterinary Entomology 26: 168–177.
26. Sanders CJ, Shortall CR, Gubbins S, Burgin L, Gloster J, et al. (2011) Influence of season and meteorological parameters on flight activity of Culicoides biting midges. Journal of Applied Ecology 48: 1355–1364.
27. Searle KR, Purse BV, Blackwell S, Falconer D, Sullivan M, et al. (2013) Environmental drivers of insect phenology across space and time: *Culicoides* in Scotland as a case study. Bulletin of Entomological Research 103: 155–170.
28. Morton D, Rowland C, Wood C, Meek L, Marston C, et al... (2011) Final Report for LCM2007 - the new UK Land Cover Map. Countryside Survey Technical Report No. 11/07 NERC/Centre for Ecology & Hydrology 112 pp. (CEH Project Number: C03259).
29. Fisher NI (1993) Statistical Analysis of Circular Data. Cambridge University Press, Cambridge.
30. Lunn D, Spiegelhalter D, Thomas A, Bes N (2011) The BUGS project: Evolution, critique and future directions. Statistics in Medicine 28: 3049–3067.
31. R-Development_Core_Team, *R: a language and environment for statistical computing*, 2009, www.r-project.org/: Vienna, Austria.
32. Spiegelhalter DJ, Best NaG, Carlin BP, Van Der Linde A (2002) Bayesian measures of model complexity and fit. Journal of the Royal Statistical Society: Series B (Statistical Methodology) 64: 583–639.
33. Gelman A, Hill J (2007) Data Analysis Using Regression and Multilevel/Hierarchical Models. Cambridge University Press, Cambridge, UK.
34. Harrup LE, Purse BV, Golding N, Mellor PS, Carpenter S (2013) *Larval development and emergence sites of farm-associated Culicoides in the*

United Kingdom. Medical and Veterinary Entomology, 2013. Online early.

35. Garros C, Gardèsa L, Allènea X, Rakotoarivonya I, Vienneta E, et al. (2011) Adaptation of a species-specific multiplex PCR assay for the identification of blood meal source in Culicoides (Ceratopogonidae: Diptera): applications on Palaearctic biting midge species, vectors of Orbiviruses. Infection, Genetics and Evolution 11: 1103–1110.
36. Bartsch S, Bauer B, Wiemann A, Clausen P-H, Steuber S (2009) Feeding patterns of biting midges of the Culicoides obsoletus and Culicoides pulicaris groups on selected farms in Brandenburg, Germany. Parasitology Research 105: 373–380.
37. Ninio C, Augot D, Delecolle J-C, Dufour B, Depaquit J (2001) Contribution to the knowledge of Culicoides (Diptera: Ceratopogonidae) host preferences in France. Parasitology Research 108: 657–663.
38. Takken W, Verhulst N, Scholte E-J, Jacobs F, Jongema Y, et al.. (2008) *The phenology and population dynamics of Culicoides spp. in different ecosystems in The Netherlands.* Preventive Veterinary Medicine, 2008. 87(1–2): p. 41–54.
39. Purse BV, Tatem AJ, Caracappa S, Rogers DJ, Mellor PS, et al.. (2004) *Modelling the distributions of Culicoides bluetongue virus vectors in Sicily in relation to satellite-derived climate variables.* Medical and Veterinary Entomology, 2004. 18(2): p. 90–101.
40. Capela R, Purse BV, Pena I, Wittman EJ, Margarita Y, et al.. (2003) *Spatial distribution of Culicoides species in Portugal in relation to the transmission of African horse sickness and bluetongue viruses.* Medical and Veterinary Entomology, 2003. 17(2): p. 165–177.
41. Carpenter S, Szmaragd C, Barber J, Labuschagne K, Gubbins S, et al. (2008) An assessment of Culicoides surveillance techniques in northern Europe: have we underestimated a potential bluetongue virus vector? Journal of Applied Ecology 45: 1237–1245.
42. Lassen SB, Nielsen SA, Skovgard H, Kristensen M (2011) Molecular identification of bloodmeals from biting midges (Diptera: Ceratopogonidae: *Culicoides Latreille*) in Denmark. Parasitology Research, 2011. 108: p. 823–829.
43. Lassen SB, Nielsen SA, Kristensen M (2012) Identity and diversity of bloodmeal hosts of biting midges (Diptera: Ceratopogonidae: *Culicoides Latreille*) in Denmark. Parasites and Vectors, 5: p. 153–161.

44. Martínez-de la Puente, Martínez J, Ferraguti M, Morales-de la Nuez A, Castro N, et al. (2012) Genetic characterization and molecular identification of the bloodmeal sources of the potential bluetongue vector *Culicoides obsoletus* in the Canary Islands, Spain. Parasites and Vectors 5: 147–155.
45. Viennet E, Garros C, Lancelot AXR, Gardes L, Rakotoarivony I, et al. (2011) Assessment of vector/host contact: comparison of animal-baited traps and UV-light/suction trap for collecting Culicoides biting midges (Diptera: Ceratopogonidae), vectors of Orbiviruses. Parasites and Vectors 4: 119.
46. Kluiters G, Sugden D, Guis H, McIntyre KM, Labuschagne K, et al... (2012) Modelling the spatial distribution of Culicoides biting midges at the local scale. Journal of Applied Ecology. Doi:10.1111/1365-2664.12030.
47. Garcia-Saenz A, McCarter P, Baylis M (2010) The influence of host number on the attraction of biting midges, Culicoides spp., to light traps. Med Vet Entomol. 25: 113–5.
48. Hartemink NA, Purse BV, Meiswinkel R, Brown HE, de Koeijer A, et al. (2009) Mapping the basic reproduction number (R-0) for vector-borne diseases: A case study on bluetongue virus. EPIDEMICS 1: 153–161.
49. Gubbins S, Carpenter S, Baylis M, Wood JLN, Mellor PS (2008) Assessing the risk of bluetongue to UK livestock: uncertainty and sensitivity analyses of a temperature-dependent model for the basic reproduction number. Journal of the Royal Society Interface 5: 363–371.
50. Mathieu B, Delecolle JC, Garros C, Balenghien T, Setier-Rio ML, et al.. (2011) *Simultaneous quantification of the relative abundance of species complex members: Application to Culicoides obsoletus and Culicoides scoticus (Diptera: Ceratopogonidae), potential vectors of bluetongue virus (vol 182, pg 297, 2011)*. Veterinary Parasitology, 2012. 187(3–4): p. 578–578.
51. Elbers ARW, Meiswinkel R, van Weezep E, Sloet van Oldruitenborgh-Oosterbaan MM, et al.. (2011) *Schmallenberg virus in Culicoides spp. biting midges, the Netherlands, 2011*. Emerging infectious diseases, 2013. 19(1): p. 106–9.

52. De Regge N, Deblauwe I, De Deken R, Vantieghem P, Madder M, et al.. (2012) *Detection of Schmallenberg virus in different Culicoides spp. by real-time RT-PCR.* Transboundary and Emerging Diseases, 2012. 59(6): p. 471–475.

15

Culicoides Midge Bites Modulate the Host Response and Impact on Bluetongue Virus Infection in Sheep

Nonito Pages[1], Emmanuel Bre´ard[2], Ce´line Urien[3], Sandra Talavera[1], Cyril Viarouge[2], Cristina LorcaOro[1], Luc Jouneau[3], Bernard Charley[3], Ste´phan Zientara[2], Albert Bensaid[1], David Solanes[1], Joan Pujols[1], Isabelle Schwartz-Cornil[3]

[1]Centre de Recerca en Sanitat Animal CReSA, Universitat Auto`noma de Barcelona – Institut de Recerca i Tecnologia Agroalimentaries, Bellaterra, Spain,
[2]Virologie, Unite´ Mixte de Recherche UMR1161, Agence Nationale de Se´curite´ Sanitaire de l'Alimentation, de l'Environnement et du Travail – Institut National de la Recherche Agronomique – Ecole Nationale Ve´te´rinaire d'Alfort, Maisons-Alfort, France,
[3]Virologie et Immunologie Mole´culaires, Unite´ de Recherche UR892, Institut National de la Recherche Agronomique, Jouy-en-Josas, France

Citation: Pages N, Bréard E, Urien C, Talavera S, Viarouge C, Lorca-Oro C, et al. (2014) Culicoides Midge Bites Modulate the Host Response and Impact on Bluetongue Virus Infection in Sheep. PLoS ONE 9(1): e83683. https://doi.org/10.1371/journal.pone.0083683

ABSTRACT

Many haematophagous insects produce factors that help their blood meal and coincidently favor pathogen transmission. However nothing is known about the ability of *Culicoides* midges to interfere with the infectivity of the viruses they transmit. Among these, Bluetongue Virus (BTV) induces a hemorrhagic fever- type disease and its recent emergence in Europe had a major economical impact. We observed that needle inoculation of BTV8 in the site of uninfected *C. nubeculosus* feeding reduced viraemia and clinical disease intensity compared to plain needle inoculation. The sheep that developed the highest local inflammatory reaction had the lowest viral load, suggesting that the inflammatory response to midge bites may participate in the individual sensitivity to BTV viraemia development. Conversely compared to needle inoculation, inoculation of BTV8 by infected *C. nubeculosus* bites promoted viraemia and clinical symptom expression, in association with delayed IFN-induced gene expression and retarded neutralizing antibody responses. The effects of uninfected and infected midge bites on BTV viraemia and on the host response indicate that BTV transmission by infected midges is the most reliable experimental method to study the physio-pathological events relevant to a natural infection and to pertinent vaccine evaluation in the target species. It also leads the way to identify the promoting viral infectivity factors of infected *Culicoides* in order to possibly develop new control strategies against BTV and other *Culicoides* transmitted viruses.

INTRODUCTION

Several biting midges of the *Culicoides* genus (Diptera: Ceratopogonidae), that are among the smallest haematophagous insects, are transmission vectors of viruses affecting human and animals, such as the Oropouche virus in human [1], the African Horse Sickness and Equine Encephalosis viruses [2], the zoonotic human/swine Vesicular Stomatitis Virus [3] and three ruminant viruses with the Bluetongue virus [4], the Epizootic Haemorrhagic Disease virus [2] and the Schmallenberg virus [5]. Bluetongue virus (BTV) - the most studied virus transmitted by *Culicoides* midges so far - belongs to the *Orbivirus* genus of the *Reoviridae* family, and includes 26 serotypes [6] determined by the outer capsid VP2 and, to a lesser extent VP5 proteins [4]. Several BTV serotypes, mainly BTV8 and 1, have been recently introduced in Europe where they caused major economical damages in the cattle and sheep industry with loss of production and mortality and trade restrictions [4]. The clinical manifestations of Bluetongue (BT) can include fever, salivation, hyperaemia/oedema, mucosae ulcerations, and abortion [4]. BTV is transmitted by many *Culicoides* species, with dominant species that vary in different parts of the world, such as *C. imicola* in Africa, *C. sonorensis* in North America, *C. brevitarsis* in Australia and *Culicoides* of the so called Obsoletus group in Northern Europe [7].

Blood sucking arthropods are often more than just delivery systems for the pathogens they carry. Indeed the arthropod bite triggers vasoconstriction, the immediate onset of acute inflammation and of innate immune responses that may interfere on the pathogen transmission. The saliva of haematophagous arthropods contains a cocktail of pharmacologic proteins that modulate vascular constriction, blood coagulation, inflammation and immunity [8]. These compounds, mainly studied in the cases of ticks, sand flies, and mosquitoes, contribute to the success of the blood meal and can also promote pathogen infectivity [8]. Regarding mosquito- transmitted viral infections, West Nile Virus infection in mice was enhanced by co-administration of mosquito saliva [9]; West Nile Virus [10] or Cache-Valley Virus [11] infections were exacerbated when the viruses were inoculated in the skin area subsequently to the feeding of mosquitoes in mice. However no effect of mosquito bite was observed in the case of Saint Louis Encephalitis Virus and Western Equine Encephalitis Virus transmission in chicken [12]. Thus the effect of insect bites on infection may depend on the pathogen, on the insect type, and possibly on the vertebrate species.

Culicoides are "pool" feeders: they lacerate the skin to feed on the effusion into this injury, which includes blood, skin cells and lymph. Nothing is known about the effect of *Culicoides* bite on viral transmission, an information of importance for the development of control strategies. *Culicoides* secrete molecules such as vasodilatator [13], anticoagulant [14] and saliva components that are able to inhibit mouse lymphocyte response and nitric oxide production by macrophages [3]. The *Culicoides* bite provokes the recruitment of inflammatory cells that might be targets for viruses such as BTV [15]. In addition, BTV produced by insect cells *in vitro* seems more infectious for sheep than BTV produced by mammalian cells [16]. However, *Culicoides* saliva cleaves the BTV VP2 outer capsid protein, resulting in processed viruses that are less infectious for mammalian Baby Hamster Kidney cells compared to the native particles [17]. Thus *Culicoides* bites may positively or negatively affect BTV transmission by the secreted saliva components, the host response to the mechanical action of their mouth part in the skin and to the saliva, the viral processing by the insect cells in the salivary gland and the sub-anatomical site where the virus is inoculated. In order to evaluate whether and how *Culicoides* bites may modulate BTV transmission, we compared the extent of viraemia, clinical signs and host responses induced by 3 infection modes with a European isolate of BTV8: 1) simple intradermal needle inoculation; 2) intradermal needle inoculation in the feeding site of C. *nubeculosus* as done in the case of West Nile Virus [10] or Cache-Valley Virus [11], and 3) transmission by intra-thoracically- infected C. *nubeculosus* midges. *C. nubeculosus* was chosen as it is the only European *Culicoides species* that can be properly maintained in laboratory conditions using standardized techniques [18]. Moreover, *C. nubeculosus* supported BTV multiplication both after virus ingestion and inoculation, and infected midges were able to transmit the virus in laboratory conditions [19]. We performed the experiments in sheep, the real target of BTV, in order to evaluate the tri-partite insect/virus/vertebrate interactions in the pertinent host species. Previous studies of pathogen transmission by insects have used mice as a model system and have very rarely involved natural target species. These experiments revealed that *C. nubeculosus* bites have a dual effect on BTV viraemia and clinical BT disease, with an inhibitory effect of control bites on needle- transmission and an enhancing effect of infected bites compared to needle- transmission, associated to differentially modulated innate host responses.

MATERIALS AND METHODS

Ethics Statement

All experiments were performed in the BSL3 facilities of the Centre de Recerca en Sanitat Animal (CReSA, Spain) under the guidelines of the European Community (86/609) and were approved by the Committee on the Ethics of Animal Experiments of *Universitat Autònoma de Barcelona* (UAB) # 1425 and Animal Experimentation Commission of DAAM # 6330. Specifically, sheep were housed under temperature, humidity and light control with 4 stages trying to simulate the day light of the farms. Food, water and block of minerals were provided ad libitum. In the current experiment, symptoms were mild (<8 on a scale of 21), thus none of the animals required pain relief. Moreover, handling of animals was performed by experienced accredited workers, avoiding undue supplemental stress. Animals were euthanized with an overdose of sodium thiopental injected intravenously.

Virus

The baby hamster kidney (BHK21) cell line used for virus growth and titration was grown in MEM supplemented with 5% FBS, 1% pyruvate, 1% non essential amino acids, and penicillin/streptomycin (=culture medium). The wild-type field BTV8 strain was from the National Reference Laboratory collection (ANSES, Maisons-Alfort, France); it has been isolated from sheep blood in France in Ardennes in 2006, cultured on embryonated chicken eggs and then passaged only 3 times in BHK21 cells. To prepare BTV stocks, confluent BHK21 cell layers were infected and incubated at 37°C in 5% CO_2 until the appearance of cytopathic effects (3 to 4 days post infection). Culture supernatants were collected and stored at −80°C. Fifty percent tissue culture infective doses ($TCID_{50}$) were estimated by endpoint titration on BHK21 cells (method of Spearman-Karber) by using a previously described protocol [20]. The inoculation stock used in this study titrated 7.5×10^6 $TCID_{50}$/ml that represented 4.57×10^{11} BTV RNA copies/ml.

BTV Inoculation to Midges and Biting Chambers

BTV8 (0.5–1 μl viral stock, 7.5×10^3 $TCID_{50}$) was inoculated intra-thoracically to *C. nubeculosus* females. Briefly, 2 or 3 day old females were slightly anesthetized using CO_2 under a dissecting microscope. Using a glass microneedle and a manual microinjector, females were intra-thoracically inoculated with BTV8 inoculum through the suture between the epimeron and episterum with a volume that forces the cerci to slightly extend the tip of their wings (0.5–1 μl). Inoculated midges (n=356 females) were sorted in 16 groups of 15–25 midges inside primary containers and maintained for an extrinsic incubation period of 6–7 days at 24±2°C, 75%RH, photoperiod 14 : 10 (light : dark) and feed *ad libitum* with 5% sucrose. To increase the number of blood engorged individuals while feeding on sheep in the pre-sensitization procedure, midges were starved for a period of 24 h and then fed on sheep. RNA was extracted from *Culicoides* right after the intra-thoracic inoculation (4 midges) and in the pools of midges after gorging on sheep and tested for content in BTV8 RNA copies. The infectivity for mammalian cells of the *C. nubeculosus*- recovered BTV was confirmed on BHK-21 cells. However we could not get reproducible infectivity titers, possibly related to *Culicoides*- derived compounds interfering on BHK-21 receptivity.

Sheep and Experimental Infection

Thirty two 6 month old Lacaune female sheep that had been checked to be BTV free and BTV seronegative were bought from Gregori Farm, Vilamajor d'Ager, Lleida (Spain). They had been treated with Deltametrine insecticide (Butox 7,5%; Intervet Schering-Plough AH) every four weeks from February to May and kept in protected area to avoid exposure to environmental midges. After a week acclimation in the CReSA BLS3 facilities (Bellaterra, Spain), the sheep were randomly allocated in 5 groups: (1) the control group (4 sheep) was intradermally inoculated with culture medium; (2) the control bite group (4 sheep, CB group) was treated by application, on the right and left inguinal areas, of *Culicoides* biting chambers with uninfected midges for 30 minutes just prior to intradermal inoculation with culture medium in the bitten zone; (3) the intradermal group (8 sheep, ID group) was infected by the intradermal route with a total of 3.75×10^{6} $TCID_{50}$ (2.28×10^{11} BTV RNA copies) per sheep; (4) the CB + ID group (8 sheep) consisted of sheep treated by application of *Culicoides* biting chambers with uninfected midges for 30 minutes followed by intradermal inoculation with a total of 3.75×10^{6} $TCID_{50}$ BTV8 per sheep (2.28×10^{11} BTV RNA copies), in the feeding site of the *Culicoides*; (5) finally, the infected bite group (8 sheep, IB group) was infected by application of two *Culicoides* biting chambers for 30 minutes with intra-thoracically infected midges, simultaneously placed on the right and left inguinal areas; as discussed in the result section, the mean RNA copy numbers available in the *Culicoides* was $2.11\pm1.4\times10^{10}$ per sheep. However the exact amount of BTV delivered to each sheep was not known in this group.

In the groups that were needle- injected, the inoculum (medium or virus) was given intradermally in the inguinal zones in 10 spots (50 μl each), with 5 spots on the right side and 5 spots on the left side. In the groups that were exposed to control or infected *Culicoides* midges, 2 biting chambers were used, one applied in the right inguinal area while the second in the left one. Biting chambers consisted on cylindrical polypropylene containers (55 mm diameter) sealed with fine mesh (1 mm wide). One of the sides was opened to introduce anesthetized *Culicoides* midges and after that immediately sealed.

Sample Collection, Local Inflammation Monitoring and Clinical Scoring

EDTA-blood samples were collected daily during the first 9 days post inoculation (dpi) and at day 11, 15 and 21 dpi. Serum samples were collected at 0, 5, 8 15 and 21 dpi. Body temperature was recorded daily for 16 days. Clinical signs were recorded daily. The clinical score table for the BTV8 animal trial and was established based on previous reports [16], [21], [22]. An integrated clinical score was calculated from the area under the curve for each sheep from 6 to 17 dpi using the Prism 5.0 software; no clinical signs were expressed before 6 dpi. The local inflammatory reaction at the inoculation point on both inguinal sides was monitored each day from 0 to 7 dpi and graded from 0 to 3 based on the intensity and area size of the local hyperaemia (score 0 to 3 per inguinal side). The integrated inflammatory reaction score per sheep was calculated from the area under the curve over the 0 to 7 dpi period.

RNA Extraction

For RNA extraction of midges, midges were homogenized in 500 µl of PBS using microbeads- containing tubes and a Ribolysor apparatus (Bio-rad, Marne-la Coquette, France). After clarification by centrifugation (2 minutes, 1000 g), RNA was extracted using the Qiamp viral RNA kit (Qiagen).

Viral RNA was extracted from 100 µl whole blood using the Qiamp viral RNA kit. Cellular RNA was extracted from EDTA blood using Trizol-LS extraction. The extracted cellular RNA was checked for quality with an Agilent 2100 Bioanalyzer using RNA 6000 Nano Kits (Agilent Technologies).

Viral RNA Detection

Viral RNA detection was performed with a commercial pan-BTV rt-RT-PCR (ADIAVET® BTV Real-time A352, ADIAGENE, France) targeting the segment 10 of BTV and GAPDH as an internal control. The quantification was done using a viral dsRNA corresponding to the BTV8 segment 10; briefly, total RNA were extracted from an infected culture cell by trizol extraction and the viral dsRNA were remove from ssRNA by lithium chloride (liCl) precipitation [23]. After migration and gel purification from a 1% agarose gel, the ds segment 10 was quantified by spectrophotometer. The number of BTV RNA copies per ml of blood was calculated at 0 to 9, 11, 15 and 21 dpi. The area under the curve over the 21 day-period was calculated using the Prism 5.0 software. An area under the curve below 2 log10 corresponded to non-viraemic sheep.

Sheep cytokine and IFN- induced gene mRNA Detection

For quantitation of cytokine and IFN-induced gene expression, RNA (200 ng) was reverse transcribed using random primers and the Multiscribe reverse transcriptase (Applied Biosystem). Real-time PCR (qPCR) was carried out using 5 ng cDNA with 300 nM primers in a final reaction volume of 25 µl of 1× SYBR Green PCR Master Mix (Applied Biosystem). The primers used to amplify ovine cDNA were designed with the Primer Express software (v2.0) using publically available GenBank sequences. PCR cycling conditions were 95°C for 10 min, linked to 40 cycles of 95°C for 15 sec and 60°C for 1 min. Real-time qPCR data were collected by the Mastercycler® ep realplex - Eppendorf system. The expression of the different genes (arbitrary units) compared to T0 and relatively to RPS24 (ribosomal protein 24) or GAPDH was calculated by the $2^{-\Delta\Delta Ct}$ method with the Realplex software.

Serological Responses to BTV

The collected sera were analyzed using a commercial BTV VP7 competitive ELISA kit (ID-VET, Montpellier, France) according to manufacturer's protocols. The development of neutralizing antibody response was assessed in BHK21 cells by serum neutralization with 50 $TCID_{50}$ BTV8 as described [24]. All dilutions were performed in triplicates.

Statistical Analyses

The qPCR viral load, Mx1 and CXCL10 gene expressions, the temperature and clinical scores, and the antibody response measurements were analyzed for statistical differences between sheep groups using the non-parametric one-sided *Mann-Whitney U* for independent groups.

RESULTS

Contrasted effects of Uninfected and Infected *C. nubeculosus* bites on Induction of BTV viraemia and Clinical Disease

We set up our goal to evaluate the effect of uninfected and infected *Culicoides* bites on BTV infection in sheep. In order to test the effect of uninfected bites, BTV was inoculated intradermally in sheep within the feeding sites of uninfected *C. nubeculosus* (control bite + intradermal, CB + ID group), and within the skin of unbitten sheep (intradermal,

ID group). The number of bites per sheep in the CB + ID group varied between 11 and 50 (data not shown). We used BTV8, the BTV serotype that spread in Europe in 2006–2008, at a dose per sheep of 3.75×10^6 TCID 50 corresponding to 2.28×10^{11} BTV RNA copies that is similar to conditions that previously induced viraemia and clinical symptoms in sheep [16], [25]. For testing the effect of infected bites, the sheep skin was bitten by BTV8-intra-thoracically inoculated *C. nubeculosus* (between 15 and 49 bites per sheep, IB group, data not shown). The same BTV8 preparation was used to infect the sheep and the *C. nubeculosus*. Four days post intra-thoracic inoculation, the mean number of BTV8 RNA copies reached 1.2×10^9/midge, whereas the initial quantity of BTV8 RNA copies detected after inoculation was 2.43×10^7/midge, indicating that viral replication occurred in *C. nubeculosus* after intra-thoracic inoculation, in the same order of magnitude as previously reported for C. *soronensis* [26]. The BTV8 RNA load in the total engorged *Culicoides* per sheep varied between 6×10^9 and 4.95×10^{10} BTV8 RNA copies per exposed sheep (mean $2.11\pm1.4\times10^{10}$). Therefore, the total quantity of BTV8 RNA available in the *C. nubeculosus* was about 10 times lower than in the intradermally- injected BTV8 inoculum.

The BTV8 RNA load in blood peaked between 5 and 7 dpi in the ID and CB + ID groups (Fig. 1A and B), whereas the peak time was more variable in the IB group (Fig. 1C), occurring between 6 and 15 dpi. When the BTV8 RNA load per sheep was integrated for the whole monitoring period (Fig. 1D), only 3 out of 8 sheep in the CB + ID group presented an integrated viral load above 3 log10 copies/per ml, whereas 6 out of 8 sheep in the ID group did. This indicates that uninfected *C. nubeculosus* bites rather inhibited BTV viraemia onset. Conversely, despite the fact that the sheep from the IB group likely received far less BTV than the CB + ID and ID sheep, all IB sheep showed integrated viral loads above 3 log10, with 4 out of 8 sheep presenting high loads that were above 6 log10, a value not reached in the ID group (Fig. 1D). Furthermore the viral load values tended to be higher in the IB group compared to the ID group ($p=0.06$). In addition, 3 sheep from the IB group developed rapid viraemia that was detected at day 3, earlier than for any sheep of the other groups (Fig. 1C). Control sheep groups (injected with medium after exposure to uninfected *Culicoides* bites (control CB) or unexposed (control)) did not present any detectable BTV RNA in their blood. Thus we show that infected bites of *C. nubeculosus* transmit BTV infection to sheep, as previously shown for *C. soronensis* in experimental conditions [26], and even appear to promote BTV transmission and high viraemia compared to needle transmission. Conversely uninfected bites combined to needle transmission rather tended to reduce BTV viraemia compared to plain needle transmission.

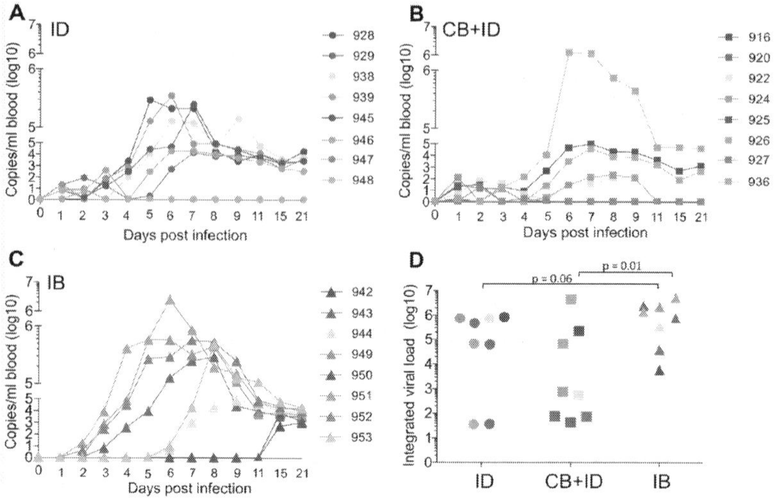

Figure 1. Measurement of viraemia in needle and *C. nubeculosous*- infected sheep. BTV viral RNA was detected in blood by qRT-PCR at 0–9, 11, 15, 21 dpi. The number of BTV copies per ml of blood is reported for each sheep with a specific color and symbol shape. The 3 infected group are represented by a distinct symbol, i.e. a circle for the intradermal needle- inoculation group (ID, **A**), a square for the control *C.* bite + intradermal needle- inoculation group (CB + ID, **B**), and triangle for the infected C. bite group (IB, **C**). In **D**, the viral load over the 21- day monitoring period was integrated for each sheep using a calculation of area under the curve (color and shape symbol as in A, B, C). Statistical comparisons between 2 groups were done with *a Mann-Whitney U* test and the p values are reported to show significance and major tendencies.

https://doi.org/10.1371/journal.pone.0083683.g001

The body temperature monitoring indicated that 5 sheep out of 8 in the ID group and 6 sheep out of 8 in the CB + ID group presented an early temperature rise at 1 dpi, that was not observed in any sheep of the IB group (Fig. 2A), nor in the control sheep. In the ID group, a clear temperature rise, that corresponded to an integrated increase of temperature >2°C, was observed between 5 and 9 dpi in 7 out of 8 sheep (Fig. 2A and B). Conversely only 3 sheep in the CB + ID group presented a body temperature rise (Fig. 2A and B). In the IB group, the temperature rise was observed in 6 of 8 sheep, from 6–11 dpi, starting a slightly later but lasting longer than in the

ID group (Fig. 2A and B). No temperature rise was observed in the control sheep. Clinical signs were observed starting from day 6 in all infected groups (Fig. 2C). The integrated clinical scores tended to be lower in the CB+ID group and were higher in the IB group compared to the scores in the ID group (p=0.07 and 0.04 respectively, Fig. 2D). The control sheep showed neither symptoms nor body temperature rise (data not shown). There was a significant correlation between the RNA viral load measured in Fig. 1 and the clinical symptom scores reported in Fig. 2 . Altogether, these data show that infected bite transmission tended to increase the clinical symptoms severity of BTV infection, compared to needle transmission. Conversely, uninfected bites reduced the clinical disease and temperature rise induced by needle transmission.

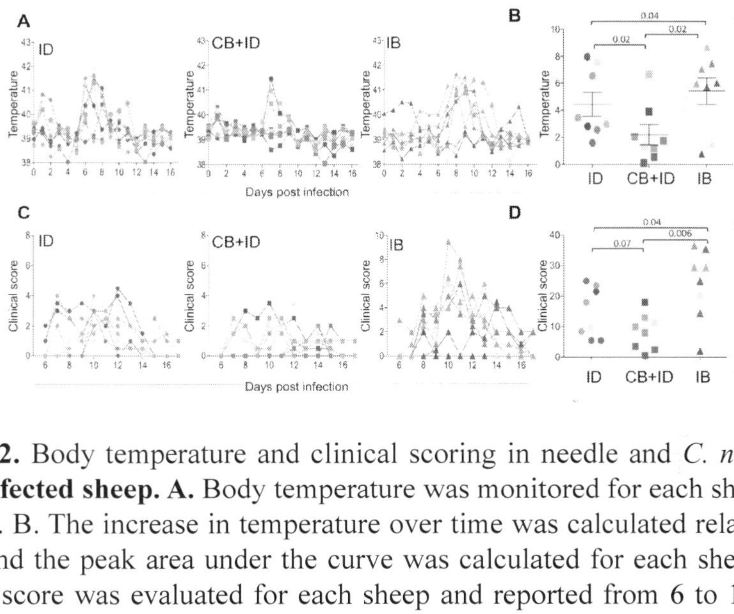

Figure 2. Body temperature and clinical scoring in needle and *C. nubeculosous*- **infected sheep. A.** Body temperature was monitored for each sheep over 17 days. B. The increase in temperature over time was calculated relatively to day 0 and the peak area under the curve was calculated for each sheep. **C.** A clinical score was evaluated for each sheep and reported from 6 to 17 dpi in B. **D.** The integrated area under the curve of the clinical scores from 6 to 17 dpi was calculated for each sheep. The same color and shape are used for each sheep as in Fig. 1. Statistical comparisons between 2 groups were done with *a Mann-Whitney U* test and the p values are reported to show significance and major tendencies.

https://doi.org/10.1371/journal.pone.0083683.g002

The Intensity of Local Inflammatory Reaction in uninfected *C. nubeculosus* bitten Sheep is Associated to Low Viral Load

During the course of the clinical monitoring, we observed that many sheep developed inflammatory reactions at the inoculation points (Fig. 3A). The highest global local reaction, measured by the integrated local inflammatory score over the 0–7 dpi period, was encountered in the CB + ID group, and the intensity of the local reaction was independent of the bite numbers. The ID and IB groups also presented significant local inflammatory reactions compared to the control group inoculated with medium (Fig. S2A). Interestingly in the CB + ID group, the sheep that showed the highest local reaction (integrated inflammatory score above 17) presented the lowest viral loads, below 3 log10 copies/ml (Fig. 3C). Even in the IB group in which the local inflammatory reaction was weaker than in the CB + ID group, the sheep with the lowest inflammatory reaction (integrated inflammatory score below 5) had the highest viral loads, reaching >6 log10 copies/ml (Fig. 3D). In the ID group, the intensity of local inflammation was very weak and could not be related to the viraemia level. These clinical observation data suggest that the uninfected *C. nubeculosus* plus the intradermal BTV8 injection potentiate each other to trigger strong local inflammation, preventing BTV viraemia. These data also lead to propose the hypothesis that the local inflammation, triggered by *Culicoides* skin laceration, may have a negative effect on BTV viraemia onset.

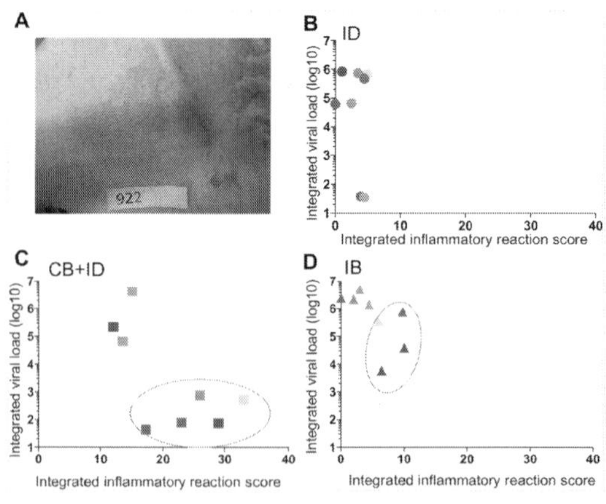

Figure 3. An elevated local inflammatory reaction appears associated to lower viraemia in *C. nubeculosous*- **bitten sheep. A.** A local inflammation at the inoculation point was observed especially in the CB + ID group, with an example of sheep 922 (score=3). **B, C, D.** The local inflammation of the

inoculation points was monitored from 0 to 7 dpi (the scores from the right and left inguinal sides were added) and the calculated area under the curve for each sheep is reported with the same color and shape symbol as in Fig. 1. B: ID group, C: CB + ID group, D: IB group. The sheep with higher inflammation and lower viral loads in the CB + ID and IB groups are circled.

https://doi.org/10.1371/journal.pone.0083683.g003

The Systemic IFN- Induced Gene Response is delayed in sheep bitten by infected C. *nubeculosus* compared to Needle- Infected Sheep

Expression of several innate immunity genes was detected in the blood cells of the control and infected sheep at 0–5 dpi (Fig. S3). There was no consistent induction of IL-1β, TNFα, XCL1, and CCL5 mRNA in infected versus control sheep. In the CB + ID group, the IL-8 mRNA expression was enhanced in sheep 920 and 925 (8 fold) and the IL-6 mRNA expression was enhanced in sheep 922 (5 fold), these sheep showing low to no viraemia and also high local inflammatory reaction; these cytokine genes were not induced to the same extent in the other sheep (Fig. S3). Sheep 953 (IB group) appeared as an over-reacting sheep, as it was the only sheep showing upregulation of most cytokine genes in addition to high fever. Interestingly, 2 IFN- induced gene mRNA, CXCL10 and MX1, were augmented at 1 or 2 dpi compared to day 0 in the 2 needle- inoculated groups (ID and CB + ID, Fig. 4) whereas in the IB group, the CXCL10 and MX1 mRNA levels were not augmented at early time points but raised at day 4 in sheep 951, 949 and 952 after the onset of viraemia. The increase in CXCL10 at day one was significantly higher in the ID and CB + ID groups than in the IB group ($p=0.001$ and $p=0.0009$ respectively). The increase in MX1 at day 1 was significantly higher in the ID and CB + ID groups than in the IB group ($p=0.02$ and $p=0.01$, respectively). Uninfected *C. nubeculosus* bite did not affect IFN- induced gene induction (data not shown). Unfortunately, type I IFN could not be detected in the serum of the infected sheep by using a biological assay, possibly due to too low amounts or inappropriate detection time points. These data indicate that IFN- induced gene induction is delayed when BTV8 is transmitted by infected *Culicoides* compared to needle transmission.

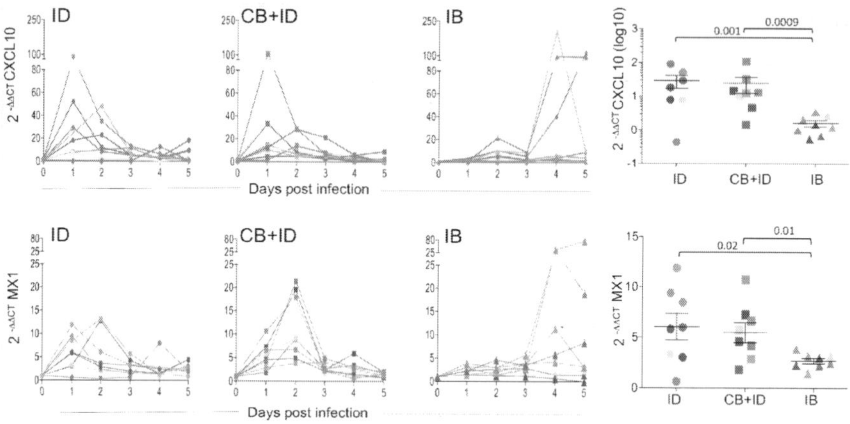

Figure 4. IFN- induced gene expression in blood cells is delayed in *C. nubeculosous-* **infected sheep.** Total RNA was extracted from blood cells of each sheep from 0–5 dpi. qPCR was performed on reverse transcribed RNA using the primers for CXCL10 and MX1 genes. The $2^{-\Delta\Delta CT}$ method was used to calculate the cytokine gene expression relatively the T0 time point level using the ribosomal RPS24 RNA as an internal control. The blood cells from control and control CB group did not show any induction of CXCL10 and MX1 gene expression during the observation period (not shown). The same color and shape symbol are used as in Fig. 1. Statistical comparisons between 2 groups were done with *a Mann-Whitney U* test and the p values are reported to show significance.

https://doi.org/10.1371/journal.pone.0083683.g004

The Humoral anti-BTV Response is delayed in sheep bitten by infected *C. nubeculosus* Compared to Needle- Infected Sheep

The anti-BTV antibody response in sera was assessed by anti-VP7 antibody detection using competition ELISA and by neutralizing antibody titration (Fig. 5). Anti-VP7 antibodies were detected as soon as by 5 dpi in the needle-infected sheep and plateaued at 8 dpi in these 2 groups, independently of the viraemia level. Conversely the anti-VP7 serological response was barely detectable at 8 dpi in the IB group and reached plateau values by 15 dpi. The neutralizing antibodies were already detectable at 8 dpi in all the needle-inoculated groups whereas it was not detectable in the IB group. Similar levels of neutralizing antibodies were reached at 15 dpi in all infected groups, independently of the viraemia level. These serological analyses show that the adaptive response to BTV occurred later in the IB group than in the ID and CB + ID groups.

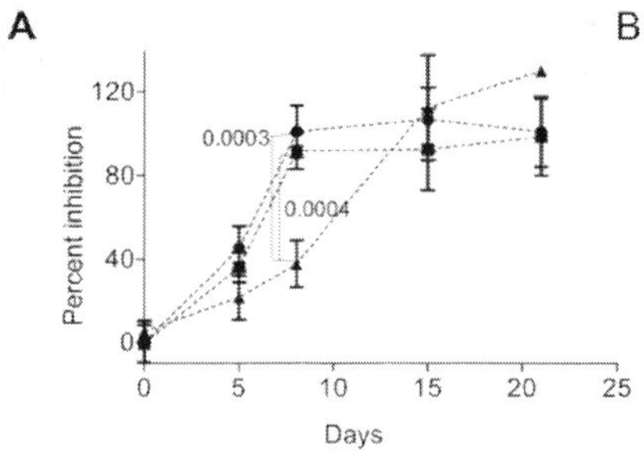

Figure 5. Anti-VP7 and neutralizing antibody response in serum is delayed in *C. nubeculosous-* **infected sheep.** A. Anti-VP7 antibodies were detected by competitive ELISA at 0, 5, 8, 15 and 21 dpi. **B.** Neutralizing antibodies were assessed at 0, 5, 8, 15 and 21 dpi in BHK21 cells by serum neutralization with 50 $TCID_{50}$ BTV8 (triplicates per sheep and per day). Statistical differences between groups were calculated by *a Mann-Whitney U* test on each day and the p values are reported for day 8.

https://doi.org/10.1371/journal.pone.0083683.g005

Altogether our results suggest that the systemic innate and adaptive responses to BTV are delayed when BTV is transmitted by infected *C. nubeculosus* bites compared to needle administration, corresponding to higher viremia and more severe clinical disease.

DISCUSSION

We showed here that *C. nubeculosus* bites impacted on the infection of sheep by BTV. Uninfected bites prevented the onset of viraemia and occurrence of symptoms in the majority of the sheep that were needle- inoculated with BTV8, possibly related to the intensity of local inflammatory response. Conversely infected *C. nubeculosus* bites promoted viraemia and disease compared to needle transmission. Both systemic innate and adaptive host responses were delayed in the sheep exposed to infected *C. nubeculosus* bites compared to the needle- inoculated sheep. These data suggest that infected *Culicoides* bites are the optimal way for BTV transmission, by avoiding the development of an early host defense responses.

Inoculation of BTV8 in the *Culicoides* bitten sites rather suppressed the development of viraemia and disease in most sheep, in contrast with Cache Valley virus and West Nile virus administration in the feeding site of mosquitoe that promoted viral infection [10], [11]. *Culicoides* lacerate the skin and trigger local inflammation that was further amplified by the subsequent needle transmission of the virus. In this certainly artificial infection method, we observed that the sheep with the highest local inflammatory response had the lowest viral RNA load in the blood, indicating that the local inflammation may prevent systemic infection, viraemia and disease. In addition, in some of these low viraemic sheep, the IL-6 and IL-8 gene expression was upregulated at early time points in the blood cells, suggesting occurrence of systemic inflammatory response. It would have been interesting to test the cytokine response within the skin lesions but we favored avoiding possible artifacts induced by taking biopsies. Besides, the sheep of the CB + ID group did all develop anti-VP7 and neutralizing antibody responses, suggesting that sufficient viral amount was detected by the host immune system to trigger high humoral response. It is possible that the local inflammation confined BTV8 to the skin, and prevented dissemination to the blood stream while still allowing immune response onset. The inhibitory action of the control *C. nubeculosus* bite may be explained by the composition of the uninfected *Culicoides* saliva that may trigger inflammation, further potentiated by the intradermal BTV inoculum. However collection of sufficient saliva from this small insect was not feasible in our experimental set up for directly evaluating the effect of the saliva *in vivo*. It remains possible that the local inflammation induced by *C. nubeculosus* corresponds to a delayed type hypersensitivity reaction (DTH), resulting from a previous exposure of the sheep *to Culicoides* bites before the experiment in BSL3. Indeed in the case of *Leishmania* transmission by sandflies, DTH to insect salivary components was shown to be detrimental to *Leishmania* establishment [27]. In the case of *Culicoides*, DTH reactions have been detected in horses [28], but not in ruminants to our knowledge. In our setting, cautions were taken to avoid exposure to midges during previous housing (insecticide treatment), but midge antigen- directed DTH can not be formally ruled out. In any event, the extent of the local inflammation triggered by uninfected *C. nubeculosus* bites varied between the different sheep and may be brought as an hypothesis for the differential individual receptivity to BTV viraemia. Finally in the IB group, there was also a trend for lower viral load in the sheep with the highest local inflammation, suggesting that individuals with high inflammatory response to *Culicoides* bites may be less susceptible to development of BTV viraemia.

The high efficacy of infected *C. nubeculosus* compared to needle administration to promote high viraemia and more pronounced clinical symptoms is striking. Many factors differ between needle and insect administration that can interfere on each other and promote viral transmission. BTV processed by the insect cells and further modified in the saliva may present biochemical alterations compared to BTV produced in mammalian cells *in vitro*, favoring infection of the sheep. Indeed the VP2 protein from 3 different BTV strains was shown to be cleaved by treatment with saliva from adult *C. soronensis in vitro* [17]. The ratio of defective particles versus infectious particles may be different between the virus produced by whole insect and mammalian cell cultures, thus affecting the balance and kinetics of immunity and infection. As uninfected *C. nubeculosus* displayed an inhibitory effect on viral transmission, it remains possible that uninfected and infected *Culicoides* saliva have different composition. Indeed the salivary gland function and virus transmission efficiency changed during the course of WNV infection due to pathological changes in the mosquito [29], resulting in a differential salivary gland transcript profile [30]. In that scenario, infected *C. nubeculosus* saliva may modulate the host response, promote the observed delay in IFN- gene and antibody responses and prevent the early peak of body temperature (Fig. 2, 4 and 5). Indeed in the case of vesicular stomatitis New Jersey virus [31], insect saliva was shown to modulate type I IFN production and to affect host resistance to the virus. Finally the mode of virus inoculation by *Culicoides* may be favorable to hit or recruit the best cells for viral transmission and dissemination.

The viral doses received by the needle- and infected *Culicoides*-inoculated sheep were most probably very different. It is generally believed that the higher the dose, the shorter the incubation period and the time from inoculation to onset of viraemia and clinical signs [32], [33], [34]. In the case of BTV infection in cattle and sheep, the size of the inoculum did not seem to influence the kinetics of viraemia [35], [36], [37]. We used here a high dose of virus for challenge, representing a reliable challenge, as done previously in other studies [16], [25]. Whereas it was expected that the lower viral inoculum delivered by the infected *Culicoides* would have been an handicap for efficient infection, it is possible that on the contrary, the high viral load in the needle- inoculum may have induced a rapid counteracting immunity that interfered with the infection efficacy. A escalating dose of BTV inoculum in parallel with evaluation of the host immunity response may help to solve this issue.

Altogether our study indicates that infected *C. nubeculosus* are more efficient than the needle- inoculation of high viral dose of BTV to induce high viraemia and BT disease in sheep. *C. nubeculosus* bites had a dual effect in our experiment, with infected bites promoting infection, and uninfected bites inhibiting infection by needle- inoculation. These results should be extended to other *Culicoides* species in order to be generalized. However our results unravel that needle- inoculation of high viral load induce the rapid onset of systemic innate and adaptive immune responses, whereby probably inhibiting the viral infection, and modifying the physio-pathological steps of BTV infection occurring in the natural infection. Uninfected bites appeared to enhance the local inflammation and further impede the infection by needle- inoculated BTV. Comparatively, infected *Culicoides* bites induced delayed immune responses that were detected after the onset of viraemia. The factors implicated in the efficacy of viral transmission by infected *Culicoides*, which are possibly related to the insect effect on the virus and/or on the host, remain to be identified. The understanding of the midge/host/virus interaction will result in the development of pertinent and novel control strategies. Indeed the insect could be targeted by vaccines, such as in the case of the SAP1 protein of sand fly that is considered as a promising target against their pathogen transmission [8]. In addition, inoculation of BTV by infected *Culicoides* stands as an optimal and reliable way to evaluate the physiopathological mechanisms of BTV infection and to test BTV targeted vaccine strategies, that may pertain to the other viruses such as Schmallenberg and African Horse Sickness viruses that are transmitted by *Culicoides* midges.

ACKNOWLEDGMENTS

We are grateful to IAH Pirbright for providing the *C. nubeculosus* individuals originating a permanent colony at CReSA facilities, and to Elena Ibáñez and Anna Mayol (UAB) for their help in microcapillary preparation. We are very grateful to CReSA personnel for their excellence in the animal care and experimental conduct. We are grateful to Claudia Bevilacqua and Lucas Lefevre (platform ICE, Jouy-en-Josas) for their help in blood RNA extraction. We warmly thank Bertrand Schwartz, Pauline Maisonnasse and Frédérick Arnaud for critical reading of the manuscript.

AUTHOR CONTRIBUTIONS

Conceived and designed the experiments: NP SZ AB JP ISC. Performed the experiments: NP EB CU ST CV CL BC DS JP ISC. Analyzed the data: NP EB CU CL AB JP LJ ISC. Wrote the paper: ISC NP JP EB.

REFERENCES

1. Mercer DR, Castillo-Pizango MJ (2005) Changes in relative species compositions of biting midges (Diptera: Ceratopogonidae) and an outbreak of Oropouche virus in Iquitos, Peru. Journal of medical entomology 42: 554–558.
2. Maclachlan NJ, Guthrie AJ (2010) Re-emergence of bluetongue, African horse sickness, and other orbivirus diseases. Veterinary research 41: 35.
3. .Bishop JV, Mejia JS, Perez de Leon AA, Tabachnick WJ, Titus RG (2006) Salivary gland extracts of Culicoides sonorensis inhibit murine lymphocyte proliferation and no production by macrophages. Am J Trop Med Hyg 75: 532–536.
4. Schwartz-Cornil I, Mertens PP, Contreras V, Hemati B, Pascale F, et al. (2008) Bluetongue virus: virology, pathogenesis and immunity. Vet Res 39: 46.
5. Beer M, Conraths FJ, van der Poel WH (2013) ‘Schmallenberg virus’–a novel orthobunyavirus emerging in Europe. Epidemiology and infection 141: 1–8.
6. Maan NS, Maan S, Belaganahalli MN, Ostlund EN, Johnson DJ, et al. (2012) Identification and differentiation of the twenty six bluetongue virus serotypes by RT-PCR amplification of the serotype-specific genome segment 2. PloS one 7: e32601.
7. Wilson AJ, Mellor PS (2009) Bluetongue in Europe: past, present and future. Philos Trans R Soc Lond B Biol Sci 364: 2669–2681.
8. Fontaine A, Diouf I, Bakkali N, Misse D, Pages F, et al. (2011) Implication of haematophagous arthropod salivary proteins in host-vector interactions. Parasites & vectors 4: 187.
9. Schneider BS, Higgs S (2008) The enhancement of arbovirus transmission and disease by mosquito saliva is associated with modulation of the host immune response. Trans R Soc Trop Med Hyg 102: 400–408.
10. Schneider BS, Soong L, Girard YA, Campbell G, Mason P, et al. (2006) Potentiation of West Nile encephalitis by mosquito feeding. Viral immunology 19: 74–82.
11. Edwards JF, Higgs S, Beaty BJ (1998) Mosquito feeding-induced enhancement of Cache Valley Virus (Bunyaviridae) infection in mice. J Med Entomol 35: 261–265.

12. Reisen WK, Chiles RE, Kramer LD, Martinez VM, Eldridge BF (2000) Method of infection does not alter response of chicks and house finches to western equine encephalomyelitis and St. Louis encephalitis viruses. Journal of medical entomology 37: 250–258.
13. Perez de Leon AA, Ribeiro JM, Tabachnick WJ, Valenzuela JG (1997) Identification of a salivary vasodilator in the primary North American vector of bluetongue viruses, Culicoides variipennis. The American journal of tropical medicine and hygiene 57: 375–381.
14. Perez de Leon AA, Valenzuela JG, Tabachnick WJ (1998) Anticoagulant activity in salivary glands of the insect vector Culicoides variipennis sonorensis by an inhibitor of factor Xa. Experimental parasitology 88: 121–130.
15. Takamatsu H, Mellor PS, Mertens PP, Kirkham PA, Burroughs JN, et al. (2003) A possible overwintering mechanism for bluetongue virus in the absence of the insect vector. J Gen Virol 84: 227–235.
16. Moulin V, Noordegraaf CV, Makoschey B, van der Sluijs M, Veronesi E, et al. (2012) Clinical disease in sheep caused by bluetongue virus serotype 8, and prevention by an inactivated vaccine. Vaccine 30: 2228–2235.
17. Darpel KE, Langner KF, Nimtz M, Anthony SJ, Brownlie J, et al. (2011) Saliva proteins of vector Culicoides modify structure and infectivity of bluetongue virus particles. PLoS One 6: e17545.
18. Boorman J (1974) Maintenance of laboratory colonies of Culicoides variipennis (Coq), Culicoides nubeculosus (Mg) and Culicoides riethi Kieff. (Diptera: Ceratopogionidae). Bulletin of Entomological Research 64: 371–377.
19. Jennings DM, Mellor PS (1988) The vector potential of British Culicoides species for bluetongue virus. Veterinary microbiology 17: 1–10.
20. Hemati B, Contreras V, Urien C, Bonneau M, Takamatsu HH, et al. (2009) Bluetongue virus targets conventional dendritic cells in skin lymph. J Virol 83: 8789–8799.
21. Huismans H, van der Walt NT, Cloete M, Erasmus BJ (1987) Isolation of a capsid protein of bluetongue virus that induces a protective immune response in sheep. Virology 157: 172–179.
22. Darpel KE, Batten CA, Veronesi E, Shaw AE, Anthony S, et al. (2007) Clinical signs and pathology shown by British sheep and cattle infected

with bluetongue virus serotype 8 derived from the 2006 outbreak in northern Europe. Vet Rec 161: 253–261.

23. Attoui H, Billoir F, Biagini P, de Micco P, de Lamballerie X (2000) Complete sequence determination and genetic analysis of Banna virus and Kadipiro virus: proposal for assignment to a new genus (Seadornavirus) within the family Reoviridae. The Journal of general virology 81: 1507–1515.
24. Stewart M, Bhatia Y, Athmaran TN, Noad R, Gastaldi C, et al. (2010) Validation of a novel approach for the rapid production of immunogenic virus-like particles for bluetongue virus. Vaccine 28: 3047–3054.
25. .Stewart M, Dubois E, Sailleau C, Breard E, Viarouge C, et al. (2013) Bluetongue virus serotype 8 virus-like particles protect sheep against virulent virus infection as a single or multi-serotype cocktail immunogen. Vaccine 31: 553–558.
26. .Baylis M, O'Connell L, Mellor PS (2008) Rates of bluetongue virus transmission between Culicoides sonorensis and sheep. Med Vet Entomol 22: 228–237.
27. Andrade BB, Teixeira CR (2012) Biomarkers for exposure to sand flies bites as tools to aid control of leishmaniasis. Frontiers in immunology 3: 121.
28. Langner KF, Darpel KE, Drolet BS, Fischer A, Hampel S, et al. (2008) Comparison of cellular and humoral immunoassays for the assessment of summer eczema in horses. Veterinary immunology and immunopathology 122: 126–137.
29. .Girard YA, Schneider BS, McGee CE, Wen J, Han VC, et al. (2007) Salivary gland morphology and virus transmission during long-term cytopathologic West Nile virus infection in Culex mosquitoes. The American journal of tropical medicine and hygiene 76: 118–128.
30. Girard YA, Mayhew GF, Fuchs JF, Li H, Schneider BS, et al. (2010) Transcriptome changes in Culex quinquefasciatus (Diptera: Culicidae) salivary glands during West Nile virus infection. Journal of medical entomology 47: 421–435.
31. .Limesand KH, Higgs S, Pearson LD, Beaty BJ (2003) Effect of mosquito salivary gland treatment on vesicular stomatitis New Jersey virus replication and interferon alpha/beta expression in vitro. J Med Entomol 40: 199–205.

32. .Alexandersen S, Quan M, Murphy C, Knight J, Zhang Z (2003) Studies of quantitative parameters of virus excretion and transmission in pigs and cattle experimentally infected with foot-and-mouth disease virus. Journal of comparative pathology 129: 268–282.

33. .Howey R, Quan M, Savill NJ, Matthews L, Alexandersen S, et al. (2009) Effect of the initial dose of foot-and-mouth disease virus on the early viral dynamics within pigs. Journal of the Royal Society, Interface/the Royal Society 6: 835–847.

34. .Quan M, Murphy CM, Zhang Z, Alexandersen S (2004) Determinants of early foot-and-mouth disease virus dynamics in pigs. Journal of comparative pathology 131: 294–307.

35. Di Gialleonardo L, Migliaccio P, Teodori L, Savini G (2011) The length of BTV-8 viraemia in cattle according to infection doses and diagnostic techniques. Research in veterinary science 91: 316–320.

36. Eschbaumer M, Wackerlin R, Rudolf M, Keller M, Konig P, et al. (2010) Infectious blood or culture-grown virus: a comparison of bluetongue virus challenge models. Veterinary microbiology 146: 150–154.

37. Martinelle L, Dal Pozzo F, Sarradin P, De Leeuw I, De Clercq K, et al. (2011) Two alternative inocula to reproduce bluetongue virus serotype 8 disease in calves. Vaccine 29: 3600–3609

16

A Bayesian Hierarchical Model for Estimation of Abundance and Spatial Density of Aedes aegypti

Daniel A. M. Villela[1], Claudia T. Codeço[1], Felipe Figueiredo[1], Gabriela A. Garcia[2], Rafael Maciel-de-Freitas[2], Claudio J. Struchiner[1]

[1]Fundação Oswaldo Cruz, Programa de Computação Científica—Rio de Janeiro, Brazil, [2]Fundação Oswaldo Cruz, Departamento de Entomologia, Laboratório de Transmissores de Hematozoários—Rio de Janeiro, Brazil

ABSTRACT

Strategies to minimize dengue transmission commonly rely on vector control, which aims to maintain *Ae. aegypti* density below a theoretical threshold. Mosquito abundance is traditionally estimated from mark-release-recapture (MRR) experiments, which lack proper analysis regarding

Citation: Villela DAM, Codeço CT, Figueiredo F, Garcia GA, Maciel-de-Freitas R, Struchiner CJ (2015) A Bayesian Hierarchical Model for Estimation of Abundance and Spatial Density of Aedes aegypti. PLoS ONE 10(4): e0123794. https://doi.org/10.1371/journal.pone.0123794

accurate vector spatial distribution and population density. Recently proposed strategies to control vector-borne diseases involve replacing the susceptible wild population by genetically modified individuals' refractory to the infection by the pathogen. Accurate measurements of mosquito abundance in time and space are required to optimize the success of such interventions. In this paper, we present a hierarchical probabilistic model for the estimation of population abundance and spatial distribution from typical mosquito MRR experiments, with direct application to the planning of these new control strategies. We perform a Bayesian analysis using the model and data from two MRR experiments performed in a neighborhood of Rio de Janeiro, Brazil, during both low- and high-dengue transmission seasons. The hierarchical model indicates that mosquito spatial distribution is clustered during the winter (0.99 mosquitoes/premise 95% CI: 0.80–1.23) and more homogeneous during the high abundance period (5.2 mosquitoes/premise 95% CI: 4.3–5.9). The hierarchical model also performed better than the commonly used Fisher-Ford's method, when using simulated data. The proposed model provides a formal treatment of the sources of uncertainty associated with the estimation of mosquito abundance imposed by the sampling design. Our approach is useful in strategies such as population suppression or the displacement of wild vector populations by refractory *Wolbachia*-infected mosquitoes, since the invasion dynamics have been shown to follow threshold conditions dictated by mosquito abundance. The presence of spatially distributed abundance hotspots is also formally addressed under this modeling framework and its knowledge deemed crucial to predict the fate of transmission control strategies based on the replacement of vector populations.

INTRODUCTION

Dengue fever is the most prevalent arbovirus infection in the world with 2.5 billion people living in areas under the risk of transmission [1]. The temporal and spatial pattern of dengue distribution is influenced by mosquito, human, viral and environmental factors. The abundance, survival and dispersal of its vector, the *Aedes aegypti* mosquito, are important factors to describe the ecology involving the mosquito. In particular, spatial distributions of mosquito populations at fine scales [2, 3] can help understand the impact on dengue transmission. Estimation of parameters that describe these components is still a complex challenge for field entomologists and modelers [4–7].

Knowledge about these key components is crucial [8] in guiding vector control strategies based on population suppression in areas at transmission risk. Several dengue endemic countries, including Brazil, plan their vector control strategies based on the assessment of infestation indices through larval surveys, most commonly House and Breteau Indices (HI and BI, respectively) [9, 10]. These traditional infestation indices show low correlation with adult mosquito abundance, as HI and BI do not consider container productivity and larval mortality [10, 11]. Alternatively, trapping of adults with a variety of devices has been proposed as a more efficient approach to monitor *Aedes aegypti* populations and many initiatives are already in place worldwide. Traps allow the development of more standardized protocols, provide indices faster and require less effort than the traditional searching approach. One drawback of a trap based infestation index, however, is that it is a relative measure of population density, with unit mosquito/trap. For comparative purposes, this may suffice. But there are situations when absolute measures of population abundance (unit: mosquitoes/area or mosquitoes/person) are of interest. For example, vector thresholds for transmission are defined in terms of mosquito/person [8]; population thresholds for *Aedes aegypti* + *Wolbachia* invasion is defined in mosquito/area [12, 13].

Traditionally, the estimation of animal population size is performed via experiments of mark, release and recapture (MRR). In MRR experiments, subjects are typically captured from the environment, marked either uniquely or as a cohort using a piece of identification (tags, colors etc.), released back into the environment and later recaptured, possibly multiple times. For this kind of experiment, mosquitoes are challenging subjects due to their small size and short life span, making recapturing difficult, which led to important modifications of the standard protocols, mainly, that marked mosquitoes are released and subsequently recaptured only once [7]. Models for estimating adult population size from MRR datasets, either deterministic or stochastic ones, include Lincoln, Jolly-Seber, and Fisher and Ford [6, 7, 14–16]. The Lincoln Index, due to its simplicity, is often the method of choice. However this simplicity comes at a price, since the strong assumptions required for the proper utilization of this index are difficult to meet in most field conditions, for example, that the population is closed and that there is no heterogeneity in capture rates. The Fisher—Ford method relaxes some of these assumptions allowing the loss of individuals by mortality. These models are not probabilistic and do not treat sampling uncertainty properly. The Jolly-Seber family of models is very popular due to their probabilistic framework, however they are tailored for data with multiple recaptures and fitting with typical mosquito data does not show convergence (results not shown). In this work we investigate the potential advantages to analyze MRR mosquito trials of a class of stochastic models containing an ecological component introduced by Royle *et al.* [17]. Stochastic models involve more complex mathematical calculations, but generate estimates of population size along with measures of uncertainty [7]. The model presented here was developed upon the basic structure of the model of Royle *et al.* [17] to accommodate particular aspects of the mosquito ecology and its observation. We perform Bayesian analysis using this model to estimate abundance and spatial distribution of the population of female *Aedes aegypti* in a dengue endemic area in the city of Rio de Janeiro, Brazil. The proposed model is also tested with artificial data from a simulator of typical mosquitoes' MRR experiments.

In the next section we describe the MRR experiments carried out in the study area in Rio de Janeiro, Brazil. The description of the experiments is helpful here as it illustrates the particular aspects of a typical MRR dataset for mosquitoes. We then describe the structure of the hierarchical model, its layered components and the simulation environment. We present estimates of abundance, spatial density and survival probability of *Aedes aegypti* in the study area comparing the estimates from the proposed hierarchical model with those from the Fisher—Ford method. Subsequently, we show some statistical properties of the proposed model obtained from its application to simulated data.

METHODS

Mark—Release—Recapture Experiments

- ***Study area.*** We conducted mark-release-recapture studies in a 7.2 ha neighborhood called Z—10, located in the Governador Island, city of Rio de Janeiro, Brazil. Z—10 (22°52′30″*S*; 43°14′53″*W*) is an isolated suburban community, surrounded by shores and mangroves, which discourage mosquito migration through its limits. This residential area has paved streets, regular sidewalks and low-moderate vegetation coverage, with around 2,350 people living in 787 houses. Most houses have 2–3 bedrooms, regular water supply and garbage collection, lacking peridomestic areas with pools or yards.

- ***Climate and MRR periodicity.*** The climate in Rio de Janeiro is characterized by a moderately dry winter (May-August) and a wet summer season (November—March), with mean temperatures of 25.1°C and 28.8°C and mean total rainfall of 46.4 mm and 132 mm, respectively. MRR experiments were conducted during nine consecutive days in September 2012 and during another period of nine days in March 2013. In Sep/2012, temperatures ranged between 22.3°C and 30.6°C (average of 28.6°C) with no rainfall. In Mar/2013, temperatures ranged from 23.1°C to 32.5°C, with an average of 29.7°C and rainfall of 0.4 mm. Air temperature and precipitation data were obtained from a meteorological station located at approximately 5 km away from the study area.
- ***Mosquitoes.*** *Aedes aegypti* adults used in MRR experiments were derived from the F1 generation of eggs collected at Z—10 using 60 ovitraps filled with hay infusion. Larvae were fed with fish food (Tetramin, Tetra Sales, Blacksburg, VA) and reared according to Consoli and Lourenço-de-Oliveira [18]. After emergence, females were kept together at 25±3°C and 65±5% relative humidity (RH) and fed with sucrose solution until the time to release.
- ***Marking and releasing.*** Before releasing in Sept/2012, *Ae. aegypti* adults were split into four cohorts, each one composed of 500 females marked with different colors of fluorescent dust (Day-Glo Color Corp., Cleveland, OH) and placed in small cylindrical cups (12 cm x 10 cm). Each cohort was released from a different outdoor location in Z—10. In Mar/2013, only one cohort composed of 2000 females was released in a central outdoor location of Z—10. In both occasions, mated, unfed females (4—day old) were released in the morning hours (between 8:00 AM and 9:00 AM), at approximately 1 hour after dust marking. Wind direction and speed was 139.7° and 4.7 m/s in the moment of release in Sep/2012. Meanwhile, in Mar/2013, wind direction and speed was 25.1° and 2.6 m/s.
- ***Capturing.*** Dust-marked females were captured using 66 uniformly distributed BG-Sentinels traps (BioGents, Regensburg, Germany), designed to attract mosquitoes seeking a host to blood feed [19, 20]. Captures started at the following day after

mosquitoes' release and was performed daily by inspection of the 66 BGs-Traps. Daily capturing stopped when no dust-marked females were collected for 3 consecutive days. Captured mosquitoes were examined under UV light to check for the presence of fluorescent dust.

Ethics Statement

Fiocruz conducts regular research activities in the city of Rio de Janeiro, in partnership and also with permission of the Rio de Janeiro Department of Health (*Secretaria de Saúde da cidade do Rio de Janeiro*) that assist in the control of transmission of infectious diseases. Mark-release-recapture experiments were approved by Fiocruz Ethical Committee (CEP 253/04) and carried out as part of this research effort, not requiring a specific permission. Since mosquitoes are released in public open spaces such as squares, there was no need to obtain individual consent for releases. In order to recapture marked mosquitoes, inside people's dwellings, we have to obtain consent from residents to install BG-Traps and also to carry daily collections after the release of marked individuals. The use of mosquito traps for sampling *Aedes aegypti* belongs to the routine of vector surveillance in Rio de Janeiro. Due to this surveillance routine, many residents in the study area are used to have mosquito traps installed in their houses and, indeed, there were just few refusals by residents to have traps installed. Written consent was not required under the ethical committee requirements. As the effort to collect mosquitoes goes along with city surveillance, these tasks are typically done with verbal consent only and our collection effort also had verbal consent, as approved by the ethics committee, and individual consents were registered by the entomology field team in regular work files.

The release of mosquitoes does not involve directly endangered or protected species and, from our experience, it does not have any significant impact on endangered or protected species. One of our main concerns during MRR experiments is related to dengue transmission by the mosquitoes released in the field. In order to address this issue, we follow two guidelines. The first one is to not release a number of mosquitoes greater than the number of mosquitoes removed by trapping in pilot experiments and monitoring activities. Second, the procedure is to interrupt or to suspend, depending on circumstances, mosquito releases if dengue cases are notified in that neighborhood. Since both trap inspection and mosquito collection are done with the assistance of the health department in the city government, i.e., the same agency that is responsible for dengue notification, dengue cases are rapidly identified.

Hierarchical Model

The proposed model has three components. The first component is a probabilistic model describing the spatial distribution of mosquitoes in the study area, while the second component is also a probabilistic model describing the daily survival of marked and native mosquitoes. Finally, the observation model describes the sampling process.

- ***Ecological process—center of activities.*** There is some evidence that mosquitoes tend to remain close to their birth location if conditions for blood feeding and ovipositing are adequate. Specialists typically agree that *Ae. aegypti* females rarely visit more than 2–3 houses during their lifetime leading to a strong spatial correlation between the distribution of immature and adults [21]. Harrington *et al.* [4] reviewed several MRR experiments carried out in different locations, different release sites (both indoors and outdoors) and experimental protocols. Overall, these studies agree on the limited dispersal behavior of *Aedes aegypti*. When release is done indoors, most of the collections occur in the same house [22, 23]. This tendency to remain in the same location (at least during the length of the experiment) can be formalized by the concept of center of activity. An individual center of activity is a central point (centroid) of the space occupied by the individual during a time interval [24]. This concept is borrowed from the works of Royle *et al.* [17, 25], in which they model the movement of tigers [17] and birds [25].

Here, there are two important claims. First, we assume that for the duration of the MRR experiments spatial density of mosquitoes can be described by the density of activity centers, as mosquitoes tend to stay within a constrained space. This is reasonable for both marked and unmarked individuals given the empirical observations as described above. Second, the marked mosquitoes are released and go through a dispersal phase. Here, we have evidence that the dispersal is fast and subsequent captures of individuals will find them already in the whereabouts of their center of activities. In the MRR experiment studied here, marked cohorts were released outdoors, in locations that stimulate dispersal towards more suitable habitats. Under stress mosquitoes can fly more than 600 m for three days to find suitable conditions [26]. Evidence for the ability to reach the whole area comes from the observation that in the first day of capturing, marked mosquitoes were already found in the most distant traps. Therefore, we distinguish between the short distance movements close to centers of activity and the long range movements realized when dispersal happens.

In the model the center of activity is given by a pair $\mathbf{s} = (s_x, s_y)$ of coordinates that describe the center of an individual area. The prior distribution for these pairs of coordinates is generally a uniform distribution. After a procedure of inference, given the observations in the MRR experiments, the distribution of center of activities is effectively evaluated as a spatial density distribution within the study area.

- ***Ecological process—survival.*** *Aedes aegypti* life span ranges from 5 to 30 days [4, 27–31], making survival an important process to be considered in an experiment that lasts for 9 days. In the capture data (shown in the Results section), there is clearly a loss of marked individuals as days go by. This loss can be attributed to mortality by natural causes. In the hierarchical model, there is a component that describes daily survival probability ϕ, assumed equal across all marked individuals. For the unmarked individuals, however, under an assumption that abundance stays at a stable level for the relatively short length of an MRR study, we consider that recruitment cancels out mortality, *i.e.*, in terms of the model, = 1 for unmarked individuals.
- ***Observation process.*** In the model each trap has associated with it a probability of capturing mosquitoes as a function of the distance between the trap location, recorded in the experiment, and any point in the study area. Hence, each trap has a probability of capturing the mosquito as function of the distance between the trap location and its center of activity. Following the approach in [17], the probability to be attracted to a trap is described by a function that has parameters to be estimated as a *Generalized Linear Model* (GLM). We choose the complementary log—log function as a link function, thus we have for the probability $\varpi_{i,j}$ of individual i be attracted to trap j, $cloglog(\varpi_{i,j})|\mathbf{s}_i \sim \beta_0 + \beta_1 d_{i,j}$, where $d_{i,j}$ is the distance from individual i's center of activity and trap j's location, and β_0 and β_1 are parameters to be estimated.

The observation model takes into account that each individual can only be trapped once. This limitation impacts the model formulation in two main aspects. First, at each observation time point (intervals given in days and total time T) the probability of an individual being captured at a particular trap is a product of two factors: the probability $\varpi_{i,j}$ of individual i to be captured at trap j, given by a function of the distance from its center of activity to the trap, and the probability to be captured at trap j, given by a categorical random variable with parameter vector given by the ratio $\frac{\pi_i}{\sum_{j=1}^{J} \pi_{i,j}}$. Once a marked individual is captured, it is removed from the study. Therefore, it cannot be observed again. In the model, removal is introduced in the survival component as a variable that indicates presence in the study. For each individual, this presence depends on its survival and also on the individual not being captured. Captures of mosquitoes at traps are described by the observation variables $y_{i,j,t}$, indicating whether a mosquito i is observed ($y_{i,j,t} = 1$) or not ($y_{i,j,t} = 0$) at trap j, $j \le J$, at time t, $t \le T$.

We want to estimate the number N_2 of unmarked mosquitoes in the study area using the hierarchical model in which each individual is indexed, whereas the number N_1 of marked mosquitoes released in the study is known. Without any loss of generality, the first N_1 individuals are the marked ones. We use the data augmentation technique [32] as used by Royle and Dorazio in [33] to estimate the number of unmarked mosquitoes N_2. The technique involves taking a number $M > N_1 + N_2$ by adding non-existing individuals, since we do not know N_2. The non-existing individuals will not be present in the study and will not be captured (zero values in observation). This requires another layer in the model to treat the zero—inflated component, that distinguishes zeroes from zero—inflation to zeroes due to non-observations. This component uses a set of variables w_i, $1 \le i \le M$, $M > N_1 + N_2$, each of which describes whether a particular individual i is effectively in the study population (either marked or unmarked population). Finally, this technique permits us to have abundance as an indirect measure obtained from an inference procedure.

Estimation and Bayesian Inference

Inference is performed via analysis of Monte Carlo Markov Chain (MCMC) simulations. The abundance, *i.e.*, the number N_2 of unmarked individuals is estimated as a latent variable:

$$N_2 = \sum_{i=N_1+1}^{M} w_i.$$

(1)

Since the estimation of variables is performed using multiple runs of MCMC simulations, results are generally given by statistical measures that include mean, median, standard deviation and credibility intervals.

The spatial density of mosquitoes is found using the posterior samples from MCMC simulation and constructing a grid over the study area. The count of individuals in each of the grid cells is found by the sum of center of activities inside each of them, similar to counts performed by Royle *et al.* in [17].

In order to find estimates, we use JAGS [34], a Bayesian analysis tool that implements MCMC through Gibbs sampling and uses a model specification very similar to another popular tool, WinBUGS. The JAGS model is included in S1 Text. The prior distribution for survival probability ϕ, defined in the [0, 1] interval, places higher weight on values closer to one, while also weakening extreme values in the [0, 1]—interval. This distribution is in accordance with empirical estimates of the survival rates of *Aedes aegypti* [35].

- ***Fisher—Ford method.*** We also find estimates of abundance using Fisher—Ford method [36], which has been used in MRR analyses and is closely related to the Lincoln index (MRR on estimating abundance of *Aedes albopictus* by Cianci *et al.* [6] and *Anopheles gambiae* by Baber *et al.* [37]). The abundance estimator derives from the same argument of the Lincoln index, including an adjustment that takes into account the survival probability ϕ for marked individuals:

$$\hat{N}_2 = \frac{n_{tot}\phi^t N_1}{m_{tot}} - \phi^t N_1,$$

where n_{tot} and m_{tot} are, respectively, the numbers of unmarked and marked individuals, captured at time t.

Bailey in [38] propose a bias correction in the case of small number of captures (typically below 20): $\hat{N}_2 = \frac{(n_{tot}+1)\phi^t(N_1+1)}{m_{tot}+1} - \phi^t(N_1+1)$. We take a bootstrap approach in order to find a confidence interval for the Fisher—Ford index, using a number R_s of re—sampling.

- ***Simulation.*** To assess the efficiency of the MRR experiment in various scenarios, an Individual-based model (IBM) was developed to simulate the dispersal and capture of marked mosquito cohorts, as a computational tool, written in the Perl programming language that mimics the field experiments. Using this type of simulation environment it is possible to observe actual emergent phenomena from simple rules and it has been gaining attention in theoretical ecology since the last two decades [39–41].

The simulator is implemented by constructing classes that define objects for each agent, including agents for mosquitoes and traps, and it also defines a square area for the environment. For each mosquito, dispersal can be considered either random or towards a randomly chosen center of activity (sampled from a uniform distribution). Traps can be positioned at random locations, or at points comprising a centralized grid. In simulations we have the possibility of varying the area around which the trap attracts mosquitoes and also possibility to define the probability that describes a Bernoulli random variable to indicate mosquito capture at traps.

The output of the simulator is given by tables discriminating trap and mosquito data as well as a capture history file typical for MRR analysis programs, such as the program MARK [42]. Full descriptions for the objects, data structures and validation are provided. The code (open-source license—GNU PGL v3) is available from https://launchpad.net/mmrrsim.

RESULTS

Field data from MRR experiments

Table 1 shows the number of daily captures observed in each of the mark—release—recapture experiments, conducted in Sept 2012 (ST1) and March 2013 (ST2). Data are shown separated by each of the cohorts (marking color) and the number of captures of unmarked mosquitoes is also shown. The counts of marked individuals generally decrease over time. For marked individuals the capture ratio is visibly higher in the days immediately subsequent to releasing in the field and has a decreasing trend along with time. The number of captures of unmarked individuals, however, tends to remain at a constant level. For the experiment in Sept. 2012, the ratio between total number of captures per number of released individuals varied from 6% (green cohort) to 13.4% (blue cohort), and including all cohorts the return capture ratio is 9.2%. For the experiment in March 2013, the capture ratio is 7.6%. Datasets are available as supporting information.

Table 1. Numbers on captures obtained from the two MRR experiments conducted in the Z—10 region in Rio de Janeiro, Brazil. ST1 refers to the study realized in Sept. 2012 and ST2 to the study realized in March 2013. For ST1, we present data on each of the four cohorts as indicated by the marked colors. For each of the studies, unmarked refers to the number of unmarked individuals captured.

Cohort	Num. Released	Day1	Day2	Day3	Day4	Day5	Day6	Day7	Day8	Total	Ratio
ST1—b	500	19	14	7	9	9	4	3	2	67	0.134
ST1—p	500	27	11	4	2	1	2	5	0	52	0.104
ST1—y	500	10	10	3	4	3	1	1	3	35	0.07
ST1—g	500	9	7	3	5	4	0	0	2	30	0.06
ST1—unmarked	—	15	22	12	17	36	14	25	21	162	—
ST2—b	2000	52	26	20	23	4	5	13	8	151	0.076
ST2—unmarked	—	119	92	95	64	93	114	110	99	786	—

https://doi.org/10.1371/journal.pone.0123794.t001

In September 2012, we also measured size of the female wings from both populations (lab—reared and wild ones). Lab-reared *Ae. aegypti* females had 2.92 mm ± 0.085 mm (mean ± SD), meanwhile wild mosquitoes in this period had 2.74 mm ± 0.12 mm.

Analysis of field data

To study the impact of increasing sample size on the abundance estimation, data from ST1 was pooled by using a separate cohort, defined by the marking color, or a combination of cohorts: cohort b (N_1 = 500), cohorts b+p (N_1 = 1000), cohorts b+p+y (N_1 = 1500) and cohorts b+ p+ y+g (N_1 = 2000).

Table 2 summarizes the pooled data and provides the abundance estimates found using the hierarchical model and the Fisher—Ford method. Fig 1 and Table 2 show the *Aedes aegypti* abundance estimates, according to both models.

Table 2. Estimates of abundance for the populations of female *Aedes aegypti* **in Z—10, Rio de Janeiro, in Sept 2012 (ST1) and March 2013 (ST2), according to the hierarchical model and the Fisher-Ford model.** Estimation realized using samples from 16000 iterations using data from both studies. The hierarchical model results are obtained using JAGS after running 10000 iterations per chain (first 2000 were discarded) in two Markov chain simulations. Fisher—Ford estimates are obtained using a bootstrap approach (re-sampling $R_s = 1000$). For the Fisher—Ford estimation the survival rate is assumed to be $\phi = 0.8$.

	Hierarchical Model				**Fisher—Ford**	
Cohort	**Mean**	**Std. dev.**	**Median**	**95% CI**	**Mean**	**95% CI**
ST1—b	667	63	669	548–783	467	392–592
ST1—b+p	660	83	652	520–851	670	480–920
ST1—b+p+y	743	83	738	593–914	730	607–977
ST1—b+p+y+g	782	84	778	633–966	822	666–996
ST2	4118	319	4142	3423–4666	4768	3185–9198

https://doi.org/10.1371/journal.pone.0123794.t002

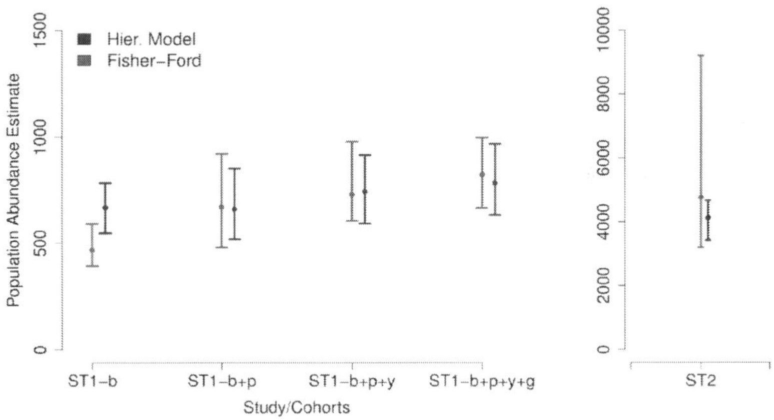

Figure 1. Results from estimation using data from field studies. Intervals colored in solid red indicate Fisher—Ford estimates. Intervals colored in solid blue indicate estimates from the hierarchical model. Labels ST1 and ST2 indicate studies conducted in September 2012 and March 2013, respectively. The September 2012 study had 4 cohorts given by 4 different colors: blue (b), pink (p), yellow (y) and green (g). Labels ST1 also include which cohorts were used in

the analysis by grouping cohorts ({b}, {b+p}, {b+p+y}, { b+p+y+g}). Estimates of population abundance are shown along with both y—axes (different scales), the one on the left side for study ST1 and on the right side for study ST2.

https://doi.org/10.1371/journal.pone.0123794.g001

Analysis using the hierarchical model provides an abundance estimate of approximately 700 female mosquitoes in Z—10 in September 2012, during the low transmission season, with point estimates ranging from 660 to 782, depending on the pooled data. The 95% credible interval varied from 548 to 966 and was the narrowest when the unpooled data (ST1-b) was used, and the widest of all when all cohorts were analyzed together. This might be due to the lowest overall capture ratio when pooling all four cohorts (about 9%) and also due to variation between cohorts. In March 2013 (ST2), during the high transmission season, the estimated abundance of *Aedes aegypti* was four times higher than in the low transmission period, with approximately 4100 female mosquitoes in Z—10. The credible interval (3423 – 4666) is wider than in the low abundance period, but the coefficient of variation is smaller (0.30 for ST2; 0.35 for ST1-b and 0.42 for ST1-b+p+y+g.).

The Fisher—Ford model tended to agree with the hierarchical model (Table 2 and Fig 1). The exception is for the ST1-b data, where the Fisher—Ford point estimate of mosquito abundance is considerably underestimated compared to the other method. Overall, the hierarchical model produced more precise estimates than the ones produced using Fisher—Ford method. This is evident observing the high abundance data (ST2) where the Fisher—Ford confidence interval is 5 times greater than the hierarchical model.

Analysis using the hierarchical model provides an estimate of the spatial distribution of mosquitoes that is not possible using the Fisher—Ford method. Figs 2 and 3 show the estimated spatial distribution of *Aedes aegypti* in Z—10 in Sept. 2012 (ST1) and March 2013 (ST2), respectively. To facilitate interpretation Figs 2 and 3 have on the left—hand side images of the study area superimposed with a bubble representation of the capture counts per trap and on the right—hand side the posterior spatial density according to the hierarchical model. First, the difference between scales should be noted. In September, local mosquito density peaks at 1.42 mosquitoes per 100 m^2 while in March, the ceiling is at 5.4 mosquitoes per 100 m^2. During the high abundance period, mosquito abundance is concentrated in the center—south region.

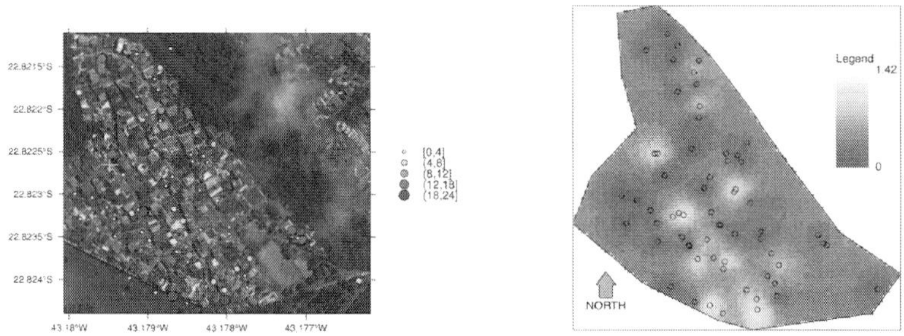

Figure 2. Spatial distribution of individuals in the Z—10 area in Rio de Janeiro, Brazil, in September 2012. Results for ST1 (Sept 2012) obtained with 16,000 iterations (8,000 iterations in each of two chains after a 2,000 burn-in period). Results from analysis using all cohort of marked individuals. Circles indicate trap locations. Bubbles as shown in the maps on the left—hand side indicate the counts of mosquitoes trapped in the MRR experiments. The release points of marked mosquitoes are depicted by crosses, each of which appear in the color of its respective cohort. The spatial density unit is number of mosquitoes per 100 m^2.

https://doi.org/10.1371/journal.pone.0123794.g002

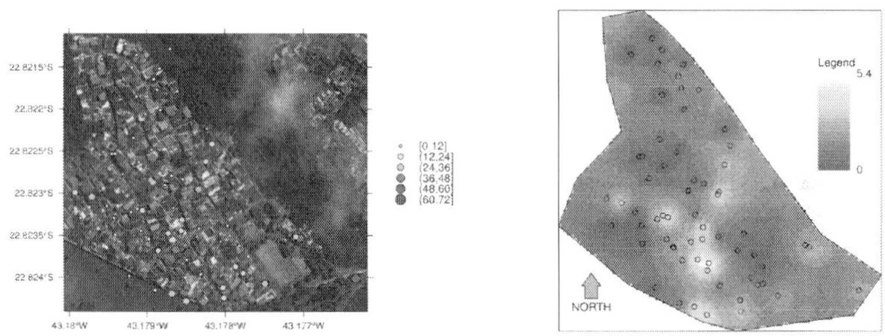

Figure 3. Spatial distribution of individuals in the Z—10 area in Rio de Janeiro, Brazil, in March 2013. Results obtained with 20,000 iterations (10,000 iterations in each of two chains after a 10,000 burn-in period). Results from analysis using the cohort of individuals marked in blue. Circles indicate trap locations. Bubbles as shown in the maps on the left—hand side indicate the counts of mosquitoes trapped in the MRR experiments. A cross indicates the point of releases of marked mosquitoes. The spatial density unit is number of mosquitoes per 100 m^2.

https://doi.org/10.1371/journal.pone.0123794.g003

The survival probabilities for the marked individuals, as estimated by analysis from the hierarchical model, are shown in Table 3. Estimates (mean values) of survival probabilities are expectedly higher for cohort combinations whose capture ratios are also higher.

Table 3. Estimates of survival probability for marked population of *Ae. aegypti* **released in Z—10, Rio de Janeiro, according to the hierarchical model.** Estimation realized using samples from 16000 iterations using data from both studies. The hierarchical model results are obtained using JAGS after running 10000 iterations per chain (first 2000 were discarded) in two Markov chain simulations. Fisher—Ford estimates are obtained using a bootstrap approach (re-sampling $R_s = 1000$).

	Hierarchical Model			
Cohort	**Mean**	**Std. dev.**	**Median**	**95% CI**
ST1—b	0.82	0.03	0.82	0.75–0.89
ST1—b + p	0.77	0.03	0.76	0.70–0.83
ST1—b + p + y	0.76	0.03	0.76	0.71–0.82
ST1—b + p + y + g	0.75	0.03	0.76	0.71–0.80
ST2	0.80	0.02	0.80	0.75–0.84

https://doi.org/10.1371/journal.pone.0123794.t003

The posterior distributions of abundance and survival probability after analysis of data from both experiments are found.

Analysis of Simulation Data

To further compare and understand the behavior of the hierarchical and Fisher—Ford estimates, artificial data from simulated scenarios were analyzed. Table 4 and Fig 4 shows the results obtained from simulations done by varying the population size of marked and unmarked individuals *i.e.*, (N_1, N_2): (200, 300), (300, 500), (400, 700), (500, 900), (2000, 3000). Besides the population size, we also compared scenarios with two attraction areas, defined as the radial spatial coverage of each trap. The larger the attraction area, the more attractive the trap is and the higher the chance of capturing mosquitoes. This is a parameter that is very difficult to measure in the field as it depends on the microenvironment and on the trap features. All simulations were done using the concept of center of activity (AC mode in simulator). Table 4 shows the number of captures and the mean and other statistics of the estimated abundance for each simulated scenario. Fig 4 shows the results graphically. In all scenarios, the credible intervals produced from the hierarchical model included the true value. The same was not true for the Fisher—Ford method, which tended to underestimate the true abundance. For the case of a larger basin attraction area the actual numbers N_2 of unmarked mosquitoes are within the intervals given by the hierarchical model, even though the capture ratio of marked individuals is small, approximately 10%. The estimated values found for survival probability and posterior distributions of abundance are found.

Table 4. Abundance estimation in simulated scenarios, using the hierarchical and the Fisher-Ford models. The total number of iterations was 12000 for each of 2 Markov chains. The first 3000 iterations are discarded as burn-in interval. The area of attraction is of size 5 (b5) and also 8 (b8), probability $p = 0.5$. The number of traps is $J = 64$. For the Fisher—Ford estimates a number of $R_s = 1000$ re-sampling was used and the daily survival probability was $\phi = 0.8$, same value used in the simulations. For the case (200, 300) with a small attraction area the sample size (2) is very small and no Fisher—Ford estimates are reported. The notation (m,u) refers to the quantity (number of captures, capture ratios) for marked and unmarked mosquitoes, respectively.

Simulated scenarios			**Hierarchical Model**				**Fisher—Ford**	
simulation	**# captures (m,u)**	**capture ratio**	**Mean**	**Std. dev.**	**Median**	**95% CI**	**Mean**	**95% CI**
b5-h / (200, 300)	(5, 29)	(0.03, 0.10)	323	85	329	152–470	—	—
b5-h / (300, 500)	(14, 48)	(0.05, 0.10)	386	128	372	169–642	217	152–332
b5-h / (400, 700)	(17, 74)	(0.04, 0.11)	583	176	566	295–953	435	246–552
b5-h / (500, 900)	(21, 85)	(0.04, 0.09)	705	185	686	414–1116	473	350–658
b8-h / (200, 300)	(17, 64)	(0.09, 0.21)	363	88	360	202–546	199	124–316
b8-h / (300, 500)	(36, 104)	(0.12, 0.21)	501	109	491	322–736	263	172–339
b8-h / (400, 700)	(41, 167)	(0.10, 0.24)	963	164	973	612–1234	539	328–811
b8-h / (500, 900)	(46, 205)	(0.09, 0.23)	818	166	790	570–1237	731	546–961
b8-h / (2000, 3000)	(186, 645)	(0.09, 0.21)	2780	286	2755	2229–3399	2895	2253–4961

https://doi.org/10.1371/journal.pone.0123794.t004

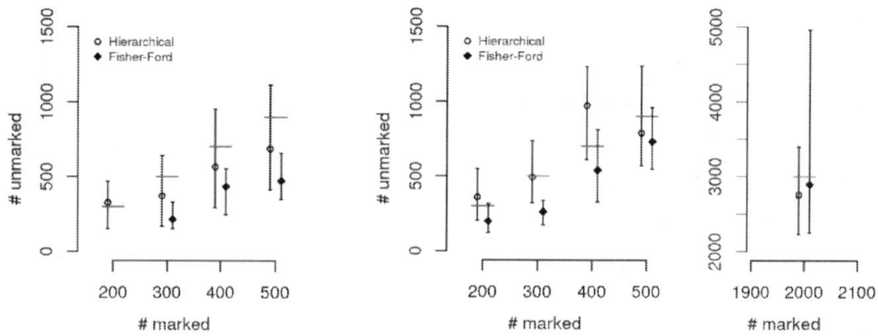

Figure 4. Estimates obtained from MRR data in the simulated scenarios. The estimated values for abundance are shown along the *y*–axis (for all cases in the *y*—axis on the left—hand side, except for the case $N_2 = 3000$ shown along the *y*—axis on the right—hand side), whereas each case is shown along the *x*-axis, described by the number (known value of) of unmarked individuals used in the simulation. The plot on the left—hand side shows results for a small basin of attraction, whereas on the right—hand side results are shown for traps' basins of attraction that have radiuses 60% greater. Red lines indicate the numbers of unmarked mosquitoes that were used in the simulations for each of the configurations.

https://doi.org/10.1371/journal.pone.0123794.g004

DISCUSSION

Recently proposed strategies [43] to control vector-borne diseases could either reduce the number of individual vector mosquitoes (population suppression) or reduce the competence of these vectors in transmitting the pathogen by replacing the susceptible wild population by genetically modified individuals refractory to the infection by the pathogen (population replacement). Evaluation of the potential success of such interventions requires detailed knowledge of the dynamics of the mosquito population, including its abundance in time and space [8]. In this paper, we present a hierarchical probabilistic model for the estimation of female *Aedes aegypti* abundance from MRR studies. The knowledge of mosquito density from estimation can help in control strategies, at least at a planning phase. The model performs well under simulation conditions and provides more precise estimates than the ones given by the Fisher—Ford method. The proposed model expands previous contributions by Royle *et al.* [17] in adding a survival component and a trap component that models the life span of individuals during MRR experiments and the trapping limitations that make difficult multiple capture events for each of the study individuals. We apply the concept of centers of activity present in [17, 25] to the distribution of a mosquito population within a given area. Clusters of human dengue cases have been identified and associated with certain locations in the community, like neighbor's homes, local meeting or gathering places and attributed primarily to variation in *Ae. aegypti* population density [44, 45]. The underlying mechanism leading to a positive association between human and mosquito infections at the level of individual houses and neighboring residences [46] is yet not entirely explained and one might speculate that this pattern expresses an indirect evidence of the presence of center of activities for the vector.

This model can be extended in order to account for difference in life—history components, for instance, if using cohorts of different origins. This could be the case of releasing cohorts of mosquitoes especially prepared for control interventions, such as *Wolbachia*—infected mosquitoes. Other extension possibilities include recruitment and survival of unmarked mosquitoes and also influence of gonotrophic cycle in the centers of activity. Such extensions would permit to study important biological relations but would also require more elaborated design in the field experiments to gather data, and by consequence, an added cost of complexity in the analysis.

The method was illustrated with data collected from Z—10, a study area located in a dengue endemic area. According to the hierarchical model's results, Z—10 had approximately 700 female mosquitoes in the late winter-early spring of 2012, corresponding to a measure of 0.99 mosquitoes/premise (95% CI: 0.80 – 1.23 mosquitoes/premise) or 0.33 mosquitoes/person (95% CI: 0.27 – 0.41 mosquitoes/person). This corresponds to the period of low dengue transmission. In March 2013, during the high transmission season, mosquito abundance jumped to 5.2 mosquitoes/premise (95% CI: 4.3– 5.9 mosquitoes/premise) or 1.8 mosquitoes/person (95% CI: 1.5 – 2.0 mosquitoes/person). The estimated daily survival probability was similar in both seasons, despite the difference in temperature, being between 0.71 and 0.80. This provides an estimate of average life span of 7.5 to 9 days (adding the four days in the laboratory) and is comparable to other values presented in the literature [35]. The analysis using the hierarchical model also provides an estimation of the pattern of spatial distribution of the mosquito population. One can see that in the low transmission season (Sept 2012), there are several hotspots distributed throughout the area, but each one holds an average of approximately 1 mosquito per 100 m^2. In the high transmission season, local mosquito abundance has a fourfold increase in some areas, and the highest abundance is concentrated in the southern part of the neighborhood. Considering that this is the area with more human movement, the presence of hotspots represents a critical situation for virus invasion and dissemination in this community.

The possibility of estimating spatial density of the population of *Aedes aegypti* is an advantage over methods that consider homogeneous distribution of mosquitoes, such as Lincoln, and Fisher—Ford. The Fisher—Ford method takes survival into account, which is quite important in the case of mosquito populations that have high mortality rates. It is, however, inherently sensitive to a good estimate of survival probability, since an error introduced in the survival parameter might underestimate or overestimate the abundance as a consequence. The attempt at using another capture—recapture method, the Jolly–Seber model, in a package implemented by Laake *et al.* [47] that uses Markov chain Monte—Carlo was not successful because of lack of convergence, due to the insufficient data (low capture ratio, single capture per individual). Such results, taken in a pre—intervention phase, integrated with other tools might be useful in control policies for limiting the spread of the vector.

Aedes aegypti larvae were raised using standard rearing protocols, which include low intra- or inter- specific competition, low variations in temperature and high amount of resources. Thus, it is expected that the wing size of lab-reared mosquitoes would be higher when compared to wild mosquitoes, as shown in the results. This would clearly influence dispersal as size has been shown to impact dispersal of individuals, i.e., small females disperse further than larger individuals [27]. However, also in [27], size had no significant impact on survival rates.

Methods for estimating *Aedes aegypti* absolute abundance, other than methods using MRR data, have been proposed in the literature. Williams *et al.* [48] and Jeffery *et al.* [49] used data from comprehensive quantitative pupae surveys to estimate absolute population abundances using life-table models. Since pupal mortality is low, one can compute the mosquito standing crop from the number of containers in the area, number of pupae per container, and adult survival. The latter is generally taken from published MRR studies. The underlying assumptions are that pupal production and survival are stable over time, which rarely hold for more than a few days. An even more detailed life-table model is the classical CIMSIM [50] which requires as input, meteorological, demographic and container availability data. A stochastic model with more than one hundred parameters is required to convert input data into estimates of mosquito abundance [43]. Despite the success of this approach in some settings [50], its large scale application might have drawbacks, especially in areas where the availability and quality of containers varies temporally and/or spatially.

Vector density and survival are key parameters that enter the expression of vectorial capacity $V = \frac{ma^2 p^v}{-\log p}$, where m is the density of mosquitoes per humans, a is the biting rate, v is the average incubation period for the pathogen, and p is the survival probability [8]. This expression has guided past control efforts, such as the use of insecticides, and remains the primary benchmark against which new strategies are compared. The methods proposed here make more efficient use of data collected in the field than under current approaches familiar to entomologists and surveillance systems. These models make explicit the sources of uncertainty by allowing the estimation of credibility intervals that accompany each estimate. By casting the estimation process under a formal statistical framework, one can benefit from well established processes of model building which consist of the entertainment of several possible sensible models, parameter estimation in these models, hypothesis testing and deviance analysis, examination of residuals, graphical considerations including displays of the observed and fitted values, and model selection by means of information criteria. This procedure does not guarantee the identification of the "true" model but provides clear steps in model validation through the quantification of uncertainty and the identification of deviations from the premises that entered the formulation of the model. Such formal framework will certainly impact positively on the evaluation process of modern complex intervention strategies that intend to suppress or replace vector populations.

ACKNOWLEDGMENTS

We would like to thank the anonymous reviewers whose comments helped us to improve the quality of this paper.

AUTHOR CONTRIBUTIONS

Conceived and designed the experiments: DAMV CTC FF GAG RM CJS. Performed the experiments: GAG RM. Analyzed the data: DAMV CTC FF GAG RM CJS. Contributed reagents/materials/analysis tools: DAMV CTC FF RM CJS. Wrote the paper: DAMV CTC FF GAG RM CJS.

REFERENCES

1. Halstead SB (2008) Dengue virus-mosquito interactions. Annu Rev Entomol 53: 273–291.
2. Olanratmanee P, Kittayapong P, Chansang C, Hoffmann AA, Weeks AR, Endersby NM (2013) Population genetic structure of *Aedes (Stegomyia) aegypti* (L.) at a micro-spatial scale in Thailand: implications for a dengue suppression strategy. PLOS Neglected Tropical Diseases 7 (1) 10.1371/journal.pntd.0001913
3. Perkins TA, Scott TW, Le Menach A, Smith DL (2013) Heterogeneity, mixing, and the spatial scales of mosquito-borne pathogen transmission PLOS Computational Biology 9 (12) 10.1371/journal.pcbi.1003327
4. Harrington LC, Scott TW, Lerdthusnee K, Coleman RC, Costero A, Clark GG, et al. (2005) Dispersal of the dengue vector *Aedes aegypti* within and between rural communities. The American Journal of Tropical Medicine and Hygiene 72: 209–220.
5. Buonaccorsi JP, Harrington LC, Edman JD (2003) Estimation and comparison of mosquito survival rates with release-recapture-removal data. Journal of Medical Entomology 40: 6–17. 10.1603/0022-2585-40.1.6
6. Cianci D, Van Den Broek J, Caputo B, Marini F, Torre AD, Heesterbeek H, et al. (2013) Estimating mosquito population size from mark-release-recapture data. Journal of Medical Entomology 50: 533–542. 10.1603/ME12126
7. Service MW (1993) Mosquito—Ecology Field Sampling Methods. Chapman & Hall.
8. Smith DL, Battle KE, Hay SI, Barker CM, Scott TW, McKenzie FE (2012) Ross, Mac-Donald, and a theory for the dynamics and control of mosquito-transmitted pathogens. PLoS pathogens 8: e1002588 10.1371/journal.ppat.1002588
9. Focks DA, Chadee DD (1997) Pupal survey: an epidemiologically significant surveillance method for *Aedes aegypti*: an example using data from Trinidad. The American Journal of Tropical Medicine and Hygiene 56: 159–167.
10. Focks DA (2003) A review of entomological sampling methods and indicators for dengue vectors. Geneva: WHO.

11. Focks DA, Brenner RJ, Hayes J, Daniels E (2000) Transmission thresholds for dengue in terms of *Aedes aegypti* pupae per person with discussion of their utility in source reduction efforts. American Journal of Tropical Medicine and Hygiene 62: 11–18.
12. Keeling M, Jiggins F, Read J (2003) The invasion and coexistence of competing Wolbachia strains. Heredity 91: 382–388. 10.1038/sj.hdy.6800343
13. Barton NH, Turelli M (2011) Spatial waves of advance with bistable dynamics:cytoplasmic and genetic analogues of Allee effects Am Nat 178: E48–E75.
14. Costantini C, Li SG, Torre AD, Sagnon N, Coluzzi M, Taylor CE (1996) Density, survival and dispersal of *Anopheles gambiae* complex mosquitoes in a West African Sudan savanna village. Medical and Veterinary Entomology 10: 203–219. 10.1111/j.1365-2915.1996.tb00733.x
15. Maciel-de Freitas R, Eiras ÁE, Lourenço-de Oliveira R (2008) Calculating the survival rate and estimated population density of gravid *Aedes aegypti* (*Diptera: Culicidae*) in Rio de Janeiro, Brazil. Cadernos de Saúde Pública 24: 2747–2754. 10.1590/S0102-311X2008001200003
16. Touré YT, Dolo G, Petrarca V, Dao A, Carnahan J, Taylor CE (1998) Mark—release—recapture experiments with *Anopheles gambiae sl* in Banambani Village, Mali, to determine population size and structure. Medical and Veterinary Entomology 12: 74–83. 10.1046/j.1365-2915.1998.00071.x
17. Royle JA, Karanth KU, Gopalaswamy AM, Kumar NS (2009) Bayesian inference in camera trapping studies for a class of spatial capture-recapture models. Ecology 90: 3233–3244. 10.1890/08-1481.1
18. Consoli RA, Oliveira RLd (1994) Principais mosquitos de importância sanitária no Brasil. Fiocruz.
19. Maciel-de Freitas R, Eiras ÁE, Lourenço-de Oliveira R (2006) Field evaluation of effectiveness of the BG-sentinel, a new trap for capturing adult *Aedes aegypti* (*Diptera: Culicidae*). Memórias do Instituto Oswaldo Cruz 101: 321–325.
20. Kroeckel U, Rose A, Eiras AE, Geier M (2006) New tools for surveillance of adult yellow fever mosquitoes: comparison of trap catches with human landing rates in an urban environment. Journal of the American

Mosquito Control Association 22: 229–238. 10.2987/8756-971X(2006)22[229:NTFSOA]2.0.CO;2

21. Getis A, Morrison AC, Gray K, Scott TW (2010) Characteristics of the spatial pattern of the dengue vector, *Aedes aegypti*, in Iquitos, Peru. In: Perspectives on Spatial Data Analysis, Springer; pp. 203–225.
22. McDonald P (1977) Population characteristics of domestic *Aedes aegypti* (*Diptera: Culi-cidae*) in villages on the Kenya Coast i. adult survivorship and population size. Journal of Medical Entomology 14: 42–48. 10.1093/jmedent/14.1.42
23. Edman J, Scott T, Costero A, Morrison A, Harrington L, Clark GG, (1998) *Aedes aegypti* (*Diptera: Culicidae*) movement influenced by availability of oviposition sites. Journal of Medical Entomology 35: 578–583. 10.1093/jmedent/35.4.578
24. Royle JA, Chandler RB, Sollman R, Gardner B (2013) Spatial Capture-Recapture. Academic Press;
25. Royle JA, Kéry M, Guélat J (2011) Spatial capture-recapture models for search-encounter data. Methods in Ecology and Evolution 2: 602–611. 10.1111/j.2041-210X.2011.00116.x
26. Honório NA, da Costa Silva W, Leite PJ, Gonçalves JM, Lounibos LP, Lourenço de Oliveira R (2003) Dispersal of *Aedes aegypti* and *Aedes albopictus* (*Diptera: Culicidae*) in an urban endemic dengue area in the State of Rio de Janeiro, Brazil. Mem Inst Oswaldo Cruz 98: 191–198. 10.1590/S0074-02762003000200005
27. Maciel-de Freitas R, Codeco C, Lourenço-de Oliveira R (2007) Body size-associated survival and dispersal rates of *Aedes aegypti* in Rio de Janeiro. Medical and Veterinary Entomology 21: 284–292. 10.1111/j.1365-2915.2007.00694.x
28. Muir LE, Kay BH (1998) *Aedes aegypti* survival and dispersal estimated by mark-release-recapture in northern Australia. The American Journal of Tropical Medicine and Hygiene 58: 277–282.
29. David MR, Lourenço-de Oliveira R, Maciel-de-Freitas R (2009) Container productivity, daily survival rates and dispersal of *Aedes aegypti* mosquitoes in a high income dengue epidemic neighbourhood of Rio de Janeiro: presumed inffuence of differential urban structure on mosquito biology. Memórias do Instituto Oswaldo Cruz 104: 927–932. 10.1590/S0074-02762009000600019
30. Tsuda Y, Takagi M, Wang S, Wang Z, Tang L (2001) Movement

of *Aedes aegypti* (*Diptera: Culicidae*) released in a small isolated village on Hainan Island, China. Journal of Medical Entomology 38: 93–98. 10.1603/0022-2585-38.1.93

31. Trpis M, Hausermann W (1986) Dispersal and other population parameters of *Aedes aegypti* in an African village and their possible significance in epidemiology of vector-borne diseases. The American Journal of Tropical Medicine and Hygiene 35: 1263
32. Van Dyk DA, Meng X-L (2001) The Art of Data Augmentation. Journal of Computational and Graphical Statistics 10: 1–50. 10.1198/10618600152418584
33. Royle JA, Dorazio RM (2012) Parameter-expanded data augmentation for Bayesian analysis of capture—recapture models. Journal of Ornithology 152: 521–537. 10.1007/s10336-010-0619-4
34. Plummer M (2003) JAGS: A program for analysis of Bayesian graphical models using Gibbs sampling. In Proceedings of the 3rd International Workshop on Distributed Statistical Computing, Vienna.
35. Maciel-De-Freitas R, Codeco CT, Lourenço-de-Oliveira R (2007) Daily survival rates and dispersal of *Aedes aegypti* females in Rio de Janeiro, Brazil. American Journal of Tropical Medicine and Hygiene 76: 659–665.
36. Dowdeswell W, Fisher R, Ford E (1940) The quantitative study of populations in the lepidoptera *I. Polyommatus icarus* rott. Annals of Eugenics 10: 123–136. 10.1111/j.1469-1809.1940.tb02242.x
37. Baber I, Keita M, Sogoba N, Konate M, Doumbia S, Traoré SF, et al. (2010) Population size and migration of *Anopheles gambiae* in the Bancoumana Region of Mali and their significance for efficient vector control. PloS ONE 5: e10270 10.1371/journal.pone.0010270
38. Bailey NT (1952) Improvements in the interpretation of recapture data. The Journal of Animal Ecology: 120–127. 10.2307/1913
39. Judson OP (1994) The rise of the individual-based model in ecology. Trends in Ecology and Evolution 9: 9–14. 10.1016/0169-5347(94)90225-9
40. Grimm V (1999) Ten years of individual-based modelling in ecology: what have we learned and what could we learn in the future? Ecological Modelling 115: 129–148. 10.1016/S0304-3800(98)00188-4
41. Bonabeau E (2002) Agent-based modeling: Methods and techniques for simulating human systems. Proceedings of the National Academy

of Sciences of the United States of America 99: 7280–7287. 10.1073/pnas.082080899

42. Cooch E, White G, editors (2013) Program MARK: A Gentle Introduction. 12th edition. URL http://www.phidot.org/software/mark/docs/book/.
43. Magori K, Legros M, Puente ME, Focks DA, Scott TW, Lloyd AL, et al. (2009) Skeeter Buster: a stochastic, spatially explicit modeling tool for studying *Aedes aegypti* population replacement and population suppression strategies. PLoS Neglected Tropical Diseases 3: e508 10.1371/journal.pntd.0000508
44. Gatrell AC, Bailey TC, Diggle PJ, Rowlingson BS (1996) Spatial point pattern analysis and its application in geographical epidemiology. Trans Institute British Geographers 21:256–274 10.2307/622936
45. Morrison AC, Getis A, Santiago M, Rigua-Perez JG, Reiter P (1998) Exploratory space-time analysis of reported dengue cases during an outbreak in Florida, Puerto Rico, 1991–1992. Am J Trop Med Hyg 58:287–298
46. Yoon IK, Getis A, Aldstadt J, Rothman AL, Tannitisupawong D, Koenraadt CJ, et al. (2012) Fine scale spatiotemporal clustering of dengue virus transmission in children and *Aedes aegypti* in rural Thai villages. PLoS Neglected Tropical Diseases 2012;6(7):e1730 10.1371/journal.pntd.0001730
47. Laake JL, Johnson DS, Conn PB (2013) marked: an R package for maximum likelihood and Markov Chain Monte Carlo analysis of capture-recapture data. Methods in Ecology and Evolution 4: 885–890. 10.1111/2041-210X.12065
48. Williams C, Johnson P, Ball T, Ritchie S (2012) Productivity and population density estimates of the dengue vector mosquito *Aedes aegypti* (*Stegomyia aegypti*) in Australia. Medical and Veterinary Entomology.
49. Jeffery JA, Yen NT, Nam VS, Hoffmann AA, Kay BH, Ryan PA (2009) Characterizing the *Aedes aegypti* population in a vietnamese village in preparation for a *Wolbachia*-based mosquito control strategy to eliminate dengue. PLoS Neglected Tropical Diseases 3: e552 10.1371/journal.pntd.0000552

50. Williams CR, Johnson PH, Long SA, Rapley LP, Ritchie SA (2008) Rapid estimation of *Aedes aegypti* population size using simulation modeling, with a novel approach to calibration and field validation. Journal of Medical Entomology 45: 1173–1179. 10.1603/0022-2585(2008)45[1173:REOAAP]2.0.CO;2

17

Latitudinal Diversity of Culex pipiens Biotypes and Hybrids in Farm, Peri-Urban, and Wetland Habitats in Europe

Chantal B. F. Vogels[1], Tim W. R. Mo¨hlmann[1,2], Diede Melsen[1], Guido Favia[3], Uno Wennergren[2], Constantianus J. M. Koenraadt[1]

[1] Laboratory of Entomology, Wageningen University and Research centre, Wageningen, The Netherlands,
[2] IFM Theory and Modelling, Linko¨ping University, Linko¨ping, Sweden,
[3] School of Biosciences and Veterinary Medicine, University of Camerino, Camerino, Italy

ABSTRACT

Despite the presence of *Culex (Cx.) pipiens* mosquitoes and circulation of West Nile virus (WNV), WNV outbreaks have so far not occurred in northern Europe. The species *Cx. pipiens* consists of two morphologically identical biotypes, *pipiens* and *molestus*, which can form hybrids. Until

---

Citation: Vogels CBF, Möhlmann TWR, Melsen D, Favia G, Wennergren U, Koenraadt CJM (2016) Latitudinal Diversity of Culex pipiens Biotypes and Hybrids in Farm, Peri-Urban, and Wetland Habitats in Europe. PLoS ONE 11(11): e0166959. https://doi.org/10.1371/journal.pone.0166959

now, population dynamic studies of *Cx. pipiens* have not differentiated between biotypes and hybrids at the European scale, nor have they used comparative surveillance approaches. We therefore aimed to elucidate the relative abundance of *Cx. pipiens* biotypes and hybrids in three habitat types at different latitudes across Europe, using two different surveillance traps. BG-Sentinel and Mosquito-Magnet Liberty Plus traps were placed in three habitat types (farms, peri-urban, wetlands), in three European countries (Sweden, The Netherlands, Italy). Collected *Cx. pipiens* mosquitoes were identified to biotype with real-time PCR. Both trap types collected equal ratios of the biotypes and their hybrids. From northern to southern latitudes there was a significant decrease of *pipiens* and an increase of *molestus*. Habitat types influenced the relative ratios of biotypes and hybrids, but results were not consistent across latitudes. Our results emphasize the need to differentiate *Cx. pipiens* to the biotype level, especially for proper future WNV risk assessments for Europe.

INTRODUCTION

Global warming, increased travel and trade, and land-use changes are important drivers for the (re-)emergence of vector-borne diseases, such as West Nile virus (WNV; family: *Flaviviridae*) [1]. The potential of WNV to quickly spread to new areas is clearly illustrated by the outbreaks that occurred in the United States of America, after the initial introduction in 1999 [2–4]. WNV outbreaks have also occurred in southern and central European countries, but no outbreaks among humans have occurred in northern Europe [5–7].

WNV is maintained in an enzootic cycle between birds and mosquitoes. The main vector for WNV is the mosquito *Culex (Cx.) pipiens* [8,9]. The *Cx. pipiens* complex consists of several closely related species and biotypes, of which only the species *Cx. pipiens* (Linnaeus 1758) occurs in Europe [10]. Because of its similar morphology, the species *Cx. torrentium* (Martini 1925) is often included in taxonomic studies of the *Cx. pipiens* complex [11]. The species *Cx. pipiens* consists of two morphologically similar biotypes, named *pipiens* (Linnaeus 1758) and *molestus* (Forskål 1775), which show distinct behaviour. Biotype *pipiens* is the most important vector in the enzootic cycle because of its preference for birds [12]. During winter, biotype *pipiens* enters diapause, which provides a means of overwintering for WNV [13,14]. Biotype *molestus* prefers mammals, including humans, as hosts, and remains active year-round [15–17]. Host availability can induce a strong shift in host feeding behaviour of biotype *molestus* from mammals to birds, especially in areas with high bird densities [18]. Previously, biotype *molestus* has been described as occurring underground [8], but recent studies show that both biotypes occur sympatrically in aboveground habitats throughout Europe [19–22]. Furthermore, biotype *pipiens* and biotype *molestus* can form hybrids which show intermediate host preference [23]. As a result of this, hybrids can play an important role in bridging WNV from birds to humans [8].

Several studies elucidated the geographic distribution of the species *Cx. pipiens* and *Cx. torrentium* (Martini 1925) at the European scale [24,25]. In general, *Cx. torrentium* is relatively more abundant in northern Europe, whereas *Cx. pipiens* is more abundant in southern Europe [24]. However, these studies did not identify *Cx. pipiens* mosquitoes to the biotype level. Identification to the biotype level is important because the behavioural differences between the two biotypes of *Cx. pipiens* and their hybrids result in different vectorial capacity for WNV. Thus far, in-depth studies that differentiated between the biotypes were done at country level [19–22,26,27]. Few of these studies systematically compared biotype ratios among different habitat types [19,20,27]. However, due to differences in experimental design it is hard to make direct comparisons between *Cx. pipiens* populations in northern and southern European countries.

The aim of this study was to assess the relative abundance of the *Cx. pipiens* biotypes with two types of traps (Biogents Sentinel and Mosquito Magnet Liberty Plus), in three different habitat types (farms, peri-urban, and wetlands), and in three countries (Sweden, The Netherlands, and Italy) at different latitudes across Europe.

MATERIALS & METHODS

Ethics Statement

Permits and approval for field work in wetlands were obtained from the county board of Östergötland in Sweden, Staatsbosbeheer in The Netherlands, and the Protected Areas Service of the San Benedetto del Tronto Municipality in Italy. For farms and peri-urban habitats approval was obtained from landowners of private properties in all three countries. No protected species were sampled in this study.

Mosquito collections

Adult mosquitoes were collected with the Biogents Sentinel (BGS) trap (BioGents GmbH, Germany) and the Mosquito Magnet Liberty Plus (MMLP) trap (Woodstream Corp., USA). A mixture of 17.5 g dry instant yeast (Bruggeman, The Netherlands), 250 g white granulated sugar and 2 l of tap water in a 5 l plastic bottle was used for CO_2 production in the BGS trap [28]. Combustion of propane provided CO_2 for the MMLP trap.

Both traps were rotated among three trapping locations, in three different habitat types (farms, peri-urban, and wetlands), in Sweden (Linköping), The Netherlands (Wageningen), and Italy (San Benedetto del Tronto; Table 1). The selected farms were dairy cattle farms with a minimum of 10 cows. Traps were placed within 50 m of the open indoor stable. Peri-urban locations were at the periphery of a city (inhabitants <150,000), and within a 50 m radius of the trap, at least two occupied residential properties were present. Locations in a wetland habitat had a minimum of 50% marshy or standing water within a 100 m radius of the traps. Trapping locations were at least 100 m apart.

Table 1. Coordinates of all 27 trapping locations in the three different habitat types (farms, peri-urban, and wetlands) in three different European countries (Sweden, The Netherlands, and Italy).

Country	Habitat type	Sampling location	Coordinates
Sweden	Farms	1	58.296530, 15.584782
(Linköping)		2	58.343622, 15.602404
		3	58.330597, 15.704327
	Peri-urban	4	58.416973, 15.499516
		5	58.401515, 15.626744
		6	58.405494, 15.595035
	Wetlands	7	58.362106, 15.651861
		8	58.361585, 15.654910
		9	58.361542, 15.659072
The Netherlands	Farms	10	51.971084, 5.761455
(Wageningen)		11	51.973637, 5.773978
		12	52.013077, 5.645998
	Peri-urban	13	52.018075, 5.655372
		14	51.979257, 5.645230

		15	51.979771, 5.660278
	Wetlands	16	51.969443, 5.758940
		17	51.967693, 5.758896
		18	51.971671, 5.747826
Italy	Farms	19	42.914466, 13.854588
(San Benedetto del Tronto)		20	42.944809, 13.859857
		21	42.943098, 13.853856
	Peri-urban	22	42.883455, 13.879388
		23	42.951012, 13.850783
		24	42.934424, 13.891933
	Wetlands	25	42.896600, 13.911895
		26	42.899042, 13.909813
		27	42.903365, 13.908667

https://doi.org/10.1371/journal.pone.0166959.t001

Collections were done during six consecutive days, every month in each country. Sampling periods were from July 2014 to June 2015, except for the winter months December, January, and February (and March for Sweden). Traps were emptied and repositioned every 24 hours between sunrise and sunset of the next day. Mosquitoes were stored at -20°C in Eppendorf tubes containing small silica beads covered with cotton wool.

Mosquito Identifications

All female mosquitoes were identified to species level, following the European identification key for female mosquitoes [29]. The number of *Cx. pipiens* mosquitoes captured each month was not sufficient to statistically test for temporal differences in biotype and hybrid ratios. Therefore, all 190 *Cx. pipiens* females available for analysis from Sweden, and a selection of 300 *Cx. pipiens* females from Italy and 299 *Cx. pipiens* females from The Netherlands were used, resulting in a total of 789 mosquitoes analysed. Samples from the Netherlands and Italy were partially random selected with 100 samples per habitat, for both countries.

Selected mosquitoes were further identified to species (*Cx. pipiens* or *Cx. torrentium*) and biotype (*pipiens*, *molestus*, or hybrid) level. We followed the real-time PCR assay for differentiation between the *Cx. pipiens* biotypes as described in detail before [22]. Briefly, for *Cx. pipiens* we used forward and reverse primers Cx_ pip_F (5'-GCGGCCAAATATTGAGACTTTC-3') and Cx_pip_R (5'-ACTCGTCCTCAAACATCCAGACATA-3'). For identification of biotype *molestus* we used probe Cpp_mol_P (5'-FAM-TGAACCCTCCAGTAAGGTA-MGB-3'), and for biotype *pipiens* we used the two probes Cpp_pip_P1 (5'-VIC-CACACAAAYCTTCACCGAA-MGB-3') and Cpp_pip_P2 (5'-VIC-ACACAAACCTTCATCGAA-MGB-3'). Hybrids were identified when both probes for biotype *pipiens* and *molestus* were amplified by real-time PCR. For identifications of *Cx. torrentium* we used forward and reverse primers Cx_tor_F (5'-CTTATTAGTATGACACAGGACGACAGAAA-3') and Cx_tor_R (5'-GCATAAACGCCTACGCAACTACTAA-3'), and probe Cx_tor_P (5'-FAM-ATGATGCCTGTGCTACCA-MGB-3'). Thermocycler conditions were 95°C for 10 min, followed by 45 cycles of 95°C for 15 s and 62°C for 1 min. The PCR was run on the CFX96 Real-Time PCR system (Bio-Rad Laboratories, Hercules, CA) and data were analysed in CFX manager 2.0 (Bio-Rad Laboratories, Hercules, CA).

Statistical Analyses

Main effects (trap type, country, and habitat) and within-effects (habitats within each country, and country within each habitat) on the ratios of *Cx. pipiens* mosquitoes were tested with Pearson's Chi-square tests. Significant effects were further evaluated with pairwise comparisons and corrected with the Bonferroni correction. All data were analysed in the statistical software package R [30].

RESULTS

In total 5,202 *Cx. pipiens* females were collected of which 3,878 females were collected with the BGS trap and 1,324 females with the MMLP trap (Table 2). Of the 789 mosquitoes selected for analysis, 663 mosquitoes were identified as *Cx. pipiens*, of which 463 (69.8%) were identified as biotype *pipiens*, 127 (19.2%) as biotype *molestus*, and 73 (11.0%) as hybrids (S1 Dataset). In addition, 14 mosquitoes were identified as *Cx. torrentium*, which all originated from Sweden. The number of *Cx. torrentium* mosquitoes was too low for reliable statistical tests, and these samples were therefore excluded from further analyses. The remaining 112 mosquitoes did not amplify a PCR product.

Table 2. Total number of collected *Cx. pipiens* **females per trap type, habitat, and country.** BGS = Biogents Sentinel trap, MMLP = Mosquito Magnet Liberty Plus trap, SW = Sweden, NL = The Netherlands, and IT = Italy.

	BGS			MMLP		
	SW	NL	IT	SW	NL	IT
Farms	19	252	128	29	64	21
Peri-Urban	56	1063	111	44	451	37
Wetlands	33	969	1247	24	101	553
Total	108	2284	1486	97	616	611

https://doi.org/10.1371/journal.pone.0166959.t002

Both trap types, BGS and MMLP, trapped similar ratios of the *Cx. pipiens* biotypes and hybrids ($\chi^2 = 2.35$, df = 2, p = 0.31; Fig 1A). Thus, data from both trap types were pooled for further analyses.

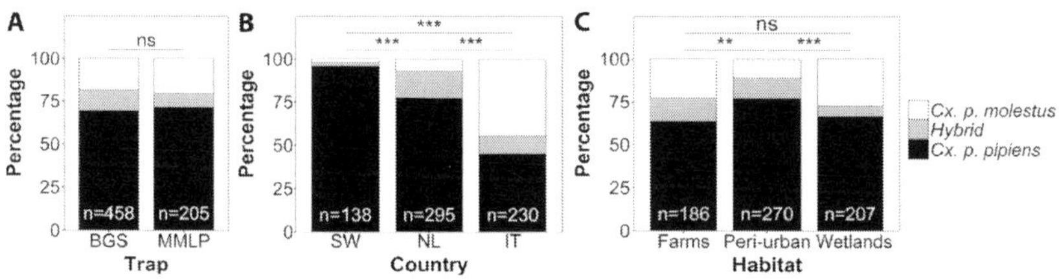

Figure 1. Main effects of (A) trap type, (B) country, and (C) habitat on the ratio of *Culex pipiens* **biotypes and their hybrids**. The total sample size (n) is indicated for each bar. Significance is displayed for each pairwise comparison, with ns = not significant, ** = $p < 0.01$, *** = $p < 0.001$. BGS = Biogents Sentinel trap, MMLP = Mosquito Magnet Liberty Plus, SW = Sweden, NL = The Netherlands, and IT = Italy.

https://doi.org/10.1371/journal.pone.0166959.g001

The ratios of *Cx. pipiens* biotypes and hybrids were significantly different between the three countries in Europe ($\chi^2 = 173.62$, df = 4, p<0.001; Fig 1B). Pairwise comparisons between countries showed that *Cx. pipiens* ratios were different between each combination of Italy, The Netherlands, and Sweden (all pairwise comparisons: p<0.001). The proportion of biotype *pipiens* was highest in Sweden (90%) and gradually decreased towards more southern latitudes, with the lowest proportion of biotype *pipiens* in Italy (40%).

The ratios of *Cx. pipiens* biotypes and hybrids were also significantly different between habitat types ($\chi^2 = 26.59$, df = 4, p<0.001; Fig 1C). Peri-urban habitats had a relatively higher proportion of biotype *pipiens* compared to both farms (p<0.01), and wetlands (p<0.001). There was no difference in ratios between farms and wetlands (p = 0.16).

In order to gain more insight in the interaction between country and habitat, pairwise comparisons were made between the habitats within each country, and the countries within each habitat type (Fig 2). Ratios of *Cx. pipiens* biotypes and hybrids were significantly different between habitats in Italy ($\chi^2 = 25.05$, df = 4, p<0.001) and The Netherlands ($\chi^2 = 26.37$, df = 4, p<0.001), but were similar within Sweden ($\chi^2 = 6.11$, df = 4, p = 0.19; Fig 2). In The Netherlands, farms were different due to the relatively high proportion of biotype *molestus* and hybrids (p<0.01), whereas in Italy wetlands were different due to the high proportion of biotype *molestus* (p<0.001).

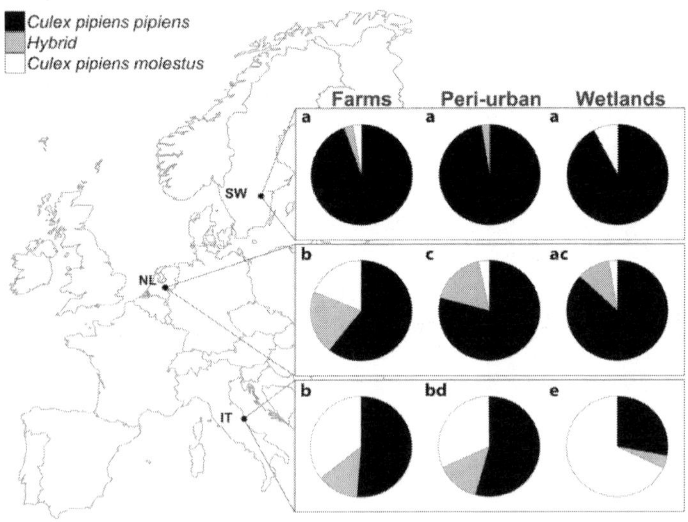

Figure 2. Within-effect of habitat within each of the three countries on the ratio of *Culex pipiens* **biotypes and their hybrids** (rows), and within-effect of country within each habitat type (columns). The sample size for each pie chart ranges between n = 26–115. Letters display significant differences between ratios shown in rows and columns, at a significance level of p < 0.05. SW = Sweden, NL = The Netherlands, and IT = Italy.

https://doi.org/10.1371/journal.pone.0166959.g002

Ratios of *Cx. pipiens* biotypes and hybrids were significantly different between countries within each of the habitat types ($p<0.001$; Fig 2). Farms in Sweden had a relatively higher proportion of biotype *pipiens* compared to Italy ($p<0.001$) and The Netherlands ($p<0.01$), which both had relatively more biotype *molestus* and hybrids. For peri-urban habitats, ratios were significantly different among all countries (pairwise comparisons: $p\leq0.001$), with a gradual increase of biotype *pipiens* towards northern latitudes. Wetlands in Italy had a relatively low proportion of biotype *pipiens* but high proportion of biotype *molestus*, compared to The Netherlands ($p<0.001$) and Sweden ($p<0.001$), which both had relatively high proportions of biotype *pipiens*.

DISCUSSION

The aim of this study was to assess the relative abundance of the *Cx. pipiens* biotypes and their hybrids in different habitats from northern to southern latitudes in Europe, using two trap types. We found a strong latitudinal effect on the ratios of the *Cx. pipiens* biotypes and hybrids, with a gradient of decreasing biotype *pipiens* from northern to southern latitudes. Habitat types also influenced the ratios of *Cx. pipiens* biotypes and hybrids, but effects were not consistent at the different latitudes.

Due to low numbers, *Cx. torrentium* was excluded from the analyses in this study. All *Cx. torrentium* females that we identified originated from Sweden, whereas no *Cx. torrentium* females were found in The Netherlands and Italy. These results are consistent with previous studies that showed a relatively high abundance of *Cx. torrentium* in northern European countries [24,31], and relatively low abundance or even absence in southern Europe [19].

No difference was found between the ratio of *Cx. pipiens* biotypes and their hybrids collected with the BGS and MMLP traps. Despite the differences in trapping mechanism between the two traps, there was no apparent difference in the attraction of the biotypes and their hybrids towards both traps. For studies focusing on relative abundance both traps can thus be used equally well, but differences in the total numbers of collected *Cx. pipiens* mosquitoes between both traps do exist. Results of studies on mosquito abundance with different traps may not be directly comparable.

Our study shows the sympatric occurrence of both *Cx. pipiens* biotypes and hybrids in aboveground habitats throughout Europe. These results are in line with previous findings from The Netherlands [22], Germany [21], Austria [27], Portugal [20,32], Italy [19], and Greece [26]. There was a clear gradient of decreasing biotype *pipiens* and increasing biotype *molestus* proportions from northern to southern latitudes. In Sweden the major part of the *Cx. pipiens* mosquitoes collected consisted of the *pipiens* biotype, whereas ratios in Italy were more equal between *pipiens* and *molestus* biotypes. This pattern was visible when ratios were determined per country without differentiating between the different habitat types, as well as when ratios for each country were split over the three habitat types. Previously, only a single record of the *molestus* biotype was known for Sweden [33]. In addition, the proportion of hybrids in Sweden was much lower compared to The Netherlands and Italy, which can be explained by the near absence of biotype *molestus* in all habitat types in Sweden. In this study we confirm that both biotypes and their hybrids occur aboveground at latitudes up to 58°24'36"N.

All peri-urban habitats combined had a relatively higher proportion of biotype *pipiens* and fewer biotype *molestus* compared to farms and wetlands. This pattern was, however, not consistent when comparing habitats within each of the three countries. In Sweden the ratios were similar for the three habitats, whereas in The Netherlands farms and in Italy wetlands ratios were different from the other two habitat types. This inconsistency could be explained by differences in, for instance, climate, microhabitat, availability of breeding sites, and hosts which all may influence the presence of the biotypes. These factors are likely to differ more between countries at different latitudes than between nearby habitats at one geographic location. Our findings also show variation at the local scale between habitats in The Netherlands and Italy. Especially the Italian wetlands stand out because of the high proportion of biotype *molestus*. Although a higher proportion of biotype *pipiens* was expected in such bird-rich habitats, a previous study showed high proportions of biotype *molestus* up to 82% in both urban and rural habitats with aboveground breeding sites in Italy [19]. The same study also showed high variation in *Cx. pipiens* biotype composition throughout Italy [19]. The ratios that we found for each country do, therefore, not represent an overall ratio of *Cx. pipiens* for the entire country, but rather for the specific sampling location. The sampling strategy used in our study is suitable for direct comparisons between locations at different latitudes, due to the consistent design over all countries. Studies that place traps at random locations throughout a country are more useful to get insight in local variation and dynamics of *Cx. pipiens* within a country [19–22,27].

Cx. pipiens populations that are dominated by biotype *pipiens* play an important role in the natural transmission cycle of WNV in birds, whereas the risk of WNV outbreaks among humans is increased in populations with high levels of hybridization [8]. Up to now, outbreaks of WNV among humans have only occurred in southern and central Europe, including Italy [34]. The overall proportion of hybrids is higher in The Netherlands than Italy, which is not consistent with the more equal proportion of biotypes in Italy than in The Netherlands. This indicates a more complex cause of hybridization than solely density dependence. If WNV would get established in The Netherlands, the higher degree of hybridization may result in a higher likelihood of bridging of WNV from birds to humans. However, other factors such as vector competence and climate determine whether transmission cycles can get established. Such factors are most likely limiting the transmission of WNV in northern Europe [35,36]. Future studies on vector competence of the *Cx. pipiens* biotypes and hybrids under different climatic scenarios are needed in order to gain more insight in the risk of the transmission of viruses by mosquitoes in Europe.

CONCLUSIONS

The BGS and MMLP traps collected equal ratios of the *Cx. pipiens* biotypes and their hybrids. A clear gradient of decreasing biotype *pipiens* and increasing biotype *molestus* proportions from northern to southern latitudes in Europe was found. Hybrids were found in all countries, but highest proportions were recorded in The Netherlands and Italy. Furthermore, *Cx. pipiens* ratios between habitat types were different. These differences were, however, not consistent when comparing habitat types within countries. Future research should focus on, (i) vector competence of *Cx. pipiens* biotypes and hybrids at different latitudes in Europe to assess the risks of WNV transmission in northern Europe, and (ii) the ecology of hybrids in order to estimate the risk of WNV being transmitted to humans.

ACKNOWLEDGMENTS

We would like to thank Claudia Damiani, Luca Bracchetti and Mahmod Hamdan from the Laboratory of Entomology and Parasitology, University of Camerino in Italy, and Rickard Pettersson and Malin Tälle from the department of Theoretical Biology, Linköping University in Sweden for their collaboration and assistance with data collection in the field. We also thank Marcel Dicke for providing comments on a previous version of this manuscript. Finally, we thank the participating farmers, nature area managers and home-owners in Sweden, The Netherlands, and Italy that provided us with the essential field locations.

REFERENCES

1. Morens DM, Folkers GK, Fauci AS. The challenge of emerging and re-emerging infectious diseases. Nature. 2004;430: 242–249. pmid:15241422
2. Davis CT, Ebel GD, Lanciotti RS, Brault AC, Guzman H, Siirin M, et al. Phylogenetic analysis of North American West Nile virus isolates, 2001–2004: evidence for the emergence of a dominant genotype. Virology. 2005;342: 252–265. pmid:16137736
3. Ebel GD, Carricaburu J, Young D, Bernard KA, Kramer LD. Genetic and phenotypic variation of West Nile virus in New York, 2000–2003. The American Journal of Tropical Medicine and Hygiene. 2004;71: 493–500. pmid:15516648
4. Snapinn KW, Holmes EC, Young DS, Bernard KA, Kramer LD, Ebel GD. Declining growth rate of West Nile virus in North America. Journal of Virology. 2007;81: 2531–2534. pmid:17182695
5. Calistri P, Giovannini A, Hubalek Z, Lonescu A, Monaco F, Savini G, et al. Epidemiology of West Nile in Europe and in the Mediterranean Basin. The Open Virology Journal. 2010;4: 29–37. pmid:20517490
6. Hubalek Z, Halouzka J. West Nile fever—a reemerging mosquito-borne viral disease in Europe. Emerging Infectious Diseases. 1999;5: 643–650. pmid:10511520
7. Sambri V, Capobianchi M, Charrel R, Fyodorova M, Gaibani P, Gould E, et al. West Nile virus in Europe: emergence, epidemiology, diagnosis, treatment, and prevention. Clinical Microbiology and Infection. 2013;19: 699–704. pmid:23594175
8. Fonseca DM, Keyghobadi N, Malcolm CA, Mehmet C, Schaffner F, Mogi M, et al. Emerging vectors in the *Culex pipiens* complex. Science. 2004;303: 1535–1538. pmid:15001783
9. Kilpatrick AM, Kramer LD, Campbell SR, Alleyne EO, Dobson AP, Daszak P. West Nile virus risk assessment and the bridge vector paradigm. Emerging Infectious Diseases. 2005;11: 425–429. pmid:15757558
10. Farajollahi A, Fonseca DM, Kramer LD, Kilpatrick AM. "Bird biting" mosquitoes and human disease: a review of the role of *Culex pipiens* complex mosquitoes in epidemiology. Infection, Genetics and

Evolution. 2011;11: 1577–1585. pmid:21875691

11. Becker N, Jöst A, Weitzel T. The Culex pipiens complex in Europe. Journal of the American Mosquito Control Association. 2012;28: 53–67.
12. Byrne K, Nichols RA. *Culex pipiens* in London Underground tunnels: differentiation between surface and subterranean populations. Heredity. 1999;82: 7–15. pmid:10200079
13. Dohm DJ, Turell MJ. Effect of incubation at overwintering temperatures on the replication of West Nile virus in New York *Culex pipiens* (Diptera: Culicidae). Journal of Medical Entomology. 2001;38: 462–464. pmid:11372976
14. Sanburg LL, Larsen JR. Effect of photoperiod and temperature on ovarian development in *Culex pipiens pipiens*. Journal of Insect Physiology. 1973;19: 1173–1190. pmid:4708144
15. Marshall JF, Staley J. Exhibition of autogenous and stenogamous characteristics by *Theobaldia subochrea*, Edwards (Diptera, Culicidae). Nature. 1936;137: 580–581.
16. Shute PG. A species of mosquito infesting deep shelters in London. The Lancet. 1941;237: 6–7.
17. Willcox W. Mosquitoes infesting deep shelters. The Lancet. 1941;237: 127.
18. Gomes B, Sousa CA, Vicente JL, Pinho L, Calderón I, Arez E, et al. Feeding patterns of *molestus* and *pipiens* forms of *Culex pipiens* (Diptera: Culicidae) in a region of high hybridization. Parasites & Vectors. 2013;6: 93.
19. Di Luca M, Toma L, Boccolini D, Severini F, La Rosa G, Minelli G, et al. Ecological distribution and CQ11 genetic structure of *Culex pipiens* complex (Diptera: Culicidae) in Italy. PLoS ONE. 2016;11: e0146476. pmid:26741494
20. Osório HC, Zé-Zé L, Amaro F, Nunes A, Alves MJ. Sympatric occurrence of *Culex pipiens* (Diptera, Culicidae) biotypes *pipiens*, *molestus* and their hybrids in Portugal, Western Europe: feeding patterns and habitat determinants. Medical and Veterinary Entomology. 2014;28: 103–109. pmid:23786327
21. Rudolf M, Czajka C, Börstler J, Melaun C, Jöst H, von Thien H, et al. First nationwide surveillance of *Culex pipiens* complex and *Culex torrentium* mosquitoes demonstrated the presence of *Culex*

pipiens biotype *pipiens/molestus* hybrids in Germany. PLoS ONE. 2013;8: e71832. pmid:24039724

22. Vogels CBF, Van De Peppel LJJ, Van Vliet AJH, Westenberg M, Ibañez-Justicia A, Stroo A, et al. Winter activity and aboveground hybridization between the two biotypes of the West Nile virus vector *Culex pipiens*. Vector-Borne and Zoonotic Diseases. 2015;15: 619–626. pmid:26394124
23. Fritz ML, Walker ED, Miller JR, Severson DW, Dworkin I. Divergent host preferences of above- and below-ground *Culex pipiens* mosquitoes and their hybrid offspring. Medical and Veterinary Entomology. 2015;29: 115–123. pmid:25600086
24. .Hesson JC, Rettich F, Merdić E, Vignjević G, Östman Ö, Schäfer M, et al. The arbovirus vector *Culex torrentium* is more prevalent than *Culex pipiens* in northern and central Europe. Medical and Veterinary Entomology. 2013;28: 179–186. pmid:23947434
25. Weitzel T, Braun K, Collado A, Jöst A, Becker N. Distribution and frequency of *Culex pipiens* and *Culex torrentium* (Culicidae) in Europe and diagnostic allozyme markers. European Mosquito Bulletin. 2011;29: 22–37.
26. Gomes B, Kioulos E, Papa A, Almeida APG, Vontas J, Pinto J. Distribution and hybridization of *Culex pipiens* forms in Greece during the West Nile virus outbreak of 2010. Infection, Genetics and Evolution. 2013;16: 218–225. pmid:23466890
27. Zittra C, Flechl E, Kothmayer M, Vitecek S, Rossiter H, Zechmeister T, et al. Ecological characterization and molecular differentiation of *Culex pipiens* complex taxa and *Culex torrentium* in eastern Austria. Parasites & Vectors. 2016;9: 1–9.
28. Smallegange RC, Schmied WH, van Roey KJ, Verhulst NO, Spitzen J, Mukabana WR, et al. Sugar-fermenting yeast as an organic source of carbon dioxide to attract the malaria mosquito *Anopheles gambiae*. Malaria Journal. 2010;9: 292. pmid:20973963
29. Becker N, Petrić D, Boase C, Lane J, Zgomba M, Dahl C, et al. Mosquitoes and their control. Springer; 2010.
30. R Core Team (2016) R: A language and environment for statistical computing. Vienna, Austria: R Foundation for Statistical Computing.
31. Hesson JC, Östman Ö, Schäfer M, Lundström JO. Geographic

distribution and relative abundance of the sibling vector species *Culex torrentium* and *Culex pipiens* in Sweden. Vector-Borne and Zoonotic Diseases. 2011;11: 1383–1389. pmid:21923273

32. .Gomes B, Sousa CA, Novo MT, Freitas FB, Alves R, Côrte-Real AR, et al. Asymmetric introgression between sympatric molestus and pipiens forms of *Culex pipiens* (Diptera: Culicidae) in the Comporta region, Portugal. BMC Evolutionary Biology. 2009;9: 262. pmid:19895687

33. Schäfer M, Lundström JO. Comparison of mosquito (Diptera: Culicidae) fauna characteristics of forested wetlands in Sweden. Annals of the Entomological Society of America. 2001;94: 576–582.

34. Rizzoli A, Jiménez-Clavero MA, Barzon L, Cordioli P, Figuerola J, Koraka P, et al. The challenge of west nile virus in Europe: Knowledge gaps and research priorities. Eurosurveillance. 2015;20.

35. Fros JJ, Geertsema C, Vogels CB, Roosjen PP, Failloux AB, Vlak JM, et al. West Nile Virus: High transmission rate in north-western European mosquitoes indicates its epidemic potential and warrants increased surveillance. PLoS Neglected Tropical Diseases. 2015;9: e0003956. pmid:26225555

36. Vogels CBF, Fros JJ, Göertz GP, Pijlman GP, Koenraadt CJM. Vector competence of northern European *Culex pipiens* biotypes and hybrids for West Nile virus is differentially affected by temperature. Parasites & Vectors. 2016;9: 1–7.

SECTION 4:
TICKS

18

Bowman A, Nuttall P: Ticks: Biology, Disease and Control

Alan R Walker

Royal (Dick) School of Veterinary Studies, Summerhall Place, Edinburgh, EH9 1QH, UK

REVIEW

This fine book demonstrates how far the field of tick biology has progressed in the last 40 years. I remember discussions with colleagues who were studying ticks and associated diseases, whilst I was studying insects as virus vectors, all in a laboratory in Kenya. They claimed their job was much tougher because the textbooks they had to help them consisted of one each from Yuri Balashov, Don Arthur and Harry Hoogstraal; whilst entomology texts could then fill a small library. Apart from a dozen recent texts on various aspects of tick biology we now have Daniel Sonenshine's comprehensive explanatory

Citation: Walker, A.R. Bowman A, Nuttall P: Ticks: Biology, Disease and Control. Parasites Vectors 2, 1 (2009). https://doi.org/10.1186/1756-3305-2-1

textbook and this wide multi-author compilation of specialist topics edited by Alan Bowman and Pat Nuttall. It originated in a commissioned special issue of *Parasitology* in 2004, now expanded by further authors and revisions. So unlike many books derived from conference proceedings, 51 well established researchers have written 21 substantial review articles, all supported by long lists of recent literature. The coverage is good, so it is easier to indicate topics not covered directly: gut and digestion, and behaviour are missing; and tick associated pathogens comprise one review on viruses, two on one bacterium each, and two on one genus of protozoa each. Of course, any attempt to cover the gamut of pathogens requires another full textbook, so some generic examples are appropriate for this book. Furthermore, the coverage is a good reflection of current research interests and so it must be emphasized here that this is a book intended for researchers rather than non-specialists seeking facts about the basics of tick biology.

Why study ticks? The standard answer is because they are economically and socially important as agents of disease. The corollary being that studying them is important. But ask a farmer in Uganda what she does about ticks on her Friesian diary herd – once a week dip in organophosphate! Or try sitting on a public health advisory board writing pamphlets about Lyme disease – wear long trousers tucked into socks! And if that is not discord enough, consider the increasing ethical problem of studying ticks. Even understanding their ecology needs information derived from feeding ticks on experimental animals – a nasty business indeed. Resolutions to this dissonance can be found in this book. Improvements in avoiding, preventing and curing vector associated diseases are at least as likely to derive, very unpredictably, from a broad advance in deep understanding, as they are from any specific applied study.

Ticks are not insects, and the great clarification of their population dynamics in relation to pathogen transmission is a major advance for vector biology in general. Tick salivary glands are wonderful subjects for anatomical, physiological and biochemical study and as a source of pharmacological agents they are a deep and rich mine. The ability of ticks to evade host immune resistance provides routes for advances in fundamental immunology as well as inspiration for pharmaceuticals for human and animal treatment. Similarly the extraordinary adaptations of tick transmitted pathogens to exploit these immune evasions, or the normally functioning dermal hypersensitivity reactions to ticks, has greatly increased our understanding of the biology of viruses, bacteria and protozoa. The natural nidality of vector associated pathogens is strongly exemplified by tick spatial distributions. So the advent of geographical information systems, coupled with studies on microclimatic factors on tick survival, has consolidated understanding of a field of such public concern that we now get news of it on television and in newspapers. That is: climate change and its threats. These topics are a selection from this book.

But, inevitably and sadly for those romantics amongst us, the best tick is a dead tick. Thus it remains vital that the essential role of conventional acaricide use remains appreciated; and scope for improvements on it, such as biological agents, manipulations with pheromones, and anti-tick vaccines, continue to be actively pursued at the applied end of a tick research front. The anti-tick vaccine of course remains a star in the tick biologist's storybook. As a final example of what I have been promoting it could not be better. The method of manufacturing specific proteins by the recombinant nucleic acid route was first used to great commercial effect with insulin. The method was invented when a small group of university researchers realized the new understanding of plasmids and how to use them for pure genetics studies could be combined with the new understanding of how DNA works to code for proteins. Structure of DNA can be traced back many decades to studies of how to use X-rays to analyse the three dimensional folding of proteins. To summarize, this book is a must-buy for every tick biologist, and library of every veterinary and medical and biology department on the globe. It is a pleasure to see Cambridge University Press produce it with their exemplary style.

19

Multi-source Analysis Reveals Latitudinal and Altitudinal shifts in range of Ixodes Ricinus at its northern Distribution Limit

Solveig Jore[1,] Hildegunn Viljugrein[1,4], Merete Hofshagen[1], Hege Brun-Hansen[2], Anja B Kristoffersen[1,5], Karin Nygård[3], Edgar Brun[1], Preben Ottesen[3], Bente K Sævik[2] and Bjørnar Ytrehus[1]

[1]Norwegian Veterinary Institute, Oslo, Norway
[2]Norwegian School of Veterinary Science, Oslo, Norway
[3]Norwegian Institute of Public Health, Oslo, Norway
[4]Centre for Ecological and Evolutionary Synthesis (CEES), University of Oslo, Oslo, Norway
[5]University of Oslo, Oslo, Norway

ABSTRACT

Background

There is increasing evidence for a latitudinal and altitudinal shift in the distribution range of *Ixodes ricinus*. The reported incidence of tick-borne disease in humans is on the rise in many European countries and has raised political concern and attracted media attention. It is disputed which factors

Citation: Jore, S., Viljugrein, H., Hofshagen, M. et al. Multi-source analysis reveals latitudinal and altitudinal shifts in range of Ixodes ricinus at its northern distribution limit. Parasites Vectors 4, 84 (2011). https://doi.org/10.1186/1756-3305-4-84

are responsible for these trends, though many ascribe shifts in distribution range to climate changes. Any possible climate effect would be most easily noticeable close to the tick's geographical distribution limits. In Norway- being the northern limit of this species in Europe- no documentation of changes in range has been published. The objectives of this study were to describe the distribution of *I. ricinus* in Norway and to evaluate if any range shifts have occurred relative to historical descriptions.

Methods

Multiple data sources - such as tick-sighting reports from veterinarians, hunters, and the general public - and surveillance of human and animal tick-borne diseases were compared to describe the present distribution of *I. ricinus* in Norway. Correlation between data sources and visual comparison of maps revealed spatial consistency. In order to identify the main spatial pattern of tick abundance, a principal component analysis (PCA) was used to obtain a weighted mean of four data sources. The weighted mean explained 67% of the variation of the data sources covering Norway's 430 municipalities and was used to depict the present distribution of *I. ricinus*. To evaluate if any geographical range shift has occurred in recent decades, the present distribution was compared to historical data from 1943 and 1983.

Results

Tick-borne disease and/or observations of *I. ricinus* was reported in municipalities up to an altitude of 583 metres above sea level (MASL) and is now present in coastal municipalities north to approximately 69°N.

Conclusion

I. ricinus is currently found further north and at higher altitudes than described in historical records. The approach used in this study, a multi-source analysis, proved useful to assess alterations in tick distribution.

BACKGROUND

Vector-borne diseases were recently identified by 30 European Ministries of Health as the biggest health threat arising from environmental change [1]. The two most prevalent tick-borne human diseases in Europe, Lyme borreliosis (LB) and tick-borne encephalitis (TBE), were ranked first and second [1]. This highlights the importance of establishing knowledge of current and future distribution ranges of the vector of these diseases, *Ixodes ricinus* (*I. ricinus*). In recent years, there has been an undocumented view in Norway that both tick abundance and their distribution range have increased. In concordance with this, the prevalence of LB and TBE in humans has shown an increasing trend (Norwegian Surveillance System for Communicable Diseases (MSIS)). Furthermore, the tick-borne disease bovine babesiosis has also had an upsurge the later years (Norwegian Cattle Health Recording system (NCHRS).

Ticks, spending the greater part of their life cycle free within their habitat, are at the mercy of abiotic factors such as climate [2]. Changes toward a warmer and wetter climate are likely to affect the distribution and abundance of ticks and, hence, the incidence of tick-borne diseases [2, 3]. Possible climate effects would be more easily noticeable close to the ticks' geographical distribution limits [3]. Mapping tick distribution is an inherently difficult task because of the complex ecology and focal distribution of *I. ricinus*. The main approaches to determine distribution are classified as model estimates based on climate or habitat suitability, indirect evidence of ticks by presence of tick-borne infection in hosts and direct observation of ticks - by scientists or by questionnaire surveys -, in vegetation or on host animals. All these approaches have their shortcomings, and studies relying on one method should be interpreted with care.

In 1935 to 1942, Tambs-Lyche [4] surveyed the distribution of *I. ricinus* in Norway by collecting ticks from domestic animals and gathering information from veterinarians concerning local presence of ticks and bovine babesiosis. More recent information about the distribution of *I. ricinus* was published by Mehl in 1983 [5]. According to both studies *I. ricinus* was distributed along the coastline of Norway to 66°N. No up-to-date maps have been published in scientific literature since 1983. *I. ricinus* is the tick species most commonly encountered by humans in Norway, although a total of 14 different species of ticks have been identified in the country [5].

The objective of this study was to improve the accuracy of the distribution estimates of *I. ricinus* by utilizing data from several sources to describe the distribution, and to evaluate if any range shifts have occurred relative to historical descriptions from 1943 and 1983.

MATERIALS AND METHODS

Historical Data

Tambs-Lyche's distribution data

In 1935 the Norwegian Veterinary Institute (NVI) requested all veterinarians in Norway to sample ticks from domestic animals and send the specimens to the Institute. In the mid thirties there were 362 authorized veterinarians in clinical practice representing all municipalities in Norway. This material consisted of around 1400 ticks collected from 97 different locations in 14 out of 19 counties, and was handed over to Tambs-Lyche (see Additional file 1). All ticks were identified as *I. ricinus*. Information on the occurrence of bovine babesiosis was collected by written requests to veterinarians. All together, information was collected from 519 municipalities, representing 76% of the 682 rural municipalities at that time. According to the information received, *I. ricinus* was restricted to: 1) the lowlands (below 150-160 metres above sea level (MASL)) in the south-eastern part of Norway, 2) below 350-500 MASL on the western coast of South Norway, and 3) below 100-150 MASL in the northernmost part of the distribution range up to 66°N (Figure 1). Tambs-Lyche concluded the most outlying districts of the western coast as tick-free, and that ticks were not found on red deer in Norway [4].

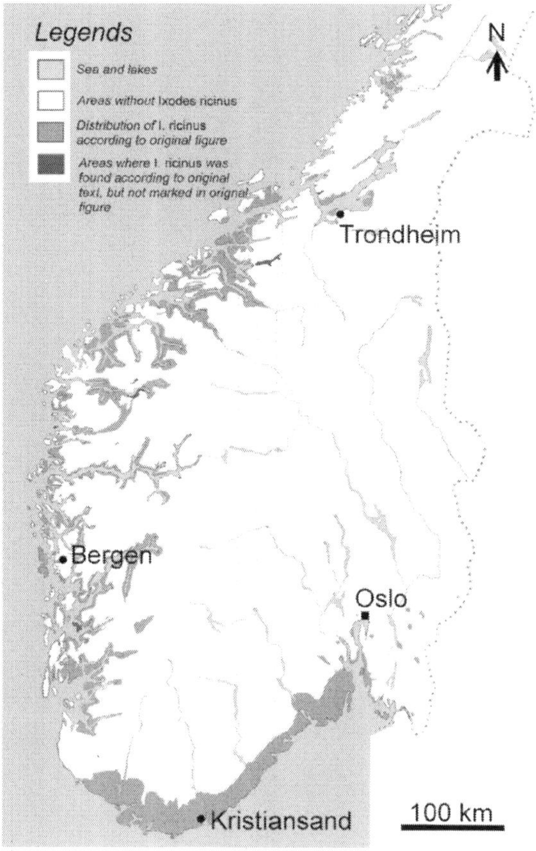

Figure 1. The distribution of *Ixodes ricinus* in South Norway 1935-43 according to Tambs-Lyche[4]. The green colour depicts areas where *I. ricinus* and/or bovine besiosis occurs. The figure is adjusted and modified from Tambs-Lyche's Figure 10 to improve conformity between the map and Tambs-Lyche's written description. (Courtesy of Norsk veterinærtidsskrift)

Mehl's distribution data

Mehl summarized his and his co-workers' findings through 15 years of extensive ectoparasite investigations in a paper that was published in 1983 [5]. They collected ticks from birds' nests, mammals, birds, reptiles and humans in Norway, identified species of ticks, and described the geographical distribution and the hosts connected to the tick species. A total of 8600 ticks, 900 nests and 7640 host animals from 220 species were examined. Mehl concluded that the distribution of *I. ricinus* was limited to a narrow zone along the southern coast from the Oslofjord to Jæren (50-150 MASL).

Along the western coast the distribution was wider and extended further into the country along the valleys and fjords (from sea level extending upto 400 MASL). Mehl's registrations consolidated Tambs-Lyches' and no definite changes in geographical distribution from 1943 to 1983 can be inferred on the basis of their investigations.

Data Sources

Lyme borreliosis incidence in humans

The source of the incidence data for LB was MSIS, where microbiological laboratories and general practitioners compulsorily report cases to the Norwegian Institute of Public Health (NIPH). The case definition for laboratory confirmed LB was "clinically suspected disseminated or chronic disease and demonstration of *Borrelia burgdorferi* sensu lato or definite antibody titres". LB was made notifiable in 1991. From 1991 to 1994, all cases confirmed by serology or culture were notifiable, including cases of erythema migrans. From 1995 onwards, only disseminated disease/chronic manifestations of LB were notifiable. The data used in the current study include cases per municipality reported to MSIS during 1991-2008. Only cases where assumed place of infection was known were included.

Bovine babesiosis incidence

NCHRS records the occurrence of diseases in cattle, including babesiosis caused by *Babesia divergens*. The treating veterinarian has to diagnose, describe and sign off the treatments given to the individual dairy cow on Cow Health Card. Detailed information is available at http://www.storfehelse.no NCHRS has been the national data base for cattle diseases since 1975 and covers over 90% of all dairy cattle in Norway. The diseases recorded are only on those treated by veterinarians. The registrations are reported once a month by the Dairy Herd Improvement Services into the NCHRS, and are statistically processed annually. An evaluation of the NCHRS has shown that when new disease codes are introduced, it takes about 5 years before the data are reliable [6]. Bovine babesiosis was introduced as a disease code in 1989, had only a few registrations early on and consequently only data from 1996-2008 are used in this study. Data on bovine babesiosis were obtained as number of treatments per farm per year, the identity of the cow treated, date of treatment and the treating veterinarian.

Cervid hunters' webpage registration of tick observations

In August 2007, a webpage was established as a joint effort between NVI and NIPH. Hunters of roe deer (*Capreolus capreolus*), red deer (*Cervus elaphus*) and moose (*Alces alces*) were encouraged to register felled animals in an on-line map application (http://www.flattogflue.no) and indicate to which degree the animal was infested with ticks. The hunters were asked to examine the cervid on the following locations; the ears, the nape of the neck, the muzzle, the axillae, distal legs and in the groin. They were asked to grade their findings as zero, few = 1-20, some = 21-100 and many >100. They should also register the place and date of killing, species, sex and age.

Newspaper webpage registration of tick observations

From June to September 2009, Norway's largest newspaper "Aftenposten" established a webpage where the readers could register observations of ticks (http://www.aftenposten.no/nyheter/iriks/article3139218.ece). There was a picture of *I. ricinus* on the webpage and a general description of the tick and the diseases it can transmit. The readers were asked to register municipality and the name of the location where they observed the tick. Eye-catching links to the registration page were visible on the home page of the newspaper, and the service was promoted in all articles describing ticks or tick-borne diseases, both on the web and in the printed edition of the paper. The newspaper has close to 1 080 000 subscribers and readers of the web-edition. Norway had 4 850 000 inhabitants by the end of 2009.

Veterinary survey of tick observations

In September 2009, a web-based questionnaire (Questback®) was e-mailed to members of the Norwegian Veterinary Association. The questionnaire comprised 14 closed questions with multiple choices; which municipalities the practice covered, how long the practice had been in the area, animal species seen in the practice, whether the veterinarian or other veterinarians/ pet owners had observed ticks, travel history for the animals, prophylactic use of tick repellents, when ticks were observed for the first time in current practice, how often and on which animals ticks were observed, if they had diagnosed tick-borne diseases, when diagnosed for the first time, what kind of diseases and how these diseases were diagnosed.

Demographic data

In 2009 there were 430 municipalities in Norway. The human population data were derived from Statistics Norway. The cattle population data were derived from the Register of Production Subsidies in Norway. The Norwegian Mapping Authority provided altitude (MASL) and fraction of open terrain data at the municipality level.

Data Analyses

Calculation of the cumulative incidence of LB was based on the total number of cases per municipality in the period 1991-2008 divided by the mean population for each municipality in the same period. Calculation of the cumulative incidence of bovine babesiosis was based on the total number of cases per municipality from 1996-2008 divided by the mean number of cattle in each municipality in the same period. The registrations from the cervid hunters' webpage represented less than 1% of the bagged cervid population and these data were not used in the principal component analysis (PCA) described below.

The registration proportion generated from data in the newspaper webpage registration was based on the number of tick registrations in each municipality divided by the population in each municipality in 2009.

In the veterinary survey the frequency of tick observations for each respondent (F_j) was graded as follows: 4 (daily basis), 3 (weekly basis), 2 (monthly basis), 1 (rare; <3 tick observations per year), 0 (never) and NA (no answer). When a respondent worked in several municipalities, the reported frequency of tick observations (F_j) was applied to all municipalities unless differences were quoted by the respondent. The frequency in each municipality (f_i) was calculated based on a weighted response of each answering respondent (*j*) with clinical practice in a given municipality *i* based on the formula:

$$fi = \frac{\sum_{j \in i} W_j \cdot F_j}{\sum_{j \in i} W_j}$$

Where the weight (W_j) depended on how many municipalities (N_j) each respondent covered in his/her practice:

$$W_j = \frac{1}{N_j}$$

A respondent covering fewer municipalities was thus given higher weight in those municipalities.

A PCA was used to identify the main spatial pattern of tick abundance as represented by the two disease registers, the newspaper web registrations and the veterinary survey. Due to the different nature of these four data sources and the skewed distribution of the disease incidences, data from each source was re-scaled, dividing by their maximum value before the PCA was performed. The standard deviation of the re-scaled data of LB, bovine babesiosis, newspaper webpage registration and veterinary survey was 0.07, 0.13, 0.10 and 0.24, respectively. The first principal component (PC1) of a PCA is given by the linear combination of the data, calculated such that it accounts for the greatest possible variance in the dataset. Weights of the data are calculated with the constraint that the sum of the squared weights is equal to one [7]. By definition, PC1 corresponds to a weighted mean of the four data sources, for which the data source with the highest variance (i.e. data from the veterinary survey) contributes most. The sensitivity of re-scaling the data was tested by comparing the resulting PC1 with the principal component obtained from standardized (mean zero and variance equal to one) data, i.e. PC1 then corresponding to the mean of standardized data.

All mapping was performed in ESRI ArcGIS 9.2 (ESRI, Redlands, CA, USA) and all analyses were performed in R version 2.6.2 [8]. Spatial structure in the data was also investigated by comparing the areas of frequent tick observations with the presence/absence of tick-borne diseases. Pair - wise correlations were calculated and reported as Kendall's correlations.

RESULTS

Lyme borreliosis Incidence in Humans

The assumed place of infection was known for 1611 (39%) notifications in 188 municipalities ranging from 1-124 notifications (median 4) per municipality (Figure 2). The number of reported cases per year increased from 4 (1991) to 79 (2008), with a maximum of 91 recorded in 2007.

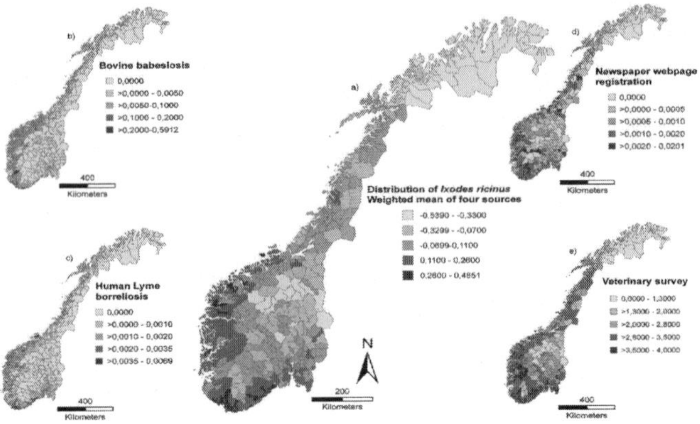

Figure 2. The present distribution of *Ixodes ricinus* in Norway depicted by (a) the weighted mean obtained by the first principal component (PC1) of a PCA-analysis of four different sources (b-e). The distribution map displays tick abundance within municipalities with increasing darkness of colour. The inserted maps show: (b) the cumulative incidence of bovine babesiosis in the Norwegian Cattle Health Recording system in the period 1996-2008 (c) the cumulative incidence of human Lyme borreliosis in Norwegian Surveillance System for Communicable Diseases during the period 1991-2008 (with known place of infection) (d) the registration proportion of tick observations registered via the newspaper webpage during summer 2009 and (e) the frequency of tick observations from the veterinary survey performed in 2009 categorized as 0: never observed, 1: rarely observed (<3 tick observations during a year), 2: monthly observation, 3: weekly observation, 4: daily observation. Each of the variables is depicted on a municipal basis.

Bovine Babesiosis Incidence

A total of 1634 cases of bovine babesiosis in 115 municipalities were reported during 1996-2008 (Figure 2). The number of cases per municipality ranged from 1 - 432 (median 8). The number of cases per farms ranged from 1 - 8 (median 2). The total number of reported cases per year increased from 8 (1996) to 161 (2008), with a maximum of 238 recorded in 2006.

Cervid hunters' Webpage Registration of tick Observations

There were 462, 426 and 255 registrations in 170 municipalities from moose, red deer and roe deer respectively, during the period 2007-2009. The number of registrations per municipality ranged from 1 - 40, with a median of 3 (Figure 3). The cervid hunters' registration gave the impression that positive observations of ticks typically were seen on the seaside and along rivers and lakes in the valleys of the inland, while negative observations most consistently were found in the hills, plateaus or mountains between the valleys (Figure 3). The low number of registrations, in relation to the total number of bagged cervids, did not allow these data to be included in the PCA.

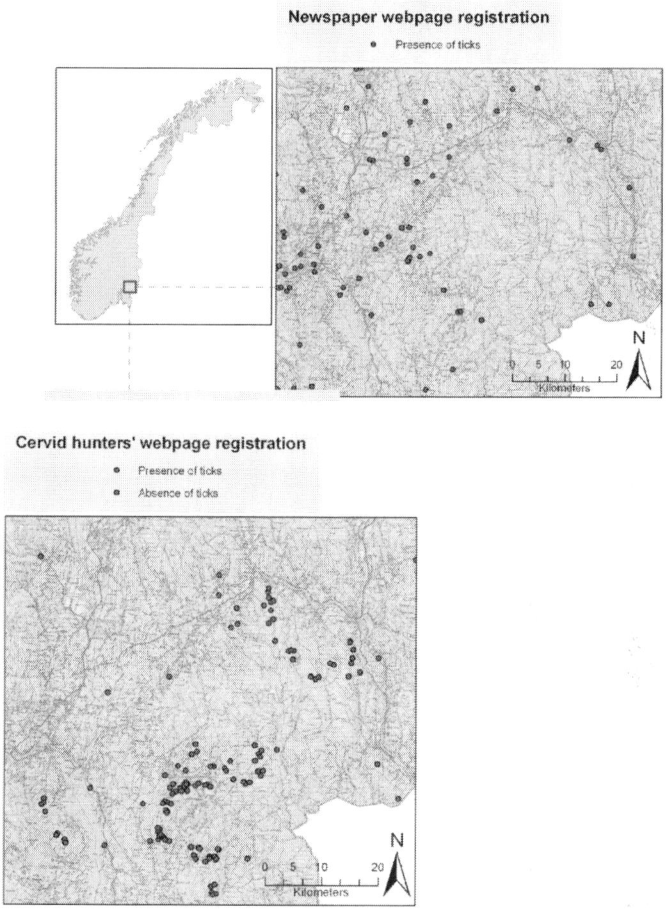

Figure 3. Distribution of *Ixodes ricinus* in a small area in south eastern Norway based on newspaper webpage registrations in 2009 and cervid hunters' webpage registrations in the period 2007-2009.

Newspaper Webpage Registration of tick Observations

A total of 3794 registrations in 338 municipalities (79% of all 430 municipalities) were done by the readers on the webpage, ranging from 1-152 registrations (median 5) per municipality (Figure 2). From June 25th to August 25th 2009 there were around 40.000 accesses at the webpage. As for the cervid hunters' registration, the point registrations made in the web page gave the impression that ticks were observed along rivers, lakes, in the valleys and on the seaside.

Veterinary Survey of Tick Observations

A total of 563 veterinarians (62% response rate) contributed with data, representing all the 430 municipalities in Norway (Figure 2). The mean number of municipalities per veterinarian was 4.2 (the 5th to 95th percentile range was (1.0, 13.6)). Fifty-five municipalities had only one respondent, and on average there were 5.4 respondents per municipality. There were 146 veterinarians with mainly large animal practice, 216 small animal practitioners, 22 horse practitioners and 179 with mixed practice. Altogether 169 veterinarians had been in the present practice since before 1990, 133 had practiced since 1990-1999, 127 since 2000-2005 and 134 since 2006. Ticks were mainly found on dogs (reported by 524 veterinarians) and cats (484) but also on cows (186), sheep (176), horses (131), cervids (14), hedgehogs (1), humans (2), hares (4) and mice (2). Those that reported ticks from hedgehogs, hares and mice, also reported ticks on animals such as cows, sheep, cats or dogs in the same municipality. The agreement between veterinarians within a municipality was in general high, as shown by the mean weighted variance being 0.36 (the 5th to 95th percentile range was (0.05, 2.2)).

Multi-Source Analysis

A visual comparison using maps of cumulative incidences of LB, bovine babesiosis, incidence data from the veterinary survey and the registration proportion from the newspaper surveys revealed spatial consistency (Figure 2). The LB, veterinary survey and newspaper registration maps show concurrent northern distribution limits at approximately 69°N, while the northern distribution limit of bovine babesiosis is at approximately 65°N. A cross-correlation analysis between the cumulative incidence of LB and bovine babesiosis respectively, and the frequency of tick observation from the veterinary survey showed significant correlations (0.42, $p <$

0.001 and 0.26, $p < 0.001$). However, a total of 60 municipalities with a tick frequency reported as daily or weekly had no registrations of LB or bovine babesiosis (Figure 4). Figure 2a shows the relative differences in tick abundance between municipalities as approximated by the weighted mean (PC1). PC1 explained 67% of the variation in the four data sources from Norway's 430 municipalities and is strongly correlated (0.86) to the average of these four re-scaled sources. The pair-wise correlations with PC1 are 0.52, 0.30, 0.44 and 0.90 for the Lyme borreliosis, bovine babesiosis, the newspaper webpage registration and veterinary survey, respectively. If the bovine babesiosis data is excluded from the PCA, the first component explained 71% of the variation in the data.

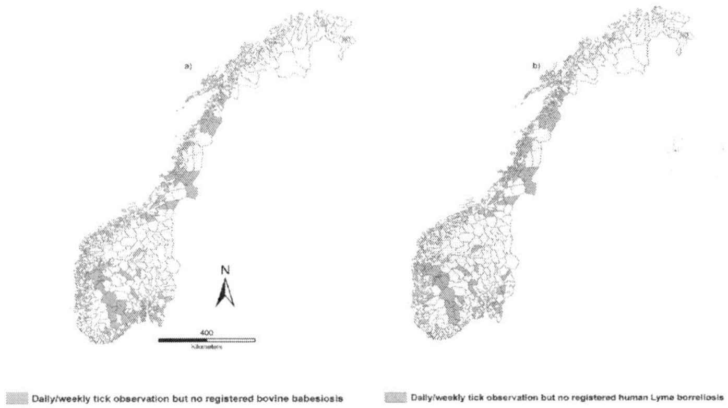

Figure 4. The municipalities (in blue) where the veterinary survey reported daily or weekly tick observations coupled with (a) absence of bovine babesiosis registrations in Norwegian Cattle Health Recording system 1996-2008 and (b) absence of Lyme borreliosis notifications in humans according to Norwegian Surveillance System for Communicable Diseases 1991-2008.

The PC1 map was relatively insensitive to the method of re-scaling data as shown by a correlation of 0.80 ($p < 0.001$) between the PC1 resulting from the two different methods (the data standardized to unit variance or by dividing by maximum value). PC1 is significantly negatively correlated to the municipalities' fraction of total area above 600 MASL (0.20, $p < 0.001$) and the fraction of uncultivated open terrain (0.28, $p < 0.001$).

Comparison between Historical and Present Distribution Data

I. ricinus is now known to be present in northern coastal municipalities to approximately 69°N, as compared with 66°N reported by Tambs-Lyche in

1943 and Mehl in 1983, and in inland municipalities where it previously was not registered. In South and Central Norway, only 13 municipalities had no reports of ticks. In these municipalities, 49-99.9% of the area has an elevation above 600 MASL. The tick has now been detected up to an altitude of at least 583 MASL, the lowest elevation in the municipality with the highest altitude reporting tick presence. In addition, the ticks are now present in outlying western coastal municipalities which Tambs-Lyche in the 1940s regarded as tick free. The registrations done by the cervid hunters clearly shows that ticks are highly prevalent on red deer (and roe deer and moose) in Norway, in contrast to what Tambs-Lyche concluded.

DISCUSSION

The updated distribution map shows shifts in latitudinal and altitudinal distribution of *I. ricinus* in Norway, compared to previously published maps of 1943 and 1983. While *I. ricinus* previously was restricted to coastal areas, it is now found in inland and mountainous areas of South and Central Norway, and has expanded its northern distribution limit approximately 400 km. This may imply that the proportion of the human and animal population in Norway potentially at risk for exposure to tick-borne pathogens is increasing and that the public awareness about possible risk in new areas needs to be considered.

The abundance of *I. ricinus* may vary greatly within short distances, and also according to season, weather conditions and time of day. This makes mapping by direct observation vulnerable to temporal and spatial variation. Mapping the distribution by use of multiple data sources as in the present study generates a cogent description. The different sources show a high degree of consistency. This study is based on «presence only» records, with the exception of the cervid hunters' webpage registration, which represents both presence and absence data. If both presence and absence data were available for each of the data sources, this would have strengthened the ability to depict the borderline areas for the distribution of *I. ricinus* in Norway. Similar approaches could be used to assess future range shifts of ticks and other disease vectors.

The latitudinal and altitudinal shift in range is in concordance with studies reporting alterations in tick distribution in Europe and North-America [9–14]. In Sweden, climate change has been associated with changes in distribution and abundance of *I. ricinus*, and the northern distribution limit expanded from below 61°N in the late 1980s to 66°N today [11, 15–18]. A shift in the altitudinal distribution of *I. ricinus* has been detected in the Czech Republic [9, 19–21] and Scotland [10], and it is suggested that the abundance of ticks at higher altitudes will increase as a response to climate change. Changes in tick distribution and abundance are however multi-factorial in cause and are likely to have multidirectional effects. Furthermore, the incidence of human disease is influenced by socioeconomic factors, making it difficult to assess the effects of climate change alone [22–24].

Norwegian series of annual mean temperature show two periods of statistically significant warming: one from 1900-1940 and then one from 1970-1994 (end of the series). The annual mean temperature has increased by about 0.7°C during the 20th century [25]. Analyses of time series from 1871-1990 of monthly mean temperature from seven weather stations located across the country, representing inland - and coastal climate, from 58°N-70°N, showed that the frost - free season length and the growth season length had increased at all stations by 10-20 days/100 years [26].

Since 1940 there has been an urbanisation trend in Norway, as in the rest of Europe. However, a dispersed pattern of settlements has been a political aim in Norway, and the country has one of the highest rural/urban breakdowns in Europe (http://www.ssb.no). This, together with the vast number of cabins and holiday-homes in the countryside, which is primarily used during the tick-season, makes it unlikely that the observed expansion in tick distribution is driven by demographic changes.

The present study analysed data on the municipality level. There is considerable local variation in tick distribution within a municipality. Typically, positive registrations were along rivers and lakes, while negative registrations were in the forested hills between them (Figure 3). This scattered and focal distribution is in concordance with previous descriptions [27].

The specificity of LB as a measurement of presence of *I. ricinus* should be high, since *I. ricinus* is the dominant vector of LB in Norway. *I. hexagonus, I. uriae* and possibly also *I. trianguliceps* may transmit *B. burgdorferi*, but due to the habitats of these species, they only occasionally bite humans. Consequently, positive disease reports caused by species other than *I. ricinus* must be regarded as exceptional. It may be argued that as tick-bites on humans easily go unnoticed, there can be considerable time between the tick-bite and the clinical diagnosis, and there can be great uncertainty about place of infection. To increase certainty, only notifications with information about assumed place of infection (39% of the cases) were used. The sensitivity of

LB as a marker of *I. ricinus* presence may be low as the incidence of LB is probably conservative estimates due to diagnostic challenges. The density of the tick population probably must exceed a certain threshold level before the pathogen is maintained in cycles, meaning that a low abundance of ticks will not be detected by disease reporting. Absence of disease therefore does not rule out the presence of ticks. There are also geographical differences in the prevalence of different strains of *B. burgdorferi* s.l. in ticks [28], providing variation in risk of infection and clinical manifestations between areas which may decrease or increase the registered incidence of disease relative to the abundance of ticks. Improvement in the notification systems, better diagnostic tests and increased public awareness may explain some of the increase in incidence and distribution range of the reported incidence of LB in humans.

Among the ticks present in Norway, only *I. ricinus* is known to transmit *B. divergens*: the cause of bovine babesiosis. Acute bovine babesiosis induces characteristic clinical signs and has few differential diagnoses ensuring reliable indications of presence of *I. ricinus*. Absence of registrations should be interpreted with care, as most cases of clinical babesiosis are seen in areas with intermediate infection pressure [29]. In the western part of South Norway, where most of the cases are distributed, the municipalities are more heterogeneous in terms of altitude range and vegetation which may induce intermediate infection pressure zones. In areas with high infection pressure, where a high percentage of ticks are infested, many animals will be exposed when young and acquire immunity without showing clinical signs. Immunity is reinforced by repeated infections in older cattle. Hence, in highly endemic areas clinical cases will seldom occur [30]. The lower correlation of bovine babesiosis to PC1 compared to the other data sources included in the PCA, probably reflects this non-linear relationship between tick abundance and clinical diagnosis, in addition to the more restricted distribution of cases.

Only licenced hunters can access and register findings on the website http://www.flattogflue.no. In Norway, no other tick than *I. ricinus* has been reported on cervids (see Additional file 1). Apart from the deer ked (*Lipoptena cervi*), which is only found in a limited region of the country and also included in this website registration, there are few other ectoparasites on cervids in Norway. Counting ticks on wild cervids is probably amongst the best way to estimate the tick-burden in the vegetation, as cervids live in tick habitat continuously, and feeding ticks are not as vulnerable to temporal variation as questing ticks are. However, animals shot in late autumn may be free from ticks, even in tick infested areas, as ticks are less active when the temperature falls. The registering system is, for the time being, limited by the sparseness of the data.

The newspaper "Aftenposten" is one of the main newspapers in the country. Although readers of the paper edition mostly are situated in urban and south-eastern part of the country, the web-edition can be freely accessed throughout the country. There are few tick species (see Additional file 1) that can be confused with *I. ricinus*. Still, the ability of an average newspaper reader to differentiate *I. ricinus* from other tick species as well as other arthropods must be taken into account. This ability must be expected to decline the further away from the core distribution range one gets, and it can thus be questioned how reliable outliers in this survey are. Despite possible recall bias, information bias and selection bias the newspaper survey had a high correlation to PC1, which indicates that this method of recording can add valuable information when assessed and used together with other sources.

The veterinary survey was representative for the population of clinical veterinarians, had a high response rate and covered all municipalities. A targeted survey amongst professionals where all the municipalities are covered should ensure validity. Nevertheless, this survey will to a certain degree describe the general public's perception of changes in tick abundance. Possible recall can be biased by the increased awareness by the veterinarians, especially in areas where the tick has appeared recently, and this may overestimate the perceived tick density. The majority of tick observations were from dogs and cats, and although the dominant tick species feeding on companion animals is *I. ricinus*[31], other tick species could take advantage of companion animals.

The significant cross-correlation between LB incidence and frequency of tick observation in the veterinary survey indicates that the occurrence of LB increases with tick density. Jaenson and Lindgren [32] propose that the density of *I. ricinus* nymphs can serve as a general indicator of risk of exposure to LB spirochaetes and thus the risk of people contracting LB. This is in line with other studies [33], although contradicting results are also reported [34].

The study identified 60 municipalities with daily/weekly tick observations, but no reported bovine babesiosis or LB in humans. One obvious explanation is that the tick densities have been overestimated in these municipalities, and that the tick populations have not reached levels where infection "spills over" into humans/cattle. It could also be that the host animals in these areas do not harbour high-virulent strains of *B. divergens* or *B. burgdorferi* s.l.. Alternatively the tick population may be too low for the maintenance of pathogen cycles or lack effective maintenance hosts for the pathogen. The pathogens could also be present, but clinical disease not observed due to resistance or subclinical manifestation.

For tick-borne infections in humans it has been argued that human behaviour determined by socio-economic conditions play a more significant role than abiotic and biotic environmental factors acting on enzootic cycles [24]. The consistency between the LB and bovine babesiosis maps (Figure 2, Figure 4) may suggest that the frequency of both diseases is associated with environmental factors and tick abundance rather than human behaviour.

CONCLUSION

The results from current study indicate that *I. ricinus* has expanded its northern distribution limit by approximately 400 km. *I. ricinus* is now present in coastal municipalities north to approximately 69°N, as compared with 66°N in 1943 and 1983. The present study does not make it possible to draw any conclusion about the factors involved in driving the expansion of the tick habitats: this will be the aim for further studies. The approach used in this study, a multi-source analysis, proved useful in the assessment of alterations in tick distribution. Similar approaches could be used to assess future range shifts of ticks and other disease vectors.

ACKNOWLEDGEMENTS

The access to the bovine babesiosis data was given by the Norwegian Dairy Herd Recording System and the Norwegian Cattle Health Services in agreement number 1/2009 by 25.02.2009. Thanks to Frøydis Hardeng (Norwegian School of Veterinary Science) for generating the babesiosis data, the Norwegian Veterinary Association, Ellef Blakstad especially, who organised the veterinary survey and to veterinarians taking the time to answer the questionnaire. Thanks to hunters who registered their findings on http://www.flattogflue.no, and to the Norwegian Mapping Authority, who provided the altitude and vegetation data at municipality level. Also thanks goes to Petter Hopp (NVI) for generating the cow population data, Attila Tarpai and Madelaine Norström (NVI) for assistance with GIS, Atle Brunvoll in "Aftenposten" for providing the data from the newspaper webpage registration and Bruce David (NVI) for language help with the paper. We are grateful to Sophie Vanwambeke (University of Louvain, Belgium) and Katsuhisa Takumi (National Institute for Public Health and the Environment, The Netherlands) for kindly taking the time to critically review the paper.

AUTHORS' CONTRIBUTIONS

The study was planned in collaboration between the authors with SJ as a project leader. SJ did the majority of manuscript writing, but all co-authors contributed with improvements, discussions and approved the final manuscript. SJ, HV and ABK carried out the data analysis. SJ, BKS and HBH did the majority of work regarding the questionnaire for the veterinary survey, but all co-authors contributed with improvements and discussions.

REFERENCES

1. Randolph SE: Tick-borne disease systems emerge from the shadows: the beauty lies in molecular detail, the message in epidemiology. Parasitology. 2009, 136: 1403-1413. 10.1017/S0031182009005782.
2. Randolph SE, Miklisova D, Lysy J, Rogers DJ, Labuda M: Incidence from coincidence: patterns of tick infestations on rodents facilitate transmission of tick-borne encephalitis virus. Parasitology. 1999, 118 (Pt 2): 177-186.
3. Gray JS, Dautel H, Estrada-Pena A, Kahl O, Lindgren E: Effects of climate change on ticks and tick-borne diseases in europe. Interdiscip Perspect Infect Dis. 2009, 2009: 593232.
4. Tambs-Lyche H: *Ixodes ricinus* og piroplasmosen i Norge (Meddelelse fra Bergens Museums zoologiske avdeling). Norsk veterinærtidsskrift. 1943, 55: 337-542.
5. Mehl R: The distribution and host relations of Norwegian ticks (*Acari, Ixodides*). Fauna norvegica Series B. 1983, 30: 46-51.
6. Osteras O, Solbu H, Refsdal AO, Roalkvam T, Filseth O, Minsaas A: Results and evaluation of thirty years of health recordings in the Norwegian dairy cattle population. J Dairy Sci. 2007, 90: 4483-4497. 10.3168/jds.2007-0030.
7. Legendre P, Legendre L: Ordination in reduced space. Developments in Environmental Modelling. Edited by: Legendre P, Legendre L. 2011, Amsterdam: Elsevier Science b.v, 387-480.
8. R development Core Team. R: A language and environment for stastistical computing. 2008, [http://www.R-project.org]
9. Danielova V, Rudenko N, Daniel M, Holubova J, Materna J, Golovchenko M, Schwarzova L: Extension of *Ixodes ricinus* ticks and agents of tick-borne diseases to mountain areas in the Czech Republic. Int J Med Microbiol. 2006, 296 (Suppl 40): 48-53.
10. Gilbert L: Altitudinal patterns of tick and host abundance: a potential role for climate change in regulating tick-borne diseases?. Oecologia. 2010, 162: 217-225. 10.1007/s00442-009-1430-x.
11. Jaenson TG, Eisen L, Comstedt P, Mejlon HA, Lindgren E, Bergstrom S, Olsen B: Risk indicators for the tick *Ixodes ricinus* and *Borrelia burgdorferi* sensu lato in Sweden. Med Vet Entomol. 2009, 23: 226-237. 10.1111/j.1365-2915.2009.00813.x.
12. Kirby AD, Smith AA, Benton TG, Hudson PJ: Rising burden of

immature sheep ticks (*Ixodes ricinus*) on red grouse (*Lagopus lagopus scoticus*) chicks in the Scottish uplands. Med Vet Entomol. 2004, 18: 67-70. 10.1111/j.0269-283X.2004.0479.x.

13. Ogden NH, Maarouf A, Barker IK, Bigras-Poulin M, Lindsay LR, Morshed MG, O'callaghan CJ, Ramay F, Walter-Toews D, Charron DF: Climate change and the potential for range expansion of the Lyme disease vector *Ixodes scapularis* in Canada. International Journal for Parasitology. 2006, 36: 63-70. 10.1016/j.ijpara.2005.08.016.
14. Scharlemann JP, Johnson PJ, Smith AA, Macdonald DW, Randolph SE: Trends in ixodid tick abundance and distribution in Great Britain. Med Vet Entomol. 2008, 22: 238-247. 10.1111/j.1365-2915.2008.00734.x.
15. Jaenson TG, Talleklint L, Lundqvist L, Olsen B, Chirico J, Mejlon H: Geographical distribution, host associations, and vector roles of ticks (Acari: *Ixodidae, Argasidae*) in Sweden. J Med Entomol. 1994, 31: 240-256.
16. Lindgren E, Talleklint L, Polfeldt T: Impact of climatic change on the northern latitude limit and population density of the disease-transmitting European tick *Ixodes ricinus*. Environ Health Perspect. 2000, 108: 119-123. 10.1289/ehp.00108119.
17. Lindgren E, Gustafson R: Tick-borne encephalitis in Sweden and climate change. Lancet. 2001, 358: 16-18. 10.1016/S0140-6736(00)05250-8.
18. Talleklint L, Jaenson TG: Increasing geographical distribution and density of *Ixodes ricinus* (*Acari*: *Ixodidae*) in central and northern Sweden. J Med Entomol. 1998, 35: 521-526.
19. Daniel M, Kriz B, Danielova V, Materna J, Rudenko N, Holubova J, Schwarzova L, Golovchenko M: Occurrence of ticks infected by tickborne encephalitis virus and Borrelia genospecies in mountains of the Czech Republic. Euro Surveill. 2005, 10: E050331.
20. Daniel M, Zitek K, Danielova V, Kriz B, Valter J, Kott I: Risk assessment and prediction of *Ixodes ricinus* tick questing activity and human tick-borne encephalitis infection in space and time in the Czech Republic. Int J Med Microbiol. 2006, 296 (Suppl 40): 41-47.
21. Materna J, Daniel M, Danielova V: Altitudinal distribution limit of the tick *Ixodes ricinus* shifted considerably towards higher altitudes in central Europe: results of three years monitoring in the Krkonose Mts. (Czech Republic). Cent Eur J Public Health. 2005, 13: 24-28.
22. Randolph SE: Evidence that climate change has caused 'emergence' of

tick-borne diseases in Europe?. Int J Med Microbiol. 2004, 293 (Suppl 37): 5-15.

23. Randolph SE: Tick ecology: processes and patterns behind the epidemiological risk posed by ixodid ticks as vectors. Parasitology. 2004, 129 (Suppl): S37-S65.

24. Randolph SE: Tick-borne encephalitis incidence in Central and Eastern Europe: consequences of political transition. Microbes and Infection. 2008, 10: 209-216. 10.1016/j.micinf.2007.12.005.

25. Hanssen-Bauer I, Førland E: Temperature and precipitation variations in Norway 1900-1994 and their links to atmospheric circulation. Int J Climatol. 2000, 20: 1693-1708. 10.1002/1097-0088(20001130)20:14<1693::AID-JOC567>3.0.CO;2-7.

26. Grimenes AA, Nissen Ø: Mathematical modelling of the annual temperature wave based on monthly mean temperatures, and comparisons between local climate trends at seven Norwegian stations. Theor Appl Climatol. 2004, 78: 229-246.

27. Estrada-Pena A, Venzal JM, Sanchez AC: The tick *Ixodes ricinus*: distribution and climate preferences in the western Palaearctic. Med Vet Entomol. 2006, 20: 189-197. 10.1111/j.1365-2915.2006.00622.x.

28. Kurtenbach K, Hanincova K, Tsao JI, Margos G, Fish D, Ogden NH: Fundamental processes in the evolutionary ecology of Lyme borreliosis. Nat Rev Microbiol. 2006, 4: 660-669. 10.1038/nrmicro1475.

29. Zintl A, Mulcahy G, Skerrett HE, Taylor SM, Gray JS: *Babesia divergens*, a bovine blood parasite of veterinary and zoonotic importance. Clin Microbiol Rev. 2003, 16: 622-636. 10.1128/CMR.16.4.622-636.2003.

30. Zintl A, Gray JS, Skerrett HE, Mulcahy G: Possible mechanisms underlying age-related resistance to bovine babesiosis. Parasite Immunol. 2005, 27: 115-120. 10.1111/j.1365-3024.2005.00748.x.

31. Jameson LJ, Medlock JM: Tick Surveillance in Great Britain. Vector-Borne and Zoonotic Diseases.

32. Jaenson TG, Lindgren E: The range of *Ixodes ricinus* and the risk of contracting Lyme borreliosis will increase northwards when the vegetation period becomes longer. TicksTick-borne Dis.

33. Tälleklint L, Jaenson TGT: Relationship between *Ixodes ricinus* density and prevalence of infection with Borrelia-like spirochetes and density of infected ticks. Journal of Medical Entomology. 1996, 33: 805-811.

34. Rizzoli A, Merler S, Furlanello C, Genchi C: Geographical information systems and bootstrap aggregation (bagging) of tree-based classifiers for Lyme disease risk prediction in Trentino, Italian Alps. J Med Entomol. 2002, 39: 485-492. 10.1603/0022-2585-39.3.485.

20

Abiotic and Biotic Factors Associated with Tick Population Dynamics on a Mammalian Host: Ixodes hexagonus Infesting Otters, Lutra lutra

Ellie Sherrard-Smith, Elizabeth Chadwick, Joanne Cable

Organisms and the Environment, Cardiff University, Cardiff, United Kingdom

ABSTRACT

The Eurasian otter, *Lutra lutra,* hosts several parasites with zoonotic potential. As this semiaquatic mammal has large ranges across terrestrial, freshwater and marine habitats, it has the capacity for wide dispersion of pathogens. Despite this, parasites of otters have received relatively little attention. Here, we examine their ectoparasite load and assess whether this is influenced by abiotic or biotic variables. Climatic phenomena such as the North Atlantic Oscillation (NAO) affect weather conditions in northern Europe. Consequently parasite distributions, particularly species with life stages exposed to the external environment, can be affected. We assessed the extent to which inter-annual variations in large-scale weather patterns (specifically

Citation: Sherrard-Smith E, Chadwick E, Cable J (2012) Abiotic and Biotic Factors Associated with Tick Population Dynamics on a Mammalian Host: Ixodes hexagonus Infesting Otters, Lutra lutra. PLoS ONE 7(10): e47131. https://doi.org/10.1371/journal.pone.0047131

the NAO and Central England (CE) temperatures) and host characteristics influenced tick prevalence and intensity. Ectoparasites consisted of a single species, the nidiculous tick *Ixodes hexagonus* (prevalence =24.3%; mean intensity =7.2; range =1–122; on n =820 otter hosts). The prevalence, but not intensity of infestation, was associated with high CE temperatures, while both prevalence and intensity were associated with positive phases of the NAO. Such associations indicate that *I. hexagonus* are most abundant when weather conditions are warmer and wetter. Ticks were more prevalent on juvenile than sub-adult or adult otters, which probably reflects the length of time the hosts spend in the holt where these ticks quest. High tick number was associated with poor host condition, so either poor condition hosts are more susceptible to ticks, or tick infestations negatively impact on host condition. Otters are clearly an important and common host for *I. hexagonus*, which has implications for vector-borne diseases. This work is the first to consider the impacts of long-term weather patterns on *I. hexagonus* and uses wild-animal cadavers to illustrate the importance of abiotic and biotic pressures impacting parasitic populations.

INTRODUCTION

Current change in climate (the long-term average meteorological conditions of a region [1]) is associated with increases in temperature and precipitation, especially in Northern temperate zones [1]. This influences parasite distributions both directly [2], [3] and indirectly, for example via impacts on host range [4], [5]. Weather (short-term variation in meteorological conditions) can cause variations in parasite distributions whilst synchronously influencing host abundance [6] but will affect specific host-parasite interactions differently [4], [7]–[9]. Weather patterns are influenced by climatic phenomena such as the North Atlantic Oscillation (NAO). The NAO affects European climate such that, when in positive phases, northern Europe experiences warmer and wetter conditions [10], [11]. Identifying associations between climate and the distribution of vectors over time (e.g. [12], [13]) is an essential pre-requisite to understanding public and wildlife health risks resulting from vector-borne infection.

Ixodid ticks are vectors for a range of pathogens causing diseases including Lyme disease, Boutonneuse fever, Q fever, tick-borne fever and tick-borne encephalitis [14]. *Ixodes hexagonus* is an efficient vector of *Borrelia burgdorferi,* the causative agent of Lyme disease [15] but in the UK, *I. ricinus* has received most attention because of its ubiquitous nature and association with transmission of pathogens to humans and livestock [14]. The distribution of *I. ricinus* is influenced by weather [16], [17] and the presence of suitable hosts and habitat [18]. Temperature increases are associated with increasing population density and geographic range of *I. ricinus,* a European tick [17], [19], and other tick species such as the North American species *I. scapularis.* [20]. The majority of ixodid ticks require >80% relative humidity for survival off the host [21]–[23] and as such, positive phases of the NAO may benefit ixodid ticks by creating suitably humid weather. Landscape, habitat use and local weather conditions have been associated with tick distributions previously [16], [17], [19], [20], [24], [25]. The impact of such environmental variables on host-parasite interactions is, however, highly variable [26]. Mustelids have been associated with the nidiculous (burrow or nest dwelling) tick *I. hexagonus* [24] but the relationship between *I. hexagonus* and weather conditions has not been examined previously.

The Eurasian otter, *Lutra lutra*, is a top predator in the UK and a sentinel of freshwater health [27]. Otters are wide ranging opportunistic predators that feed in terrestrial, freshwater and marine habitats [28]. They are therefore potentially exposed to a wide diversity of pathogens and a great deal can be learned about the distribution of parasites in UK ecosystems by screening such a generalist host. Here, we identified the tick species that use otters as a host. Next, we investigated how weather patterns and host characteristics are associated with tick infestations of otters in England and Wales. Specifically, we hypothesised that tick occurrence (prevalence and intensity) would be positively correlated with temporal variation in: i) the NAO (associated with warmer and wetter weather in the UK), and; ii) higher Central England (CE) temperatures (a long-term record of temperature in central England, see Materials and Methods). Based on these findings we hypothesised that spatial variation in tick counts among meteorologically distinct regions of the UK would correlate positively with rainfall and temperature.

RESULTS

Tick Species

Tick (*Ixodes hexagonus*) prevalence on Eurasian otters, between 2004 and 2010, was 24.3% (199 out of 820) (Figure 1). On some hosts, all post-hatch tick life stages (larva, nymph and adults) were recovered (18 cases), but almost 40% of hosts had only one life stage present at collection (Larvae =13 cases, Nymph =40 cases, Adult =26 cases) (Table 1). Infested otters were widespread across England and Wales (Figure 1) with no evidence of clustering of infestation within the otter distribution (Ripley's K analysis at the 95% confidence level using radii ranging from 1 km-130 km).

Table 1. Summary of *Ixodes hexagonus* **on otters.**

Parasite stage	Prevalence (%)	Count (/820 hosts)	Mean Intensity (95% CI)	Range
Any stage	24.3	199	7.2 (5.5–9.2)	1–122
Larva	9.3	76	7.7 (4.7–11.7)	1–112
Nymph	15.1	124	4.0 (2.9–5.2)	1–44
Adult	11.8	97	2.7 (2.1–3.5)	1–26

Ixodes hexagonus **infestations of** *Lutra lutra* **in England and Wales between 2004 and 2009 (n=820); showing prevalence, parasite count, mean intensity** with upper and lower 95% bootstrap confidence interval (10000 iterations), and maximum intensity for each tick life stage.

https://doi.org/10.1371/journal.pone.0047131.t001

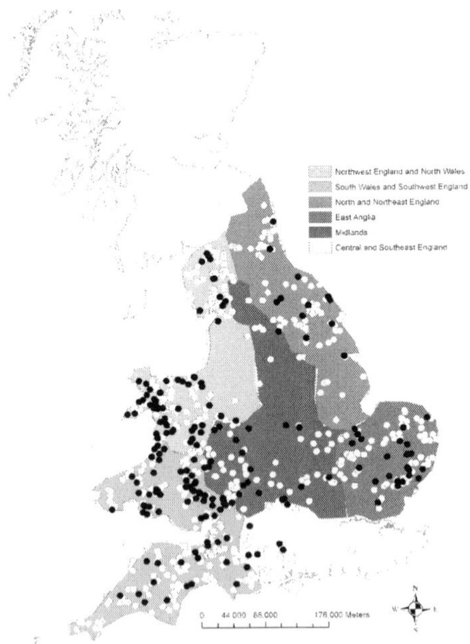

Figure 1. Tick distributions across the UK. Distribution of *Ixodes hexagonus* infested (dark circles) and uninfested (clear circles) otters in England and

Wales. Meteorologically distinct regions (East and Northeast England, East Anglia, Southeast England and Central South, Northwest England and North Wales, South Wales and Southwest England, and Midlands) defined by the Meteorological Office UK Climate Impacts Programme (data available online).

https://doi.org/10.1371/journal.pone.0047131.g001

Abiotic Factors

Tick prevalence on otters was associated with higher Central England (CE) temperatures for the 12 month period preceding host death (GLM: t =2.594, df=569, p<0.01), and more positive phases of the North Atlantic Oscillation (NAO) over the 12 month period preceding host death (GLM: t=2.099, df=569, p<0.05) (Figure 2). Tick intensity was not significantly associated with CE temperatures over any period preceding host death (GAM: t=1.445, df=155, p=0.15). Tick intensity was, however, positively associated with the NAO at month of host death (GAM: t=2.670, df=155, p<0.05) (Figure 3).

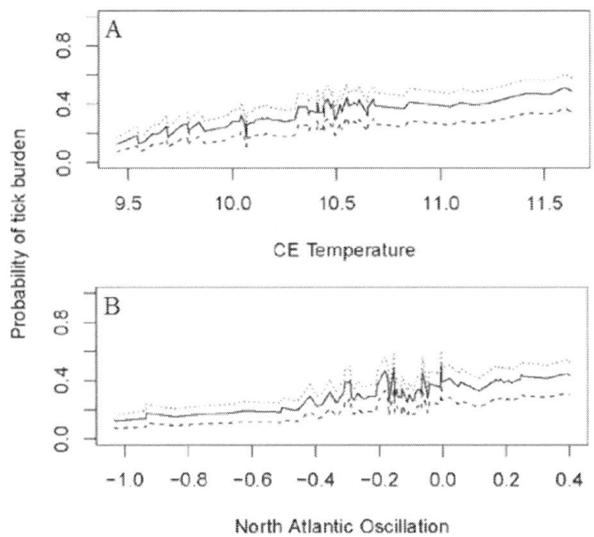

Figure 2. Abiotic impact on tick prevalence. Probability plot for a model of the association between tick prevalence and the explanatory variables A) Central England Temperature for the 12 month period preceding host death, B) North Atlantic Oscillation for the 12 month period preceding death for each host age class: Dotted line = juvenile hosts; Solid line = Adult hosts; Dashed line = Sub-adult hosts.

https://doi.org/10.1371/journal.pone.0047131.g002

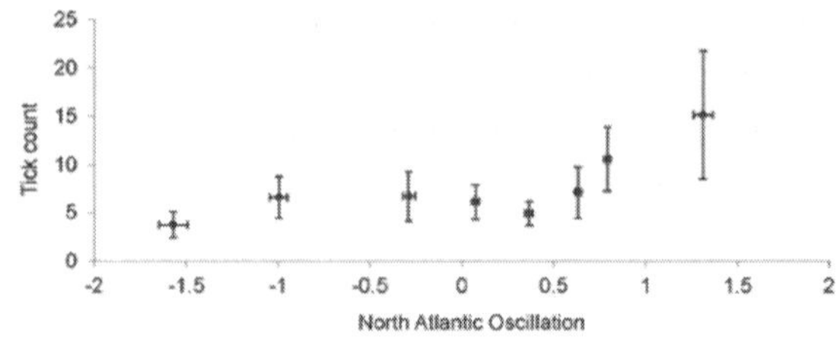

Figure 3. The North Atlantic Oscillation impacts tick counts on otters. Relationship of tick count to mean North Atlantic Oscillation at month of host death. Standard error bars shown.

https://doi.org/10.1371/journal.pone.0047131.g003

The South Wales and Southwest England region (Figure 1) had significantly higher tick counts on otters than all other regions, while East and Northeast England had significantly lower tick counts than all other regions; these two regions contributed most strongly to the statistically significant difference in counts between regions (χ^2 =302.169, df=5, p<0.001). Mean intensities for each region did not, however, correlate with maximum or minimum temperature, or mean rainfall for the long-term yearly average (1971–2000) regional data (Correlations, all p>0.1) (Figure 4). There were no seasonal associations between larval, nymph or adult stage ticks on otters (GLM, p>0.1).

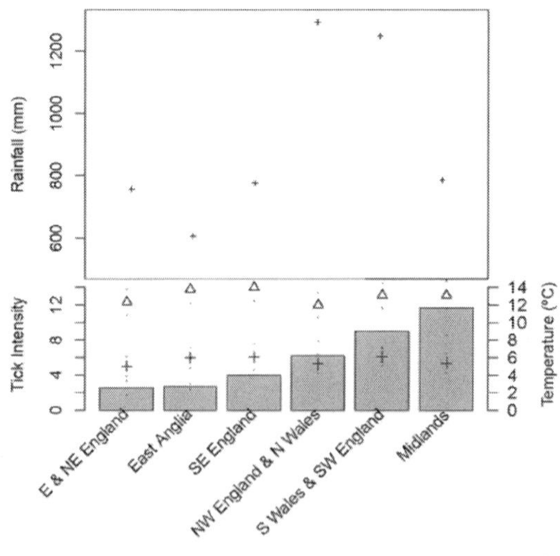

Figure 4. Abiotic impact on tick intensity. Mean tick intensity (grey bars) in each meteorologically distinct region (East and Northeast England, East Anglia,

Southeast England and Central South, Northwest England and North Wales, South Wales and Southwest England, and Midlands) and corresponding 30 year average (1971–2000) summed mean rainfall (mm) (upper Y-axis), maximum (triangle) and minimum (cross) 30 year (1971–2000) average temperature (°C) for each region (lower Y-axis). Standard error marks for rainfall, maximum and minimum temperature correspond to variability in monthly averages.

https://doi.org/10.1371/journal.pone.0047131.g004

Biotic Factors

More juvenile otters were infested than older age-classes (GLM: $p>0.01$; Figure 2). The mean host condition 'K' for the sampled population was 1.0286. Tick intensity was inversely related to otter condition so that as otter condition increased, tick intensity decreased (GAM: $t=3.137$, $df=155$, $p>0.01$) (Figure 5).

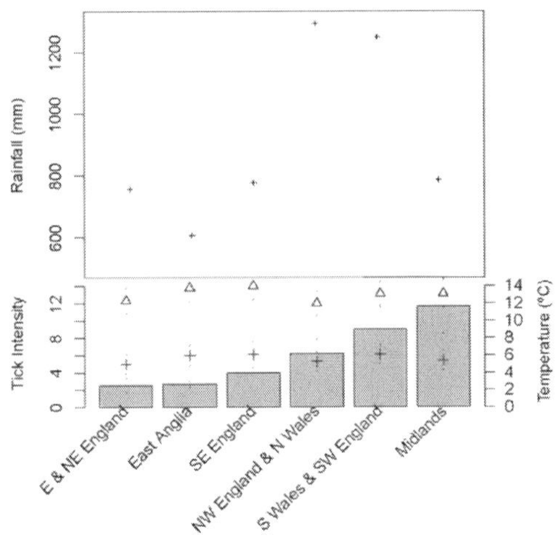

Figure 5. Biotic impact on tick count. Relationship of tick intensity to host condition (K). Standard error bars shown.

https://doi.org/10.1371/journal.pone.0047131.g005

DISCUSSION

Ixodes hexagonus is the only tick species reported from the Eurasian otter ([29], current study). *Ixodes hexagonus* can complete its life cycle on the

European hedgehog [24], fox [30] and American Mink [31]. As all three post-hatch tick life stages were found on the otter in the current study, it appears that *I. hexagonus* can also potentially complete its life cycle on this mammal. The prevalence of *I. hexagonus* on otters (24.3%) is lower than that reported on European hedgehogs, which are the preferred host for this tick [24], [32] (53.3% prevalence on hedgehogs from Western Europe, [32]). *I. hexagonus* is encountered by domestic dogs and cats in urban areas [33] illustrating the close proximity of this particular tick to human populations. The prevalence of *I. hexagonus* on otters is, however, high in comparison to its prevalence on the domestic dog in the UK (5.6%, n=3534 [34]). Further, the mean intensity (the total number of parasites of a particular species found in a sample divided by the number of infested hosts [35]) of ticks is higher on otters (7.2 per host) than on hedgehogs (3.8 [36]) despite examination of cadavers in the current study and live hosts in the hedgehog study. This suggests that otters are a noteworthy host for *I. hexagonus*. The association between otters and *I. hexagonus* populations may be important for pathogen transmission, particularly if otters act as either reservoir or amplifier hosts, or reduce pathogen abundance through the dilution effect [37]. Further, otters have large home ranges [28], [38] indicating that this host has the potential to transfer ticks between habitat islands. The nocturnal and aquatic nature of otters may deter other tick species from utilising such a resource, explaining the absence of diversity in tick species.

Positive phases of the North Atlantic Oscillation (NAO) were associated with increased prevalence and intensity in tick populations on otters. Strong positive phases of the NAO are linked with above average temperature and precipitation across northern Europe. Together with the elevated humidity produced, such weather conditions may lead to increased abundance of *I. hexagonus*, as reported for *I. ricinus* and *I. scapularis* [17], [19], [20]. This may be related to the weather conditions causing changes in the behaviour of either the parasite or the host thereby altering infestation rates (see [39]). No previous literature was found relating NAO to *I. hexagonus*. In *I. ricinus* however, the NAO did not correlate with intensity of tick infestation but negative winter NAO phases (associated with warmer and wetter winters) corresponded to increased Borreliae infections [40]. Further investigation into the underlying pathogenic infections of otters would be useful to examine whether this association holds for *I. hexagonus*.

Both NAO and CE temperatures are indices that can be used to describe temporal variation in weather; they do not provide spatially explicit weather data within the UK. The significant relationships found in the current study therefore describe an association between otter ticks (prevalence and intensity) and temporal variation in weather. We tested subsequently whether warmer and wetter regions were associated with higher tick infestations of otters, and found significantly more ticks in South Wales and Southwest England – a region associated with higher rainfall [41]. Overall, however, at the regional scale we identified no significant correlations between the mean intensity of ticks and either temperature or rainfall. This may be because at the regional scale temperature and rainfall are negatively correlated, so a more detailed analysis of local weather is necessary to clarify their interaction. Other factors such as the distribution of non-otter hosts, and variation in habitat type, may also heavily influence spatial variation in tick abundance. In preliminary investigations we explored the impact of local weather, alternative hosts, and habitat on *I. hexagonus* distribution, but subsequently removed these from our analyses because: i) Restricted availability of data meant that inclusion of both spatial and temporal variation in weather reduced the size of the dataset considerably, rendering conclusions less robust; ii) Information on the reported distribution of alternative hosts (hedgehog and fox) and of *I. hexagonus* were obtained from the National Biodiversity Network (NBN). Hedgehogs and foxes are both widespread and abundant in the UK and therefore availability of alternative hosts seems unlikely to limit *I. hexagonus* distribution at the regional scale. Further, *I. hexagonus* records from the NBN are concentrated in the London area of South East England, but because this database relies heavily on records submitted by members of the public this is likely to represent bias due to distribution of the human population. The NBN records map presence only (and not absence on screened hosts), so it was not possible to test for clustering within the host distribution as we did for *I. hexagonus* on otters. Comparisons were therefore uninformative; iii) Data on land use (arable, broadleaf and coniferous woodland, improved and semi-natural grassland, and upland habitat) were obtained from the Countryside Information System (CIS version 8, available online). ArcMap GIS (version 9.2) was used to interrogate these data and to assign percentage cover of each land-use within a 20 km radius of each otter. Significant negative associations were revealed (between tick prevalence and arable land, improved grassland, and conifer woodland), but interpretation is questionable because of the heterogenous and patchy nature of habitat data, the relatively large areas examined which may not accurately reflect the nature of real otter ranges (these tend to be linear along water courses, and vary considerably in length from a few to

40km [28]), and the difficulty in defining where, within this unknown range, an otter may have become infested.

Tick prevalence, but not intensity on otters, was associated with CE temperature. As far as we are aware, there are no previous records of temperature effects on *I. hexagonus* and the only long-term study on population dynamics of *I. hexagonus* indicates little seasonal variation and low-level abundance (on hedgehogs [36]). In general, however; temperature has a key role in driving tick development rates [17], [19] and so affects population dynamics [16], [17]. Additionally, temperature tends to be associated with length of diapause, larval activity and adult interactions [42]. Particularly strong associations are found between *I. ricinus* and temperature [17]. Stochastic temperature variations across the year are predicted to alter population dynamics of *I. ricinus* with subsequent impacts on the transmission of vector borne diseases [17]. The contrasting impact of temperatures on *I. hexagonus* and *I. ricinus* may be attributable to the ecological differences between the two species. The most important of these is likely to be habitat choice. *I. hexagonus* is nest dwelling, and so to some extent insulated from changes in ambient temperatures. In contrast, *I. ricinus* uses open areas for questing [14], so is likely to be exposed to wider fluctuations in air temperature.

Juvenile otters were more frequently infested with *I. hexagonus* than adult hosts. Host age, in general, influences the intensity of infestations, but can also affect parasite-induced mortality, and the distribution of the parasite among host individuals [43]. Several hypotheses [43], [44] predict that juveniles will carry heavier infestations than older hosts, either because: i) adult hosts develop immunity and/or behavioural adaptations to avoid or remove parasites; and/or ii) heavily infested juveniles die before adulthood (selection hypothesis [44]) although this is very unlikely as a direct cause of death. Grooming is a learned activity in otters [28] and may contribute to lower tick numbers on older otters [28]. Additionally, young otters spend the majority of their time in holts, the resting place of otters [28], so are disproportionately exposed to such parasites. Effects may be underestimated here, however, because road killed samples tend to reflect the healthier section of the population [45].

Finally, we found a relationship between host condition and tick intensity such that a better host condition is associated with decreased intensity. This is not a reflection of the elevated infestations on juvenile hosts because the host condition index used here [46] controls for size and therefore age, in addition to sexual dimorphism. This positive relationship could imply that otters in better condition are more efficient at grooming and thereby rid themselves of ticks, or that ticks have a negative impact on otter condition.

We acknowledge that data from road-killed hosts are likely to underestimate tick counts and recognise that road-kill samples are a stochastic sub-sample of a population and may lead to bias in terms of the proportion of the host population examined. For protected species, however, road-kill samples remain the only way to obtain large sample sizes for analysis. The absence of tick species other than *I. hexagonus,* in concordance with the only other report of otter tick infestations [29], could reflect differences in emigration patterns when abandoning a dead host, while tick emigration rates from dead hosts may interact with local microclimate. Our analysis of recently killed versus decomposed otters (see Materials and Methods), however, reveals no significant difference in infestation levels or species diversity, suggesting that observed associations are robust. Such data can therefore successfully illustrate associations between inter-annual variations in weather patterns, host characteristics and *I. hexagonus* populations.

To our knowledge this work is the first to consider the impacts of weather on *I. hexagonus,* and reveals that inter-annual variations in large-scale weather patterns, together with host characteristics, combine to affect the distribution, prevalence and intensity of *I. hexagonus* on Eurasian otters. Associations were identified between positive NAO phases, CE temperatures and tick prevalence, suggesting that the predicted change in climate in northern temperate zones may cause an increase in *I. hexagonus* populations. Although the associations highlighted here may not necessarily parallel what is observed on other hosts for this tick, *I. hexagonus* is common on domestic cats and dogs [33] and we suggest that tick research should, perhaps, target species other than *I. ricinus* in the future. This study illustrates how surveys of wild-animal cadavers can be hugely informative about parasitic populations.

MATERIALS AND METHODS

The Host

The Cardiff University Otter Project receives dead otters, *Lutra lutra,* from across England and Wales. Most (86% of the current study) have been killed by road traffic and are stored subsequently at −20°C. The location (British

National Grid Reference) and date of death (month and year), sex, age-class (juvenile (n=25), sub-adult (238) and adult (312) and size (weight (kg) and length (m)) were recorded for each otter collected between 2004 and 2010. A condition index K was calculated controlling for the dimorphism of otter sexes, following [46]. Such that:

$$K = \text{weight}/[a \times \text{length}^{1/n}]$$

where a =5.02 and n=2.33 for females, and a =5.87 and n=2.39 for males [46]. Seasons were defined as winter: December-February, spring: March-May, summer: June-August, and autumn: September-November. Very decomposed otters were excluded from the analysis. Remaining otters included in model (and excluding those with missing data; n=575) were distributed across seasons and years as follows: spring =137; summer =83; autumn =179; and winter =176; 2004=12; 2005=48; 2006=70; 2007=122; 2008=146; 2009=116; and 2010=61.

Parasite Identification

Ticks were removed (via pelt searching and fur combing) and stored in 90% molecular grade ethanol prior to immersion in 0.1% saline solution for microscopic examination (x30 magnification) using a Nikon dissecting microscope with fibre optic illumination, and identified to species using morphological features [14], [47]. *Ixodes hexagonus* was the only tick species present and species identification of 15 specimens (5 adults, 5 nymph and 5 larvae) was confirmed by the Natural History Museum. Occasionally, damage or desiccation prevented morphological identification so for these specimens we sought confirmation using mitochondrial DNA cytochrome oxidase sub-unit 1 (COX1) analysis as follows.

DNA was extracted from 13 specimens, three adults, three nymphs and seven larvae, from four geographically separate hosts. Ethanol was evaporated fully from each sample. Extractions were conducted using a QIAGEN kit as per the manufacturer's protocol (QIAGEN DNeasy Blood and Tissue Handbook 2006) with the additional step of manually crushing each tick body with a sterile pipette tip at the start of the process. PCR followed standard procedures (QIAGEN DNeasy Blood and Tissue Handbook 2006). Novel primers (IHEXCO1F: 5′- TCATAAAGACATTGGGACT-3′, IHEXCO1R: 5′-TGGTAAAGAATGGGGTCT-3′) were designed by alignment of COX1 mtDNA from 8 reference tick species (GenBank: *Dermacentor reticulatus* AF132829, *Haemaphysalis punctata* FN394339.1, *Hyalomma aegyptium* AF132821, *Ixodes uriae* NC006078, *I. hexagonus* AF081828.1, *I.*

lividus GU124743, *I. ricinus* FN394342 and *Rhipicephalus sanguineus* NC002074). These primers are specific to *I. hexagonus*. The PCR reaction conditions were carried out in a 50 µl final volume, with 10x PCR buffer II (Applied Biosystems, UK), 50 mM MgCl (Applied Biosystems, UK), 2.5 mM of each dNTP, 10pmol/µ l of each primer, 0.5 U Taq DNA polymerase (Invitrogen) for each 10 µl DNA template. PCR conditions (GenAmp PCR System 9700, Applied Biosystems, UK) were: 95°C for 5 min, followed by 35 cycles of 94°C for 30 sec, 53°C for 1 min and 72°C for 1 min, with a final extension of 72°C for 10 min. PCR products produced identical sized bands for all tick samples on a 1.5% agarose gel. Four larvae, one nymph and one adult were sequenced (QIAGEN, Genomic Services, Germany) using both forward and reverse primers. All 592 bp sequences from the current study were identical and showed 99% similarity to the corresponding region of GenBank *I. hexagonus* AF081828.1. This reference sequence, AF081828.1, was obtained from laboratory maintained ticks over ten years ago [48], perhaps explaining the 3 bp discrepancy (at position 130 transition T to C, and at positions 172 and 188 transversions A to C). The next closest sequence match was 82% with *Ixodes asanumai* Kitaoka 1973 (GenBank: AB231674.1).

Data Preparation

Temporal variation in weather was quantified using mean monthly temperatures (°C) for Central England (CE temperature) [49] and North Atlantic Oscillation (NAO) phases (http://www.cpc.ncep.noaa.gov/products/precip/CWlink/pna/nao.shtml, data provided by the Climate Prediction Centre of the U.S. National Oceanographic and Atmospheric Administration website). The mean of each was calculated for: i) the month of host death; ii) the sixth month period preceding host death; and iii) the year preceding host death, for each otter. These time periods were selected based on literature indicating that populations may be influenced by conditions during the previous season or year [16], [50].

Spatial variation in climate was quantified using long-term averages, which are a useful tool to describe the state of the climate in a particular region [41]. Long-term yearly average (1971–2000) temperature (maximum and minimum, °C) and rainfall (mm) measures for meteorologically distinct regions of England and Wales were collated [41]. These regions are defined as East and Northeast England, East Anglia, Southeast England and Central South, Northwest England and North Wales, South Wales and Southwest England, and Midlands (Figure 1) and are used by the Meteorological Office

UK Climate Impacts Programme (UKCIP) to summarise weather patterns in the UK (http://www.metoffice.gov.uk/climate/uk/averages/19712000/). To determine the abiotic conditions for each sampled carcass, otters were assigned the regional average for climate data depending on their geographic location at time of death. Associations between these measures of climate and tick prevalence (the number of hosts infested with specific parasitic species, in the current study ticks, divided by the total number of hosts examined [35]), intensity (the number of individuals of a particular parasite species on a single infested host [35]) and tick count (the total number of ticks within a population) were examined from otters found between 2004 and 2010.

Data Analysis

Ectoparasites are thought to abandon dead hosts [51]. We tested initially, therefore, whether there was a difference in tick abundance between fresh (collected within 24 h of death, n=610) and not fresh (otters characterised as slightly or moderately decomposed, n=210) otters. We found no significant difference in tick presence/absence (χ^2=0.515, n=820, df=1, P=0.473) or median intensity (Kruskal-Wallis H=0.35, n=195, df=1, P=0.556) and subsequently pooled all data for further analyses.

The NAO and CE temperatures, for each time period examined, and host factors (sex, age, condition, season and year of death) were combined in a general linear model fitted to the tick presence/absence data with a binomial error distribution (n=575 hosts, individuals were omitted where data was missing). A generalised additive model incorporating these explanatory variables was fitted to the tick intensity data (tick counts per otter excluding zero counts) with negative binomial error distribution. Relationships between explanatory variables and tick intensity were non-linear. A generalised additive model (GAM) was applied therefore with splines fitted appropriately. Final models were selected using Akaike Information Criterion (AIC).

Tick counts were compared between the meteorologically distinct regions of England and Wales (described above) by calculating regional mean tick intensities and testing for a correlation with the long-term yearly average maximum and minimum temperature (°C) and total rainfall (mm) in each region.

The spatial distribution of infested otter carcasses (n=199) was examined to look for clustering within the host distribution (n=820) by calculating a modified Ripley's K statistic, K[i.](r), using Ripley's isotropic edge correction [52] with a simplified border of England and Wales as a boundary (for further details of methodology [53]). All statistical analyses were conducted using R version 2.12 [54].

ACKNOWLEDGMENTS

We thank Aaron Nahil for technical assistance, and referees for comments on an earlier version of this manuscript. The Health Protection Agency confirmed the morphological identification of *Ixodes hexagonus*. Samples were provided by the Cardiff University Otter Project funded by the Environment Agency: EAC.

AUTHOR CONTRIBUTIONS

Conceived and designed the experiments: ESS EAC JC. Performed the experiments: ESS EAC. Analyzed the data: ESS. Contributed reagents/materials/analysis tools: ESS EAC JC. Wrote the paper: ESS EAC JC.

REFERENCES

1. IPCC (2007) Climate Change 2007: Synthesis Report. Contribution to working groups I, II and III to the Fourth Assessment Report of the Intergovernmental Panel on Climate Change. RK Pachauri, A Reisinger, editors. IPCC, Geneva, Switzerland.
2. Kutz SJ, Hoberg EP, Nagy J, Polley L, Elkin B (2004) "Emerging" parasitic infections in arctic ungulates. Integ Comp Biol 44: 109–118.
3. Mangal TD, Paterson S, Fenton A (2008) Predicting the impact of long-term temperature changes on the epidemiology and control of schistosomiasis: a mechanistic model. PLoS One 3: e1438.
4. Patz JA, Epstein PR, Burke TA, Balbus JM (1996) Global climate change and emerging infectious diseases. JAMA 275: 217–223.
5. Harvell CD, Mitchell CE, Ward JR, Altizer S, Dobson AP, et al. (2002) Climate warming and disease risks for terrestrial and marine biota. Science 296: 2158–2162.
6. Cattadori IM, Haydon DT, Hudson PJ (2005) Parasites and climate synchronise red grouse populations. Nature 433: 737–741.
7. Kovats RS, Campbell-Lendrum D-H, McMichel AJ, Woodward A, Cox JSH (2001) Early effects of climate change: do they include changes in vector-borne disease? Phil Trans R Soc B 356: 1057–1068.
8. Lafferty KD (2009) The ecology of climate change and infectious diseases. Ecology 90: 888–900.
9. Moller AP (2010) Host-parasite interactions and vectors in the barn swallow in relation to climate change. Glob Change Biol 16: 1158–1170.
10. Hurrell JW, Kushnir Y, Ottersen G, Visbeck M (2003) An overview of the North Atlantic Oscillation. The North Atlantic Oscillation: Climate Significance and Environmental Impact. Geophys Monogr 134: 1–36.
11. Lopez-Moreno JI, Vicente-Serrano SM (2007) Positive and negative phases of the wintertime North Atlantic Oscillation and drought occurrence over Europe: A multitemporal-scale approach. J Climate 21: 1220–1243.
12. Mills JN, Gage KL, Khan AS (2010) Potential influence of climate change on vector-borne and zoonotic diseases: A review and proposed research plan. Environ Health Persp 118: 1507–1514.

13. Jongejan F, Uilenberg G (2004) The global importance of ticks. Parasitol 2004: S3–S14.

14. Hillyard PD (1996) Ticks of North-west Europe Synopses of the British Fauna (New Series). Barnes RSK, Crothers JH, editors. pp. 178. The Natural History Museum, Field Studies Council.

15. Gern L, Toutoungi LN, Hu CM, Aeschlimann A (1991) *Ixodes* (Pholeoixodes) *hexagonus,* an efficient vector of *Borrelia burgdorferi* in the laboratory. Med Vet Entomol 5: 431–435.

16. Randolph SE, Green RM, Hoodless AN, Peacey MF (2002) An empirical quantitative framework for the seasonal population dynamics of the tick *Ixodes ricinus* . Int J Parasitol 32: 979–989.

17. Hancock PA, Brackley R, Palmer SCF (2011) Modelling the effect of temperature variation on the seasonal dynamics of *Ixodes ricinus* tick populations. Int J Parasitol 41: 513–522.

18. Gray JS, Kahl O, Janetzki C, Stein J (1992) Studies on the ecology of Lyme disease in a deer forest in county Galway, Ireland. J Med Entomol 29: 915–920.

19. Lindgren E, Talleklint L, Polfeldt T (2000) Impact of climate change on the Northern latitude limit and population density of the disease-transmitting European tick *Ixodes ricinus.* . Environ Health Persp 108: 119–123.

20. Ogden NH, Maarouf A, Barker IK, Bigras-Poulin M, Lindsay LR, et al. (2006) Climate change and the potential for range expansion of the Lyme disease vector *Ixodes scapularis* in Canada. Int J Parasitol 36: 63–70.

21. MacLoed J (1934) *Ixodes ricinus* in relation to its physical environment: the influence of climate on development. Parasitol 26: 189–305.

22. MacLoed J (1939) The ticks of domestic animals in Britain. Empire J Exp Agr 7: 97–110.

23. Medlock JM, Pietzch ME, Rice NVP, Jones L, Kerrod E, et al. (2008) Investigation of ecological and environmental determinants for the presence of questing *Ixodes ricinus* (Acari: Ixodidae) on Gower, South Wales. J Med Entomol 45: 314–325.

24. Arthur DR (1953) The host relationships of *Ixodes hexagonus* Leach in Britain. Parasitol43: 227–238.

25. Hoberg EP, Brooks DR (2008) A macroevolutionary mosaic: episodic host-switching, geographical colonization and diversification in

complex host-parasite systems. J Biogeogr 35: 1533–1550.

26. Tylianakis JM, Didham RK, Bascompte J, Wardle DA (2008) Global change and species interactions in terrestrial ecosystems. Ecol Lett 11: 1351–1363.
27. Chadwick EA, Simpson VR, Nicholls AE, Slater FM (2011) Lead levels in Eurasian otters decline with time and reveal interactions between sources, prevailing weather, and stream chemistry. Environ Sci Technol 45: 1911–1916.
28. Kruuk H (2006) Otters: ecology, behaviour and conservation, Oxford University Press, Oxford.
29. Kelly TC, Sleeman DP, Fennessy GJ, Dillon A, Walton GA, et al. (2001) The Mammal Ticks (Acari: Ixodoidae) of Ireland: Their distribution and hosts. Irish Naturalists' Journal 26: 363–370.
30. Harris S, Thompson GB (1978) Populations of the ticks *Ixodes hexagonus (Pholeoixodes) hexagonus,* and *Ixodes (Pholeoixodes) canisuga* infesting suburban foxes, *Vulpes vulpes* . J Zool 186: 83–93.
31. Page RJC, Langton SD (1996) The occurrence of ixodid ticks on wild mink *Mustela vison* in England and Wales. Med Vet Entomol 10: 359–364.
32. Pfäffle M (2010) Influence of parasites on fitness parameters of the European hedgehog. PhD Thesis, Karlsruhe Institute of Technology.
33. Ogden NH, Cripps P, Davis CC, Owen G, Parry JM, et al. (2000) The ixodid ticks species attaching to domestic dogs and cats in Great Britain and Ireland. Med Vet Entomol 14: 332–338.
34. Smith FD, Ballatyne R, Morgan ER, Wall R (2011) Prevalence, distribution and risk associated with tick infestation of dogs in Great Britain. Med Vet Entomol 25: 377–384.
35. Bush AO, Lafferty KD, Lotz JM, Shostak AW (1997) Parasitology meets ecology on its own terms: Margolis *et al.* revisited. J Parasitol 83: 575–583.
36. Pfäffle M, Petney T, Skuballa T, Taraschewski H (2011) Comparative population dynamics of a generalist (*Ixodes ricinus*) and specialist tick (*I. hexagonus*) species from European hedgehogs. Exp Appl Acarol 54: 151–164.
37. Schmidt KA, Ostfeld RS (2001) Biodiversity and the dilution effect in disease ecology. Ecology 82: 609–619.

38. Green J, Green R, Jefferies DJ (1984) A radio-tracking survey of otters *Lutra lutra* on a Perthshire river system. Lutra 27: 85–145.
39. Kerr GD, Bull CM (2006) Interactions between climate, host refuge use, and tick population dynamics. Parasitol Res 99: 214–222.
40. Hubalek Z, Halouzka J, Juricova Z (2003) Longitudinal surveillance of the tick *Ixodes ricinus* for borreliae. Med Vet Entomol 17: 46–51.
41. Perry M, Hollis D (2005) The development of a new set of long-term climate averages for the UK. Int J Climatol 25: 1023–1039.
42. Randolph SE (2004) Tick ecology: processes and patterns behind the epidemiological risk posed by ixodid ticks as vectors. Int J Parasitol 129: S37–S65.
43. Hawlena H, Abramsky Z, Krasnov BR (2006) Ectoparasites and age-dependent survival in a desert rodent. Oecologia 148: 30–39.
44. Sol D, Jovani R, Torres J (2003) Parasite mediated mortality and host immune response explain age-related differences in blood parasitism in birds. Oecologia 135: 542–547.
45. Nusser SM, Clark WR, Otis DL, Huang L (2008) Sampling considerations for disease surveillance in wildlife populations. J Wildl Manage 72: 52–60.
46. Kruuk H, Conroy JWH, Moorhouse A (1987) Seasonal reproduction, mortality and food of otters *Lutra lutra* L. in Shetland. Syrup. Zool Soc Lond 58: 263–278.
47. Snow KR (1979) Identification of Larval Ticks Found on Small Mammals in Britain. Reading, UK: Mammal Society.
48. Black WC, Roehrdanz RL (1998) Mitochondrial gene order in not conserved in Arthropods: Prostriate and metastriate tick mitochondrial genomes. Mol Biol Evol 15: 1772–1785.
49. Parker DE, Legg TP, Follan CK (1992) A new daily Central England Temperature Series, 1772-1991. Int J Climatol 12: 317–342.
50. Ruiz-Fons F, Gilbert L (2010) The role of deer as vehicles to move ticks, *Ixodes ricinus*, between contrasting habitats. Int J Parasitol 40: 1013–1020.
51. Nelder MP, Reeves WK (2004) Ectoparasites of road-killed vertebrates in northwestern South Carolina, USA. Vet Parasitol 129: 313–322.
52. Ripley BDA (1988) Statistical Inference for Spatial Processes. Cambridge University Press, Cambridge, UK.

53. Sherrard-Smith E, Cable J, Chadwick EA (2009) Distribution of Eurasian otter biliary parasites, *Pseudamphistomum truncatum* and *Metorchis albidus* (Family Opisthorchiidae), in England and Wales. Parasitol 136: 1015–1022.

54. R Development Core Team (2008) R: A Language and Environment for Statistical Computing. R Foundation for Statistical Computing. Vienna, Austria.

21

DNA of Piroplasms of Ruminants and Dogs in Ixodid Bat Ticks

Sándor Hornok[1], Krisztina Szőke[1], Dávid Kováts[2], Péter Estók[3], Tamás Görföl[4], Sándor A. Boldogh[5], Nóra Takács[1], Jenő Kontschán[6], Gábor Földvári[1], Levente Barti[7], Alexandra Corduneanu[8] and Attila D. Sándor[8]

[1] Department of Parasitology and Zoology, University of Veterinary Medicine, Budapest, Hungary

[2] Department of Evolutionary Zoology and Human Biology, Debrecen University, Debrecen, Hungary

[3] Department of Zoology, Eszterházy Károly University, Eger, Hungary

[4] Department of Zoology, Hungarian Natural History Museum, Budapest, Hungary

[5] Department of Nature Conservation, Aggtelek National Park Directorate, Jósvafő, Hungary

[6] Plant Protection Institute, Centre for Agricultural Research, Hungarian Academy of Sciences, Budapest, Hungary

[7] Romanian Bat Protection Association, Satu Mare, Romania

[8] Department of Parasitology and Parasitic Diseases, University of Agricultural Sciences and Veterinary Medicine, Cluj-Napoca, Romania

Citation: Hornok S, Szőke K, Kováts D, Estók P, Görföl T, Boldogh SA, et al. (2016) DNA of Piroplasms of Ruminants and Dogs in Ixodid Bat Ticks. PLoS ONE 11(12): e0167735. https://doi.org/10.1371/journal.pone.0167735

ABSTRACT

In this study 308 ticks (*Ixodes ariadnae*: 26 larvae, 14 nymphs, five females; *I. vespertilionis*: 89 larvae, 27 nymphs, eight females; *I. simplex*: 80 larvae, 50 nymphs, nine females) have been collected from 200 individuals of 17 bat species in two countries, Hungary and Romania. After DNA extraction these ticks were molecularly analysed for the presence of piroplasm DNA. In Hungary *I. ariadnae* was most frequently identified from bat species in the family Vespertilionidae, whereas *I. vespertilionis* was associated with Rhinolophidae. *Ixodes ariadnae* was not found in Romania. Four, four and one new bat host species of *I. ariadnae*, *I. vespertilionis* and *I. simplex* were identified, respectively. DNA sequences of piroplasms were detected in 20 bat ticks (15 larvae, four nymphs and one female). *I. simplex* carried piroplasm DNA sequences significantly more frequently than *I. vespertilionis*. In *I. ariadnae* only *Babesia vesperuginis* DNA was detected, whereas in *I. vespertilionis* sequences of both *B. vesperuginis* and *B. crassa*. From *I. simplex* the DNA of *B. canis*, *Theileria capreoli*, *T. orientalis* and *Theileria* sp. OT3 were amplified, as well as a shorter sequence of the zoonotic *B. venatorum*. Bat ticks are not known to infest dogs or ruminants, i.e. typical hosts and reservoirs of piroplasms molecularly identified in *I. vespertilionis* and *I. simplex*. Therefore, DNA sequences of piroplasms detected in these bat ticks most likely originated from the blood of their respective bat hosts. This may indicate either that bats are susceptible to a broader range of piroplasms than previously thought, or at least the DNA of piroplasms may pass through the gut barrier of bats during digestion of relevant arthropod vectors. In light of these findings, the role of bats in the epidemiology of piroplasmoses deserves further investigation.

INTRODUCTION

Bats (Chiroptera) are the second largest order of mammals, with more than 1200 species. Owing to their flying habit, feeding preference and high level of adaptability, bats are widely distributed and present on all continents except the Antarctica [1]. In this scenario, invasion of men into bat habitats and adaptation of bats to urban areas increased the chances for contact between humans, domestic animals and bats [2], thus promoting the chances of pathogen transmission.

Accordingly, during the past decades, the epidemiological significance of bats has become increasingly recognized. Bats can be reservoirs or carriers of numerous species of viruses, bacteria and parasites, among them many with zoonotic potential to infect humans [3]. This is especially important in the context of synanthropic life of several bat species, which tend to roost or breed in human settlements, even in man-made buildings such as steeples, attics, cellars and barns [4]. In comparison with bat species that prefer forested habitats, urbanized populations of chiropterans may form larger, more stable and aggregated colonies [1,4]. Thus, urban bats may reach the highest individual number in the local mammalian fauna, further increasing their epidemiological significance. In addition, bat species that are indigenous in Europe feed predominantly on arthropods, and this makes it possible for them to get into contact with vector-borne pathogens not only from their ectoparasites, but also from their food [5].

Blood-sucking ectoparasites of bats may be potential vectors of a broad range of pathogens [2]. Depending on their taxonomic group, these ectoparasites can be highly (e.g. bat flies) or less specialized to bats (e.g. soft ticks, bugs), and exceptionally can settle on other mammals, even on humans [6]. Interestingly, while ixodid ticks of bats are not known to feed on other mammals, except *Ixodes vespertilionis* on humans [7], ixodid ticks that frequently infest domestic animals (e.g. *Ixodes ricinus*, *Dermacentor reticulatus*, *Haemaphysalis* spp.) have also been collected from bats [8, 9, 10].

Three species of ixodid bat ticks (Acari: Ixodidae) occur in Europe, i.e. *I. vespertilionis*, *I. ariadnae* and *I. simplex* [11]. These are specialized for bats, i.e. all three developmental stages that need a blood meal (larvae and nymphs for moulting, females for oviposition) will typically suck blood on bats, as exemplified by *I. ariadnae* [11]. While these three tick species and their genotypes appear to be widely distributed across the Palaearctic [12], only few data are available on their vector potential. Concerning molecular investigations of vector-borne pathogens in ixodid bat ticks, bartonellae were reported from *I. vespertilionis* [13], but to the best of our knowledge none from *I. simplex* or *I. ariadnae*. Therefore the present study was undertaken to ameliorate this lack of data on pathogens and/or pathogen DNA carried by ixodid bat ticks. Recently, DNA of *Babesia canis* has been detected in bat faeces [5], therefore piroplasms were chosen as the target group of analyses. It was also within the scope of this study to examine the geographical range and host spectrum of these tick species in Hungary and Romania, by including 17 species of bats (from three families and six genera) from these two countries.

MATERIALS AND METHODS

Tick Collection and Identification

Ixodid ticks were collected from bats, caught for monitoring purposes, on 24 locations in Hungary in 2008–2015, and on seven locations in Romania in 2015 (Fig 1; data in S1 Table). These bats were caught (as part of a monitoring program) at the entrance of caves between sunset and dawn, using standard Ecotone mist-nets (Gdynia, Poland) with 12 m length, 2.5 m height and 14 × 14 mm mesh. Bats were handled without anesthesia, but released immediately after tick removal to alleviate suffering. Data (species, sex) of bats, from which the ticks were removed, were recorded. The ticks were immediately put into and stored in 96% ethanol. Morphological identification was done with a stereo microscope (SMZ-2T, Nikon Instruments, Japan) using standard morphological keys (subadults: [14]; females: [15]).

Figure 1. Sampling sites of the present study. Color of collection sites for *Ixodes ariadnae* are marked with red dots, for *I. vespertilionis* with yellow dots, and for *I. simplex* with black dots. Letters: A—Ariadne Cave System and caves in the Pilis Mountains (bats sampled at three caves), B—Bükk Highlands Cave system (bats sampled at nine caves). Numbers in black circles indicate places, where piroplasm-carrier bat ticks were collected (Table 3). Small and unseparated dots with different colour indicate the same place with two tick species. Two places close to each other in northeast Hungary (Baradla and Béke Caves) are marked with one dot. Coordinates for the individual places are shown in S1 Table.

https://doi.org/10.1371/journal.pone.0167735.g001

DNA Extraction and Molecular Analyses

DNA was extracted (except for one *I. simplex* larva from *Barbastella barbastellus*) with the QIAamp DNA Mini Kit (QIAGEN, Hilden, Germany) according to the manufacturer's instructions and including extraction controls. Ticks were dried, then washed three times (in detergent containing water, in tap water and in distilled water) and minced at the bottom of 1.5 ml Eppendorf tubes in 100 μl PBS with pointed scissors. Between each sample the scissors were washed and burned for decontamination. Samples were then incubated overnight at 56°C in tissue lysis buffer containing proteinase-K.

DNA samples were molecularly screened with a conventional PCR that amplifies an approx. 500 bp long part of the 18S rDNA gene of piroplasms, modified from Casati et al. [16]. The primers BJ1 (forward: 5'-GTC TTG TAA TTG GAA TGA TGG-3') and BN2 (reverse: 5'-TAG TTT ATG GTT AGG ACT ACG-3') were used in a reaction volume of 25 μl, which included 5 μl of extracted DNA, and 20 μl of reaction mixture containing 0.5 unit HotStarTaq Plus DNA polymerase (5U/ μl), 200 μM PCR nucleotid mix, 1 μM of each primer and 2.5 μl of 10× Coral Load PCR buffer (15 mM $MgCl_2$ included). For amplification an initial denaturation step at 95°C for 10 min was followed by 40 cycles of denaturation at 95°C for 30 s, annealing at 54°C for 30 s and extension at 72°C for 40 s. Final extension was performed at 72°C for 5 min.

All PCRs were run with appropriate positive and negative controls. During all tests positive controls showed positivity, whereas negative (non-template) controls and extraction controls remained negative (the latter indicating absence of sample contamination). PCR products were electrophoresed in a 1.5% agarose gel, stained with ethidium-bromide and visualized under ultra-violet light. Purification and sequencing (twice) were done from all PCR positive samples by Biomi Inc. (Gödöllő, Hungary). Representative sequences were submitted to GenBank (accession numbers KU958544-53). Phylogenetic analyses were conducted according to the Tamura-Nei model [17] and Maximum Composite Likelihood method by using MEGA version 5.2 [18].

Statistical Analyses

Association of tick species with bat families was assessed by Fisher's exact test. Intensities of tick infestation (i.e. number of ticks on a bat individual) were compared between bat species by using Mann-Whitney U-test and

Kruskal Wallis H-test in R program. The following bat species (in Table 1, Table 2: harboring different stages of the same tick species) were analysed: *Myotis bechsteinii* (n = 7), *My. daubentonii* (n = 28), *My. emarginatus* (n = 9), *Miniopterus schreibersii* (n = 95), *Rhinolophus ferrumequinum* (n = 16), *R. hipposideros* (n = 12). Bat species with small sample size (n<5) were excluded from the latter analysis. The COIN (Conditional Inference Procedures in a Permutation Test Framework) package was used to correct P values of linked parameters. Differences were considered significant when P<0.05.

Table 1. Tick species and stages collected in Hungary, shown according to their bat hosts. Five females and two nymphs of *I. ariadnae*, which were collected from cave walls (Ariadne Cave System), are not included.

Tick		**Bat (number of ticks per number of bats)**											
	Vespertilionidae									**Rhinolophidae**		**Miniopteridae**	
Species	**Stage**	**MALC**	**MBEC**	**MNAT**	**MEMA**	**MDAU**	**MDAS**	**MMYO**	**PAUR**	**BBAR**	**RHIP**	**REUR**	**MSCH**
Ixodes ariadnae	larva	4/2	3/3	1/1	6/6	5/3	5/2	-	1/1	-	1/1	-	-
	nymph	-	4/4	-	4/3	-	-	1/1	3/3	-	-	-	-
	female	-	-	-	-	-	-	-	-	-	-	-	-
Ixodes vespertilionis	larva		1/1	-	-	-	-	-	1/1	-	8/7	-	-
	nymph	-	-	-	-	-	-	-	-	-	6/4	-	-
	female	-	-	-	-	-	-	-	-	-	2/1	1/1	-
Ixodes simplex	larva	-	-	-	-	-	-	-	-	1/1	-	-	23/10
	nymph	-	-	-	-	-	-	-	-	-	-	-	11/10
	female	-	-	-	-	-	-	-	-	-	-	-	4/4

Abbreviations: MALC- *Myotis alcathoe*, MBEC—*My. bechsteinii*, MNAT—*My. nattereri*, MEMA—*My. emarginatus*, MDAU—*My. daubentonii*, MDAS—*My. dasycneme*, MMYO—*My. myotis*, PAUR—*Plecotus auritus*, BBAR—*Barbastella barbastellus*, RHIP—*Rhinolophus hipposideros*, REUR—*R. euryale*, MSCH—*Miniopterus schreibersii*.

https://doi.org/10.1371/journal.pone.0167735.t001

Table 2. Tick species and stages collected in Romania, shown according to their bat hosts.

Tick		Bat (number of ticks per number of bats)								
	Vespertilionidae					Rhinolophidae			Miniop-teridae	
Species	Stage	MNAT	MCAP	MDAU	MBLY	ESER	REUR	RFER	RMEH	MSCH
Ixodes vesper-tilionis	larva	1/1	9/1	28/16	4/1	2/2	-	26/9	7/4	2/2
	nymph	-	4/2	9/8	2/2	-	-	6/6	-	-
	female	-	1/1	1/1	-	1/1	1/1	1/1	-	-
Ixodes simplex	larva	-		-	-	-	-	-	-	56/33
	nymph	-		-	-	-	-	-	-	39/33
	female	-		-	-	-	-	-	-	5/5

Abbreviations: MNAT—*Myotis nattereri*, MCAP—*My. capaccinii*, MDAU—*My. daubentonii*, MBLY—*My. blythii*, ESER—*Eptesicus serotinus*, REUR—*Rhinolophus euryale*, RFER—*R. ferrumequinum*, RMEH—*R. mehelyi*, MSCH—*Miniopterus schreibersii*.

https://doi.org/10.1371/journal.pone.0167735.t002

Ethical Approval

Authorization for bat capture was provided by the National Inspectorate for Environment and Nature in Hungary (No. 14/2138-7/2011). Bat banding license numbers are TMF-14/32/2010 (DK), 59/2003 (PE) and TMF-493/3/2005 (TG), 65/2003 (SAB). Bats were handled according to the current law of animal welfare regulation (1998. XXVIII.). Permission from the Institutional Animal Care and Use Committee (IACUC) was not necessary, because bats were released in the field after tick removal (none taken to participating Institutes).

RESULTS

Tick Infestations of Bats

In the two countries 308 ixodid ticks have been collected from 200 individuals of 17 bat species (Table 1, Table 2). *Ixodes ariadnae* was represented by 45,

I. vespertilionis by 124 and *I. simplex* by 139 specimens (larvae, nymphs and females). In Hungary *I. ariadnae* was significantly more frequently found on bat species in the family Vespertilionidae, whereas *I. vespertilionis* was associated with Rhinolophidae ($P<0.00001$). *Ixodes ariadnae* was not collected in Romania, where *I. vespertilionis* occurred usually on representatives of both Vespertilionidae and Rhinolophidae (Table 2). Discounting one larva collected from *Barbastella barbastellus*, *I. simplex* was exclusively found on *Mi. schreibersii*.

In general, there was no significant difference between the intensity of tick infestation between bat species ($p = 0.4279$, $df = 5$, $\chi^2 = 4.9026$), but there was a significant difference in the intensity of infestation of bats with different tick stages (larva: $n = 155$; nymph: $n = 79$; female: $n = 13$; $p = 0.0005$, $df = 2$, $\chi^2 = 14.924$), i.e. ixodid tick larvae occurred in highest individual number on their hosts. In the case of *I. ariadnae* or *I. vespertilionis* there was no significant difference in the intensity of infestation between bat species, i.e. between *My. bechsteinii* and *My. emarginatus* ($p = 0.4497$, $W = 28$) or between *My. daubentonii*, *R. ferrumequinum* and *R. hipposideros* ($p = 0.8719$, $df = 2$, $\chi^2 = 0.27423$), respectively (Table 1, Table 2). Similarly, concerning these five bat species, there was no significant difference between intensities of their infestations with different tick stages, except for *My. daubentonii* and *R. ferrumequinum* on which larvae occurred significantly more frequently than nymphs/females ($p = 0.02013$, $df = 2$, $\chi^2 = 7.8106$). Similarly, infestation of *Mi. schreibersii* with *I. simplex* had the highest intensity when larvae were present on bats (larva: $n = 79$; nymph: $n = 50$; female: $n = 9$; $p = 0.001674$, $df = 2$, $\chi^2 = 12.786$).

DNA of Piroplasms in bat ticks

DNA sequences of piroplasms were detected in 20 bat ticks (Table 3). *Ixodes simplex* carried piroplasm DNA significantly more frequently (13 of 138 specimens), than *I. vespertilionis* (3 of 124 specimens) ($P = 0.02$). The largest variety of *Babesia* and *Theileria* DNA sequences was also shown to be present in *I. simplex* (Table 3).

Table 3. Results of molecular analyses of bat ticks for the presence of piroplasms.

***Ixodes* species**	**Tick stage or sex**	**PCR positive / all analysed ticks**	**Results of sequencing (length, % identity, sample number)**	**Bat host of PCR positive ticks[#]**	**Location(s) of PCR positive ticks in Fig 1**	**Reference sequence**	**Accession number of sequence in this study (name of isolate)**
I. ariadnae	larva	4/26	*Babesia vesperuginis* **(448** bp, 100%, 4×)	MDAS	1	AJ871610	KU958544 **(Ia-Bv-1)**
	nymph	0/14	-	-	-	-	-
	female	0/5	-	-	-	-	-
I. vespertilionis	larva	3/89	*Babesia vesperuginis* **(448** bp, 100%, 2×)	ESER, MDAU	2	AJ871610	KU958544 **(Ia-Bv-1)**
			Babesia crassa **(410** bp, 98.5%, 1×)	RHIP	3	KF791205	KU958546 **(Iv-Bcr-1)**
	nymph	0/27	-		-	-	-
	female	0/8	-		-	-	-

I. simplex	larva	8/79	*Babesia crassa* **(410** bp, 98.3%, 1×)	MSCH	4	KF791205	KU958545 **(Is-Bcr-1)**
			Babesia venatorum-like (105 bp, 100%, 1×)	MSCH	2	KC007118	KU958553 **(Is-Bv-1)**
			Babesia canis **(420** bp, 100%, 1×)	MSCH	2	JF461253	KU958552 **(Is-Bca-2)**
			Theileria capreoli **(425** bp, 99.5%, 1×)	MSCH	4	KJ188219	KU958547 **(Is-Tc-1)**
			Theileria orientalis **(432** bp, 100%, 4×)	MSCH	2, 4, 5	AB668373	KU958549 **(Is-To-1)**
	nymph	4/50	*Babesia crassa* **(410** bp, 98.5%, 1×)	MSCH	2	KF791205	KU958546 **(Iv-Bcr-1)**
			Babesia canis **(420** bp, 100%, 1×)	MSCH	2	KC902833	KU958551 **(Is-Bca-1)**
			Babesia canis **(420** bp, 100%, 2×)	MSCH	2	JF461253	KU958552 **(Is-Bca-2)**
	female	1/9	*Theileria* **sp. OT3 (432** bp, 100%, 1×)	MSCH	4	DQ866839	KU958550 **(Is-TOT3-1)**

#Abbreviations: MDAS—*Myotis dasycneme*, ESER—*Eptesicus serotinus*, MDAU—*M. daubentonii*, RHIP—*Rhinolophus hipposideros*, MSCH—*Miniopterus schreibersi*

https://doi.org/10.1371/journal.pone.0167735.t003

In *I. ariadnae* only a DNA sequence of *B. vesperuginis* (identity: 448/448 bp = 100%) was shown to be present. All four PCR-positive larvae were removed from the same bat. In *I. vespertilionis* larvae sequences of *B. vesperuginis* (identity: 448/448 bp = 100%) and *B. crassa* (identity: 404/410 bp = 98.5%) were detected (Table 3).

Adding to the presence of the latter sequence in *I. simplex* nymphs, in a larva of this tick species the sequence of another genotype of *B. crassa* (identity: 403/410 = 98.3%) was demonstrated (Table 3), which was not detected before in Hungary. From *I. simplex* a shorter sequence of the zoonotic *B. venatorum* (identity: 105/105 bp = 100%) was also amplified, showing less identity with other piroplasms (second closest to *B. occultans* and *T. equi*, with 103/105 bp = 98.1% identity). In *I. simplex* larvae/nymphs two sequences of *B. canis* (both identities: 420/420 bp = 100%) were also detected.

Results of sequencing demonstrated DNA of two *Theileria* spp. exclusively in *I. simplex* larvae. These were *T. capreoli* (identity: 423/425 bp = 99.5%) and *T. orientalis* (identity: 432/432 bp = 100%). In addition, one female *I. simplex* carried the sequence of *Theileria* sp. OT3 (identity: 432/432 bp = 100%) (Table 3).

In the phylogenetic analysis, all sequences of *Babesia* and *Theileria* spp. amplified from bat ticks in the present study clustered together with relevant genotypes available in GenBank (and published from previously known "type" hosts of these piroplasms) (Fig 2). Their separation from other piroplasms was confirmed by high bootstrap values (Fig 2). Taken together, piroplasm sequences were demonstrated in bat ticks from three places of Hungary and two places of Romania (Table 3, Fig 1); three sequences of piroplasms were detected only in samples from Hungary, three of them only in Romania and three in both countries (Fig 2).

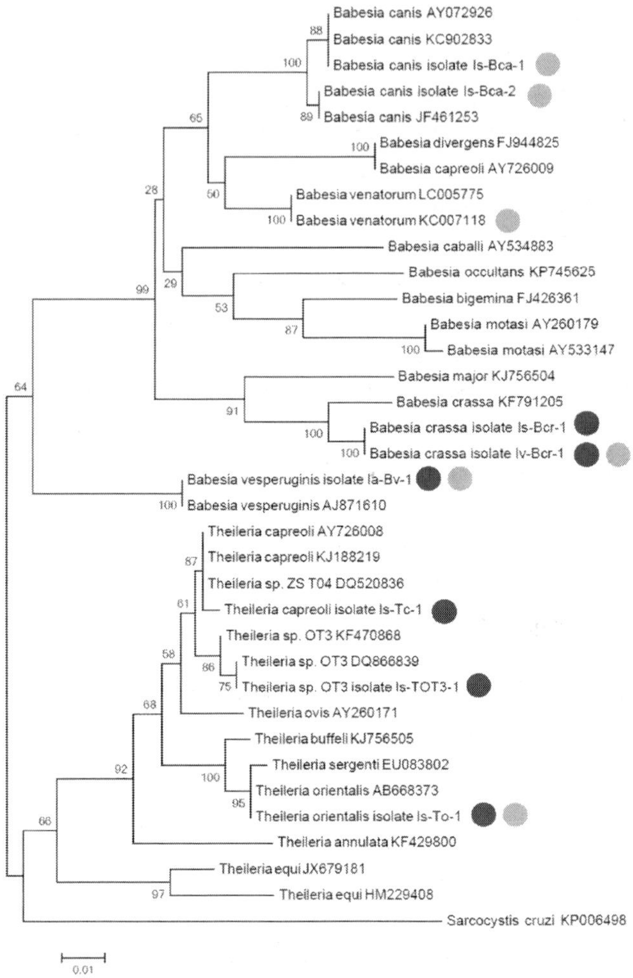

Figure 2. Phylogenetic relationships of 18S rDNA sequences of piroplasms identified in this study and relevant sequences previously deposited in GenBank. Sequences identified in the present study in Hungary and in Romania are highlighted with purple or turquoise dots, respectively. The shorter sequence of *Babesia venatorum* from this study is not included, therefore its reference sequence (to which it showed 100% identity) is marked. Branch lengths correlate to the number of substitutions inferred according to the scale shown.

https://doi.org/10.1371/journal.pone.0167735.g002

DISCUSSION

In this study all three ixodid tick species have been collected, which are specialized to bat hosts in Europe. Among them, *I. ariadnae* is known to occur in three countries of Central and Western Europe (Hungary, Germany and Belgium: [19]), whereas *I. vespertilionis* and *I. simplex* are more widespread on the continent (the latter predominantly south of latitude 49° N: [20]). In the present study *I. ariadnae* was not found in Romania, suggesting that the bat tick fauna in this country is similar to that in the Balkans, with the predominance of *I. simplex* [21]. A plausible explanation for this phenomenon is the absence of regular bat migration between the mountainous regions of Hungary and Romania, which otherwise could have caused the spread of *I. ariadnae* towards the southeast (the main route of long distance bat migration in the region is in the southwestern-northeastern direction: [22]).

The present results confirmed that the preferred hosts of *I. ariadnae* belong to Vespertilionidae [11, 19], those of *I. vespertilionis* to Rhinolophidae, while *I. simplex* is adapted to parasitize *Mi. schreibersii* (Miniopteridae) [23]. Nevertheless, to the best of our knowledge, several new host associations of ixodid bat ticks are reported here for the first time. In particular, *I. ariadnae* was found newly on four *Myotis* spp., *I. vespertilionis* on two *Myotis* spp., as well as on *Eptesicus serotinus* and *Rhinolophus mehelyi*, finally *I. simplex* on *Barbastella barbastellus* (Table 4).

Table 4. Host associations of ixodid bat ticks reported previously and in this study.

	Host species reported previously	References	New host species in this study
Ixodes ariadnae	*Myotis alcathoe*	[11, 19]	*Myotis dasycneme*
	Myotis bechsteinii		*Myotis daubentonii*
	Myotis blythii		*Myotis emarginatus*
	Myotis myotis		*Myotis nattererii*
	Plecotus auritus		

Ixodes vespertilionis	*Miniopterus schreibersii*	[9, 21, 23, 24, 25, 26, 27, 28]	*Eptesicus serotinus*
	Myotis bechsteinii		*Myotis capaccinii*
	Myotis blythii		*Myotis dasycneme*
	Myotis brandtii		*Rhinolophus mehelyi*
	Myotis daubentonii		
	Myotis emarginatus		
	Myotis myotis		
	Myotis mystacinus		
	Myotis nattererii		
	Plecotus auritus		
	Rhinolophus euryale		
	Rhinolophus ferrumequinum		
	Rhinolophus hipposideros		
Ixodes simplex	*Miniopterus schreibersii*	[21, 23, 25, 29]	*Barbastella barbastellus*
	Myotis alcathoe		
	Rhinolophus euryale		
	Rhinolophus ferrumequinum		

https://doi.org/10.1371/journal.pone.0167735.t004

The intensity of tick infestation was not significantly different between small and large size bat species (in the case of *My. daubentonii* vs. *R. ferrumequinum*, respectively), suggesting that factors depending on body size (such as the body surface area, interrelated with metabolic rate, heat emission: [30]) may not be crucial for host finding by bat ticks. Similarly, it has been reported that body size of passeriform bird species did not significantly influence the intensity of their tick infestation [31]. On the other hand, the intensity of infestation with bat tick larvae was significantly higher, than with later stages in the life cycle. This finding is consistent with the significant decrease of individual number of tick stages with the advance of tick developmental cycle [32].

Taking into account the considerable lack of data in literature on the vector potential of ixodid bat ticks, and the recent finding of *B. canis* DNA in bat faeces [5], DNA extracts of 307 specimens were molecularly analysed for the presence of piroplasms (Apicomplexa: Piroplasmida). Among piroplasms, *Babesia* species are known to be transmitted transovarially by female ticks to the next generation (i.e. to larvae prior to their blood meal), whereas *Theileria* species are transmitted transstadially [33]. The latter implies that there is no other way for tick larvae to harbor theileriae or to contain theileria DNA, than to ingest these with the blood meal from a host/reservoir which is either theileria-infected or at least theileria DNA is present in its blood stream.

Babesia vesperuginis DNA was molecularly identified here in *I. ariadnae* and *I. vespertilionis*. This piroplasm is pathogenic to bats, and was reported to infect *Pipistrellus pipistrellus* [34], several *Myotis* spp. (including *My. daubentonii*, on which bat species a PCR positive tick was collected in the present study) as well as *Plecotus auritus* [35]. Taking into account that soft ticks (*Argas vespertilionis*) have been incriminated as vectors of *B. vesperuginis* [34], the present results suggest that bat ticks carrying this piroplasm (or its DNA) ingested it with the blood meal, i.e. further *Myotis* spp. (exemplified by *My. dasycneme*) and *Eptesicus serotinus* might also be susceptible to *B. vesperuginis*.

Babesia crassa has low pathogenicity in small ruminants and its vector is unknown [36]. This piroplasm (or closely related genotypes) were reported to occur only in the Middle-East, but recently one genotype has also been identified in *Haemaphysalis inermis* ticks in Central Europe, Hungary [37]. In the present study two different DNA sequences of *B. crassa* were detected in bat ticks (*I. vespertilionis*, *I. simplex*) in both Hungary and Romania. These two bat tick species have never been reported from small ruminants, and therefore their PCR positivity can be explained by ingesting *B. crassa* DNA-containing blood meal from bats. In this context it may be epidemiologically relevant that *B. crassa* was reported to be present in *H. sulcata* [38], and this tick species was reported to infest bats in the larval stage and small ruminants in the adult stage [39].

Babesia canis is an important parasite of dogs. Wild canids are also susceptible [40]. The known vector of this piroplasm is *Dermacentor reticulatus*, which is a tick species seldom infesting bats, including *Mi. schreibersii* [8]. Recently, bats were reported to pass the DNA of *B. canis* in their faeces [5]. Taking into account that it is very unlikely that relevant lineages of *I. simplex* (found to be PCR positive here) had become infected from canids (from which hosts *I. simplex* has never been reported) in a previous stage or generation, *B. canis* or its DNA might have been present in the blood of relevant bats. This possibility is supported by recent finding of *B. canis* DNA in bat tissues [41].

Interestingly, the DNA of *B. venatorum* was amplified from one larva of *I. simplex* in Romania. Although the sequence was 100% identical with *B. venatorum* and differed from other piroplasms, because of its shortness no final conclusion can be drawn on the occurrence of *B. venatorum* DNA in bat ticks. This piroplasm (associated with cervids as hosts) is zoonotic, with *I. ricinus* as its vector. It is noteworthy that *I. ricinus* occurs on bats (e.g. [9]), and the present results suggest that this may allow bats to become carriers of *B. venatorum* or its DNA (taking into account that *I. simplex* has never been reported from cervids or from humans). On the other hand, the host of *I. simplex*, *Mi. schreibersii* may live in large colonies in the human environment (e.g. mines, man made tunnels, ruins: [20]). Therefore, this preliminary finding deserves further molecular epidemiological investigation.

Among *Theileria* spp. and genotypes, the DNA of *Theileria* sp. OT3 has been detected here in a female *I. simplex*. This piroplasms (with unknown pathogenicity) was formerly reported to infect small ruminants in Italy [42], but recently its DNA has also been reported from *Haemaphysalis punctata* in northern Hungary [37], and this tick species is known to infest bats [10].

In addition, the DNA of two *Theileria* spp. have been shown here to be present in larvae of *I. simplex* from *Mi. schreibersii*. Among them, *T. capreoli* is a mildly pathogenic parasite of cervids. The tick species *H. concinna*, in which the DNA of *T. capreoli* has been recently demonstrated in Hungary[37], is also known to occasionally infest bats [43].

Members of the *T. orientalis* complex (*T. orientalis*, *T. buffeli*) infect cattle in the tropical-subtropical regions of the globe, usually with low pathogenicity. Recently, *T. orientalis* has been shown to emerge in Central Europe [37] and Australia [44], sometimes severely affecting cattle [45]. Vectors of the *T. orientalis* complex are *Haemaphysalis* spp. [33]. *Haemaphysalis* spp. may accidentally infest bats, and in particular *H. punctata*, the most likely vector of *T. orientalis* in Europe, was synonymously called «*H. rhinolophi*" [10]. Furthermore, in South-East Asia (where species of the *T. orientalis* complex are widespread) at least one *Haemaphysalis* sp. has bats as preferred hosts [46]. These literature data attest a possible connection between large ruminants and bats via *Haemaphysalis* sp. ticks. In the present study only bat tick (*I. simplex*) larvae were PCR positive for

T. capreoli and *T. orientalis*. This means that relevant piroplasms (or their DNA) could have been acquired by the larvae exclusively from the blood of bat hosts, because there is no transovarial, only transstadial transmission in the case of theileriae [33]. The significance of the potential epidemiological role of bats in bovine theileriosis deserves further attention, as several bat species may use cattle stables for roosting [47].

In summary, competent vectors of the above piroplasms (that have been hitherto reported from hosts other than bats) are *D. reticulatus*, *I. ricinus* and *Haemaphysalis* spp. These tick species are rarely found on bats, most likely attaching to bats when roosting in nests of small mammals (e.g. in tree holes) [5], or when gleaning bat species feed on insects from the lower vegetation in meadows or forests. However, *Mi. schreibersii*, associated with most of the piroplasms identified in the present study, is not known to forage on the ground level [48]. Alternatively, blood-sucking flies have the potential to carry and transmit *Babesia* spp. [49] and *Theileria* spp. [50], and flies (Insecta: Diptera) are among the frequent food items of e.g. *Mi. schreibersii* [51]. This implies that bats may get into contact with or may have access to piroplasms or piroplasm DNA from their food.

Thus, there are two plausible explanations for the above, unexpected findings. The first is that bats get into frequent contact with the DNA of vector-borne pathogens contained in their food. During digestion this DNA may pass through the gut wall (barrier) un- or only partly digested, thus appearing in the circulation (or perhaps other tissues) from where bat ticks can take it up with their blood meal. In support of this possibility, it has recently been verified that meal-derived DNA fragments (even long ones) can avoid degradation and through not-yet-known mechanisms enter the circulation, at least in humans [52].

Another, although less likely explanation is that bats are susceptible to a broader range of piroplasms than previously thought. The phylogeny of piroplasms has recently been shown to reflect considerable host diversity and limited host specificity [53], suggesting that these tick-borne protozoa have undergone frequent host switches during their evolution. In this context the present results may imply that bats may share piroplasms with a broad range of mammals (from various orders). Similarly, several *Babesia* and *Theileria* spp. are known to infect hosts from different mammalian orders (e.g. *B. caballi*, *B. canis*, *B. divergens*, *B. microti*, *T. equi*: [53]).

CONCLUSIONS

Bat ticks are not known to infest dogs or ruminants, i.e. typical hosts and reservoirs of piroplasms molecularly identified in *I. vespertilionis* and *I. simplex*. Therefore, DNA sequences of piroplasms detected in these bat ticks most likely originated from the blood of their respective bat hosts. This may indicate that either bats are susceptible to a broader range of piroplasms than previously thought, or at least the DNA of piroplasms may pass through the gut barrier of bats during digestion of relevant insect vectors. In light of these findings, the role of bats in the epidemiology of piroplasmoses deserves further investigation

ACKNOWLEDGMENTS

The survey was organized in the framework of the EurNegVec COST Action TD1303. Financial resources were provided by OTKA 115854. The research was also supported by the 11475-4/2016/FEKUT grant of the Hungarian Ministry of Human Resources. GF was supported by the János Bolyai Research Scholarship of the Hungarian Academy of Sciences. PCE 236/2011 grant and the framework of Domus Hungarica, Hungarian Academy of Science provided financial help to ADS. Jére, Cs., Csősz, I., Bücs, Sz., Varga, Á., D'Amico, G. and Ionică, A. provided help in different stages of sample collection in Romania.

REFERENCES

1. Dietz C, Nill D, von Helversen O. Bats of Britain, Europe and Northwest Africa. A & C Black, London, 2009. 406 pp.
2. Szőke K, Hornok S. Epidemiological importance of bats (Chiroptera) in Europe, with emphasis on their blood-sucking ectoparasites as potential transmitters of vector-borne pathogens. Magy Állatorvos. 2016;138:15–29.
3. Calisher CH, Childs JE, Field HE, Holmes KV, Schountz T. Bats: important reservoir hosts of emerging viruses. Clin Microbiol Rev. 2006;19:531–545. 10.1128/CMR.00017-06
4. Kunz TH, Lumsden LF. Ecology of cavity and foliage roosting bats In: Kunz TH, Feneon MB (eds): Bat ecology. The University of Chicago Press, Chicago, 2003;3–89.
5. Hornok S, Estók P, Kováts D, Flaisz B, Takács N, Szőke K et al. Screening of bat faeces for arthropod-borne apicomplexan protozoa: *Babesia canis* and *Besnoitia besnoiti*-like sequences from Chiroptera. Parasit Vectors. 2015;8:441 10.1186/s13071-015-1052-6
6. Estrada-Peña A, Jongejan F. Ticks feeding on humans: a review of records on human- biting Ixodoidea with special reference to pathogen transmission. Exp Appl Acarol. 1999;23:685–715
7. Piksa K, Nowak-Chmura M, Siuda K. First case of human infestation by the tick *Ixodes vespertilionis* (Acari: Ixodidae). Int J Acarol. 2013;38:1–2.
8. Neumann LG. Ixodidae: *Dermacentor reticulatus*. Das Tierreich. 1911;26:98–100.
9. Ševčík M, Krištofík J, Uhrin M, Benda P. New records of ticks (Acari: Ixodidae) parasitising on bats in Slovakia. Vespertilio 2010;13–14:139–147.
10. Estrada-Peña A. *Las garrapatas* (Acari: Ixodoidea) parásitas de *murciélagos* (Mammalia, Chiroptera). (II). Rev Ibér Parasitol. 1989;49:165–175.
11. Hornok S, Kontschán J, Kováts D, Kovács R, Angyal D, Görföl T et al. Bat ticks revisited: *Ixodes ariadnae* sp. nov. and allopatric genotypes of *I. vespertilionis* in caves of Hungary. Parasit Vectors. 2014;7:202 10.1186/1756-3305-7-202
12. Hornok S, Estrada-Peña A, Kontschán J, Plantard O, Kunz B, Mihalca AD et al. High degree of mitochondrial gene heterogeneity in the bat

tick species *Ixodes vespertilionis*, *I. ariadnae* and *I. simplex* from Eurasia. Parasit Vectors. 2015;8:457 10.1186/s13071-015-1056-2

13. Hornok S, Kovács R, Meli ML, Kontschán J, Gönczi E, Gyuranecz M et al. First detection of bartonellae in a broad range of bat ectoparasites. Vet Microbiol. 2012;159:541–543. 10.1016/j.vetmic.2012.04.003
14. Feider Z. Ixodoidea In: Fauna of the Popular Republic of Romania. 1965, volume 5/2 Published by Academiei Republicii Populare Romane, Bucuresti. [in Romanian].
15. Hornok S, Kontschán J, Estrada-Peña A, Fernández de Mera IG, Tomanović S, de la Fuente J. Contributions to the morphology and phylogeny of the newly discovered bat tick species, *Ixodes ariadnae* in comparison with *I. vespertilionis* and *I. simplex*. Parasit Vectors. 2015; 8:47 10.1186/s13071-015-0665-0
16. Casati S, Sager H, Gern L, Piffaretti JC. Presence of potentially pathogenic *Babesia* sp. for human in *Ixodes ricinus* in Switzerland. Ann Agric Environ Med. 2006;13:65–70.
17. Tamura K, Nei M. Estimation of the number of nucleotide substitutions in the control region of mitochondrial DNA in humans and chimpanzees. Mol Biol Evol. 1993;10:512–26.
18. Tamura K, Peterson D, Peterson N, Stecher G, Nei M, Kumar S. MEGA5:molecular evolutionary genetics analysis using maximum likelihood, evolutionary distance, and maximum parsimony methods. Mol Biol Evol. 2011;28:2731–9. 10.1093/molbev/msr121
19. Hornok S, Krawczyk A. First record of *Ixodes ariadnae* in Western Europe, Belgium. Acta Vet Hung. 2016;64:467–471.
20. Hutson AM, Aulagnier S, Benda P, Karataş A, Palmeirim J, Paunović M. Miniopterus schreibersii. The IUCN Red List of Threatened Species 2008: e.T13561A4160556.
21. Burazerović J, Ćakić S, Mihaljica D, Sukara R, Ćirović D, Tomanović S. Ticks (Acari: Argasidae, Ixodidae) parasitizing bats in the central Balkans. Exp Appl Acarol. 2015;66:281–291. 10.1007/s10493-015-9891-6
22. Hutterer R, Ivanova T, Meyer-Cords C, Rodrigues L. Bat Migrations in Europe. A Review of Banding Data and Literature. Naturschutz und BiologischeVielfalt 28. Federal Agency for Nature Conservation, 2005, Bonn.

23. Arthur DR. The *Ixodes* ticks of Chiroptera (Ixodoidea, Ixodidae). J Parasitol. 1956;42:180–196.

24. Rupp D, Zahn A, Ludwig P. Actual records of bat ectoparasites in Bavaria (Germany). Spixana 2004;27:185–190.

25. Siuda K, Stanko M, Piksa K, Górz A. Ticks (Acari: Ixodida) parasitizing bats in Poland and Slovakia. Wiad Parazytol. 2009;55:39–45.

26. Mihalca AD, Dumitrache MO, Magdaş C, Gherman CM, Domşa C, Mircean V et al. Synopsis of the hard ticks (Acari: Ixodidae) of Romania with update on host associations and geographical distribution. Exp Appl Acarol. 2012;58:183–206. 10.1007/s10493-012-9566-5

27. Frank R, Kuhn T, Werblow A, Liston A, Kochmann J, Klimpel S. Parasite diversity of European *Myotis* species with special emphasis on *Myotis myotis* (Microchiroptera, Vespertilionidae) from a typical nursery roost. Parasit Vectors. 2015;8:101 10.1186/s13071-015-0707-7

28. Piksa K, Stańczak J, Biernat B, Górz A, Nowak-Chmura M, Siuda K. Detection of *Borrelia burgdorferi* sensu lato and spotted fever group rickettsiae in hard ticks (Acari, Ixodidae) parasitizing bats in Poland. Parasitol Res. 2016;115:1727–31. 10.1007/s00436-016-4936-2

29. Krištofík J, Danko Š. Arthropod ectoparasites (Acarina, Heteroptera, Diptera, Siphonaptera) of bats in Slovakia. Vespertilio 2012;16:167–189

30. Hock RJ. The metabolic rates and body temperatures of bats. Biol Bull. 1951;101:289–299.

31. Hornok S, Flaisz B, Takács N, Kontschán J, Csörgő T, Csipak Á et al. Bird ticks in Hungary reflect western, southern, eastern flyway connections and two genetic lineages of *Ixodes frontalis* and *Haemaphysalis concinna*. Parasit Vectors. 2016;9:101 10.1186/s13071-016-1365-0

32. Korotkov IuS. Methods for estimating the demographic structure of the taiga tick (Ixodidae) based on results of standard parasitological observations. Parazitologiia 2004;38:492–502. [in Russian]

33. Fujisaki K, Kawazu S, Kamio T. The taxonomy of the bovine *Theileria* spp. Parasitol Today. 1994;10:31–33.

34. Gardner RA, Molyneux DH. *Babesia vesperuginis*: natural and experimental infections in British bats (Microchiroptera). Parasitology 1987;95 (Pt 3):461–9.

35. Sebek Z, Sixl W, Rosicky B. Ein Beitrag zur Charakteristik der Naturherde der Piroplasmose und zur Kenntnis der Wirtstiere mit Daten zur Rinderplasmose in der Steiermark und von Kleinsäugeruntersuchungen in der CSSR (Sporozoa, Haemosporina). Mitteilungen der Abteilung für Zoologie am Landesmuseum Joanneum Graz 1975;4:67–80.
36. Hashemi-Fesharki R, Uilenberg G. *Babesia crassa* n.sp. (Sporozoa, Babesiidae) of domestic sheep in Iran. Vet Q. 1981;3:1–8. 10.1080/01652176.1981.9693787
37. Hornok S, Takács N, Kontschán J, György Zs, Micsutka A, Iceton S et al. Diversity of *Haemaphysalis*-associated piroplasms of ruminants in Central-Eastern Europe, Hungary. Parasit Vectors. 2015;8:627 10.1186/s13071-015-1236-0
38. Aktas M. A survey of ixodid tick species and molecular identification of tick-borne pathogens. Vet Parasitol. 2014;200:276–283. 10.1016/j.vetpar.2013.12.008
39. Filippova NA, Neronov V M, Farhang-Azad A. Data on the ixodid fauna (Acarina, Ixodidae) of small mammals in Iran. Entomol Obozrenie. 1976;55:467–479. [in Russian]
40. Kuttler KL. Canine babesiosis In: Ristic M (ed.): Babesiosis of domestic animals and man. CRC Press, Boca Raton, Florida; 1988. pp. 12–13.
41. Corduneanu A, Sándor AD, Mihalca AD, Hrazdilová K, Modry D, Hornok S: Molecular evidence of canine pathogens in tissues of European bats–Proceedings of the 17th International Bat Research Conference, 2016, Durban, South Africa, pp. 50–51. Available: http://www.harrison-institute.org/IBRC2016_FullProgram.pdf, Accessed November 16, 2016
42. Giangaspero A, Marangi M, Papini R, Paoletti B, Wijnveld M, Jongejan F. *Theileria* sp. OT3 and other tick-borne pathogens in sheep and ticks in Italy: molecular characterization and phylogeny. Ticks Tick Borne Dis. 2015;6:75–83. 10.1016/j.ttbdis.2014.09.007
43. Lebedeva NN, Korenberg EI. Distribution of *Haemaphysalis concinna* Koch in the Soviet Union and some general features of its ecology. Folia Parasitol. 1981;28:249–261.
44. Kamau J, de Vos AJ, Playford M, Salim B, Kinyanjui P, Sugimoto C: Emergence of new types of *Theileria orientalis* in Australian cattle and possible cause of theileriosis outbreaks. Parasit Vectors. 2011;4:22 10.1186/1756-3305-4-22

45. Izzo MM, Poe I, Horadagoda NJdVA, House. Haemolytic anaemia in cattle in NSW associated with *Theileria* infections. Aust Vet J. 2010;88:45–51. 10.1111/j.1751-0813.2009.00540.x
46. Hoogstraal H: Studies on Southeast Asian *Haemaphysalis* ticks (Ixodoidea, Ixodidae). Redescription, hosts, and distribution of *H. traguli* Oudemans. The larva and nymph of *H. vidua* W. and N. Identity of *H. papuana toxopei* Warburton (n. comb.). J Parasitol. 1964;50:765–782.
47. Dekker JJA, Regelink JR, Janssen EA, Brinkmann R, Limpens HJGA. Habitat use of female Geoffroy's bats (*Myotis emarginatus*) at it's two northernmost maternity roosts and the implications for their conservation. Lutra 2013;56:111–120.
48. Vincent S, Nemoz M, Aulagnier S. Activity and foraging habitats of *Miniopterus schreibersii* (Chiroptera, Miniopteridae) in southern France: implications for its conservation. Hystrix It J Mammal. 2011;22:57–72.
49. Friedhoff KT. Transmission of *Babesia*: Other means of transmission In: Ristic M (ed.): Babesiosis of domestic animals and man. CRC Press, Boca Raton, Florida; 1988. p. 45.
50. Hammer JF, Jenkins C, Bogema D, Emery D. Mechanical transfer of *Theileria orientalis*: possible roles of biting arthropods, colostrum and husbandry practices in disease transmission. Parasit Vectors. 2016;9:34 10.1186/s13071-016-1323-x
51. Presetnik P, Aulagnier S. Diet of *Miniopterus schreibersii* in northeastern Slovenia (Central Europe). Mammalia 2013;77:297–305.
52. Spisák S, Solymosi N, Ittzés P, Bodor A, Kondor D, Vattay G, et al. Complete Genes May Pass from Food to Human Blood. PLoS ONE 2013;8(7):e69805 10.1371/journal.pone.0069805
53. Lack JB, Reichard MV, Van Den Bussche RA. Phylogeny and evolution of the Piroplasmida as inferred from 18S rRNA sequences. Int J Parasitol. 2012;42:353–363. 10.1016/j.ijpara.2012.02.005

22

Molecular Ecological Insights into Neotropical Bird–Tick Interactions

Matthew J. Miller[1,2], Helen J. Esser[1,3], Jose R. Loaiza[1,4,5], Edward Allen Herre[1], Celestino Aguilar[1,4], Diomedes Quintero[6], Eric Alvarez[1,5] and Eldredge Bermingham[1,7]

[1] Smithsonian Tropical Research Institute, Panama City, Republic of Panama
[2] Department of Biology, Villanova University, Villanova, Pennsylvania, United States of America
[3] Department of Environmental Sciences, Wageningen University, Wageningen, the Netherlands
[4] Centro de Biodiversidad y Descubrimiento de Drogas, Instituto de Investigaciones Científicas y Servicios de Alta Tecnología (INDICASAT-AIP), Panama City, Republic of Panama
[5] Centralamerican Master's Program in Entomology, University of Panama, Panama City, Republic of Panama
[6] G. B. Fairchild Invertebrate Museum, University of Panama, Panama City, Republic of Panama
[7] Patricia and Phillip Frost Museum of Science, Miami, Florida, United States of America

Citation: Miller MJ, Esser HJ, Loaiza JR, Herre EA, Aguilar C, Quintero D, et al. (2016) Molecular Ecological Insights into Neotropical Bird–Tick Interactions. PLoS ONE 11(5): e0155989. https://doi.org/10.1371/journal.pone.0155989

ABSTRACT

In the tropics, ticks parasitize many classes of vertebrate hosts. However, because many tropical tick species are only identifiable in the adult stage, and these adults usually parasitize mammals, most attention on the ecology of tick-host interactions has focused on mammalian hosts. In contrast, immature Neotropical ticks are often found on wild birds, yet difficulties in identifying immatures hinder studies of birds' role in tropical tick ecology and tick-borne disease transmission. In Panama, we found immature ticks on 227 out of 3,498 individually–sampled birds representing 93 host species (24% of the bird species sampled, and 13% of the Panamanian land bird fauna). Tick parasitism rates did not vary with rainfall or temperature, but did vary significantly with several host ecological traits. Likewise, Neotropical–Nearctic migratory birds were significantly less likely to be infested than resident species. Using a molecular library developed from morphologically–identified adult ticks specifically for this study, we identified eleven tick species parasitizing birds, indicating that a substantial portion of the Panamanian avian species pool is parasitized by a diversity of tick species. Tick species that most commonly parasitized birds had the widest diversity of avian hosts, suggesting that immature tick species are opportunistic bird parasites. Although certain avian ecological traits are positively associated with parasitism, we found no evidence that individual tick species show specificity to particular avian host ecological traits. Finally, our data suggest that the four principal vectors of Rocky Mountain Spotted Fever in the Neotropics rarely, if ever, parasitize Panamanian birds. However, other tick species that harbor newly–discovered rickettsial parasites of unknown pathogenicity are frequently found on these birds. Given our discovery of broad interaction between Panamanian tick and avian biodiversity, future work on tick ecology and the dynamics of emerging tropical tick-borne pathogens should explicitly consider wild bird as hosts.

INTRODUCTION

Wild birds are increasingly recognized as playing an important role in human and animal health. Emerging zoonotic diseases such as avian influenza, West Nile Virus, and Lyme disease often have wild birds in their transmission cycle [1–3], and wild birds can act as reservoir hosts in endemic areas [4]. Birds also have the potential to introduce diseases into previously naïve populations by spreading pathogens and/or their vectors over large distances via migratory flyways [5,6]. Hard ticks (Ixodidae) are haematophagous ectoparasites regularly found on wild birds, providing them habitat and blood-meal resources, and vector more human pathogens than any other arthropod group [7]. Migratory birds captured in temperate regions upon their return from wintering grounds have been observed infected with larval and nymphal stages of tropical tick species [5]. The ability of migratory birds to move tropical ticks over long distances, in concert with global climate change [8], potentially exposes extra-tropical regions to novel tropical tick-borne pathogens and vice versa [7,9,10]. Therefore, empirical studies are needed that evaluate which tick species are frequently involved in bird parasitism and which ecological characteristics of bird species are related to increased tick infestation levels.

In the New World tropics, the greatest risk of tick-borne disease for human comes from *Rickettsia rickettsii*, the etiological agent of Rocky Mountain Spotted Fever (RMSF), which has a current fatality rate of 20–40% [11]. In the Neotropics, the principle vector of RMSF are members of the *Amblyomma cajennense sensu lato* species complex [12], although other species of ticks have been show to harbor *R. rickettsii*. In Panama, four confirmed RMSF vectors occur: *A. mixtum*, which is the local taxon in the *A. cajennense* species-complex [13], along with three other species which are widely found in the Neotropical region: *Dermecantor nitens*, *Haemaphysalis leporispalustris*, and *Rhipicephalus sanguineus* [14]. Although the first clinical cases of RMSF were reported in Panama in the 1950s, the disease went unreported for over 50 years until a fatal case in 2004 [15]. In the past decade, additional fatal cases have been reported in Panama [16] as well as adjacent Costa Rica [17] and Colombia [18]. Improvements in diagnosis are clearly responsible for at least part of the surge in recent cases, although climate change, habitat modification, and/or increases in human–wildlife contact may also be responsible [19]. While adults of the genus *Amblyomma* and several other Neotropical tick species typically exploit mammals, or reptiles and amphibians to a lesser degree [20], immature forms are routinely found on birds [7,20–26] including, rarely, nymphs of the *A. cajennense* species complex [27].

At the same time, improvements in molecular detection of tick endosymbionts has uncovered a diversity of novel *Rickettsia* strains of unknown pathogenicity in Neotropical ticks [21,28,29], and wider distributions of other *Rickettsia* species known to cause RMSF-like symptoms. Recently, *R. rickettsii* has been found in a variety of tick species throughout the Americas and is believed to have caused many overlooked cases of rickettsial disease in South America [19]. Rickettsial bacteria have been found in several species of ticks recovered from Neotropical birds [21,25,26].

Our understanding of Neotropical bird-tick associations and hence their role in disease transmission has been hampered by species identification problems; most immature Neotropical ticks, especially those of the genus *Amblyomma*, are not readily identifiable to species by morphology alone. *Amblyomma* species diversity peaks in the Neotropics, where taxonomic keys serve to identify the nymphal stage of few *Amblyomma* species [30]. For example, in a recent survey of ticks found parasitizing humans in Panama, only 38% of the recovered specimens of *Amblyomma* could be identified to species [31]. Similarly, in a study documenting tick infestation patterns in wild birds from southeastern Brazil, nearly 48% of the specimens of *Amblyomma* could not be identified to species [27]. That study and others [21,32] demonstrate that new tools and approaches are essential to properly assess the role of wild birds in tick ecology and tick-borne disease transmission in the Neotropics.

Museum specimens are often valuable resources for broad studies of ecological patterns [33]. Here, we exploit seven years of extensive bird specimen collection across Panama to clarify the ecological interactions between the diverse tick and bird fauna of this Neotropical country. The collecting program of the STRI Bird Collection visited over 100 field sites, sampling nearly 3500 terrestrial birds from 384 species, almost half of the roughly 800 non-aquatic bird species recorded from Panama. Our study provides an unparalleled insight into the ecological relationships between birds and ticks in a Neotropical setting. Specifically, our goals were to: 1) identify how avian ecological traits influence the frequency of tick parasitism of Panamanian bird species; 2) produce a robust DNA barcode library capable of identifying most of the commonly encountered immature ticks in Panama; and 3) use the DNA barcode library to identify to species immature ticks collected off of Panamanian wild birds to determine host specificity patterns and the role of parasite ecological filtering in shaping bird-tick associations in Panama. Our results should provide insights into the relation between avian and ixodid biodiversity that may better inform our understanding of Neotropical tick ecology and may provide insights for our understanding of emerging tropical tick-borne diseases.

RESULTS

Ecological traits of Panamanian Wild Birds Parasitized by Ticks

We evaluated patterns of tick parasitism in 3,498 bird specimens from the STRI bird collection, representing 384 species, i.e. nearly half of the roughly 800 non-aquatic bird species recorded from Panama. Ticks parasitized a total of 227 specimens of Panamanian birds (6.5% of all individuals; S1 Table) representing 93 avian species and 24 families. Among the 227 infested birds, both the median and modal number of ticks recovered was 1. However, 15% of the infested birds had 4 or more ticks; one bird had 101 ticks, which was the most recovered from any bird in the study. Resident birds (i.e. species that breed in Panama) were 3.8 times as likely to be parasitized by ticks compared to non-breeding Nearctic–Neotropical migratory species (6.8% vs. 1.8%; Fisher's exact test, $P = 0.001$). 90 of 343 resident species had at least one bird sampled infested with ticks, whereas only 3 of 40 Nearctic–Neotropical migratory species had an infested individual, although sample sizes were lower for migratory species relative to resident species (mean $N_{resident} = 9.5$, mean $N_{migratory} = 5.4$; two-tailed unequal variance t-test, $P < 0.001$).

Among avian families with at least 25 specimens evaluated, the families with the greatest proportion (%) of specimens parasitized by ticks were: Thamnophilidae (antbirds: 18%; 12 of 20 species); Furnariidae (ovenbirds and woodcreepers: 15%; 11 of 20 species); Polioptilidae (gnatcatchers: 15%; 1 of 3 species); Turdidae (thrushes: 15%; 7 of 15 species); and Troglodytidae (wrens: 14%; 9 of 16 species). Because of behavioral differences between males and females (including time spent in the nest, where infestation may be more likely), we anticipated a difference in infestation rates between males and females. Although females had a slightly higher prevalence of tick parasitism than males (7% vs. 6%), the difference was not significant (Fisher's exact test, $P = 0.23$). Among resident birds, a logistic model indicated that only forest habitation, terrestrial foraging, bark insectivory, and lowland residency were significantly positively associated with tick parasitism (Table 1).

Table 1. Ecological traits associated with tick parasitism among Panamanian resident wild birds (N = **3274).**

Ecological Trait	**Odds Ratio**	***P* value (Wald's)**
Bark insectivores*	8.3	0.004
Terrestrial foragers*	4.0	$< 10^{-15}$
Forest vs. non-forest dwellers*	3.3	$< 10^{-9}$
Tree hole nesters	2.5	0.20
Ground cavity nesters	2.0	0.14
Lowland vs. montane*	1.7	0.01

Traits significant for a positive association with tick parasitism marked with (*). Significance determined by multiple logistic regression.

https://doi.org/10.1371/journal.pone.0155989.t001

Among 26 locations where we sampled a minimum of 20 birds, we found no relationship between the frequency of tick parasitism and annual mean temperature, temperature seasonality, annual precipitation, or precipitation seasonality. However we did recover an effect of taxonomic composition, specifically, the proportion of the sampled avifauna belonging to the five families most frequently parasitized by ticks (S1 Table).

DNA Identifications of Adult Ticks Agree with Morphological Taxonomy

The 96 individuals in the adult reference library of morphologically identified ticks from central Panama formed 20 clusters with pairwise Kimura-2 parameter (K2P) genetic distances greater than 5% (S1 Fig). All 96 could be placed in clusters in agreement with the original morphological identification of the voucher with bootstrap support values of at least 99%. Among the 20 species—clusters, average nearest-neighbor K2P distance to another cluster was 15.6% and the minimum nearest-neighbor K2P distance was 12.5% (range: 12.5%– 20.5%).

Two species contained DNA barcode sub-clusters with between-cluster sequence divergence well below the 12.5–20.5% difference observed between named species, but greater than 3.0%. *Haemaphysalis juxtakochi* comprised two clusters that differed by 3.0% pairwise K2P distance, with each cluster supported by 99% bootstrap support, while *Amblyomma ovale* contained one cluster of four individuals supported by 97% bootstrap and a fifth individual that varied by an average K2P distance of 3.2%. In both cases, individuals from both clusters were collected at a shared location, suggesting that these might represent cryptic biological species. As a consequence, a numerical DNA barcoding taxonomy (BIN barcode identification number) based on genetic distances among barcodes clusters recovered 22 unique BINs in our adult dataset; representing the 20 clusters that agree with our morphological named species as described above, as well as second BINs for both *H. juxtakochi* and *A. ovale*.

DNA Identifications of Immature ticks from Birds

We generated useable DNA barcode sequences from 130 immature ticks out of a total of 172 samples attempted (76% success rate). Two failures were due to double peaks in the electropheragram recovered in multiple amplification and sequencing attempts (MJM2941-T01 and MJM4264-T01); we removed these individuals from further analyses. One individual (MJM 7015) amplified its avian host (*Poecilotriccus sylvia*) DNA. The remaining 39 individuals failed in either the PCR or the sequencing step (S1 Table).

Sequences from the immature ticks formed 13 DNA barcode clusters. When we merged the immature barcode dataset with the adult reference library (Fig 1), 122 of 130 (94%) taxa formed 11 DNA barcode clusters which included an adult reference. Thus, we can confirm that immature ticks of the following species parasitize wild birds in Panama: *H. juxtakochi*, *A. dissimile*, *A. ovale*, *A. longirostre*, *A. geayi*, *A. sabanerae*, *A. varium*, *A. calcaratum*, and *A. nodosum* (S1 Table). Our data establish that at least 8 of the 18 species of *Amblyomma* ticks found in Panama parasitize birds. This includes the second global record of *A. dissimile*–a reptile and anuran specialist–parasitizing a wild bird (Mangrove Cuckoo, *Coccyzus minor*, [20]). In the case of *H. juxtakochi*, which is represented in the adult reference library by two DNA barcode clusters (and two unique BINs in the molecular taxonomy), we recovered both clusters from bird samples. The two remaining clusters of immature ticks did not include an adult reference, so they can only be identified using molecular taxonomy (see below).

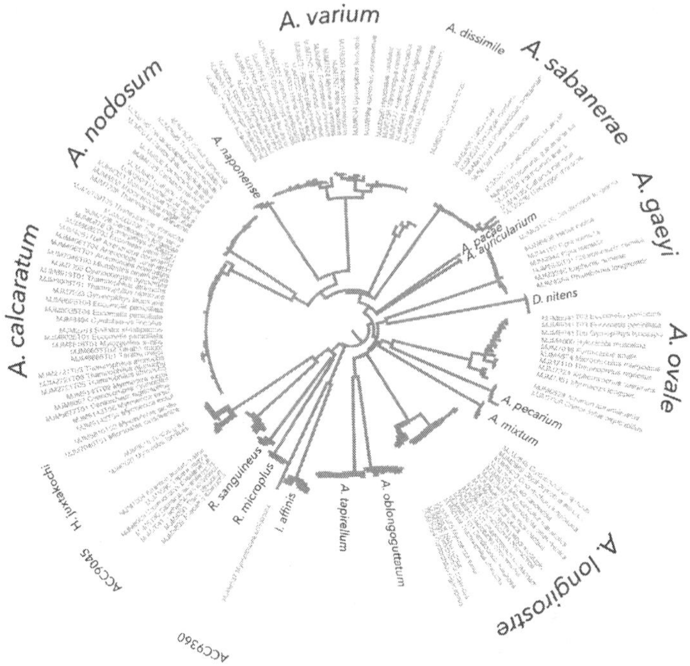

Figure 1. DNA barcoding neighbor-joining tree of combined data matrix of immature ticks and adult reference library ticks**.** Thin gray tip labels refer to specimen number and host species. Clade labels in blue refer to tick species recovered from birds, clade labels in gray refer to ticks unobserved on birds. Two clades where not represented by in the adult reference library so they are labeled in red with their molecular taxonomy Barcode Identification Number (BIN). The distribution of Panamanian ticks recovered from wild birds is unbalanced towards certain species of *Amblyomma*. Importantly, we recovered no immature ticks on birds from species known to vector human disease [6].

https://doi.org/10.1371/journal.pone.0155989.g001

Using the BIN molecular taxonomy, our sample of immature ticks clustered into 13 BINs, including 11 of the 22 BINs recovered in the adult reference library. Hence, we were able to assign a BIN to 100% of the samples that provided DNA sequences and to 76% (130 of 172) of all samples for which we attempted DNA barcoding. Eight other immature ticks (6%) formed two novel clusters on our phylogenetic trees; and for these we were only able to assign a BIN numerical taxonomic identification. The first BIN (ACC9360) was formed by just one immature tick that had a nearest neighbor-distance of 14.4% to the reference library cluster of *Ixodes affinis*. A second cluster (BIN: ACC9045) was formed by seven immatures whose nearest-neighbor cluster on the BOLD database did not include ticks from this study. Instead, they were most closely related to *H. leporispalustris* collected in Canada, with a nearest-neighbor distance of 6.0%. *H. leporispalustris* also occurs in Panama, but without a Panamanian adult reference sequence and given the large sequence variation, we are unable to determine whether our sample represents a genetically-divergent Panamanian *H. leporispalustris* population or distinct species of *Haemaphysalis* yet to be recorded in Panama. Thus, we refer to these two taxa as probable members of *Ixodes* and *Haemaphysalis* genera, respectively.

Using either BIN or traditional taxonomy, species accumulation curves for our sample of immature ticks recovered from wild birds were essentially asymptotic, as were species richness estimators designed to account for unobserved species. Using the BIN taxonomy, the Chao1 species richness estimate was 13.3 (95% confidence interval: 13.0–19.0), compared to an observed BIN species richness of 13 (Fig 2). Likewise, the species accumulation curve using the traditional taxonomy recovered a mean Chao1 estimate of 11.5 (95% confidence interval: 11.0–19.3), compared to an observed species richness of 11 (S2 Fig). These results suggest that at most only a few more tick species would be recovered from wild birds in Panama given considerably greater sampling effort, and that their occurrence on wild birds would be exceedingly rare.

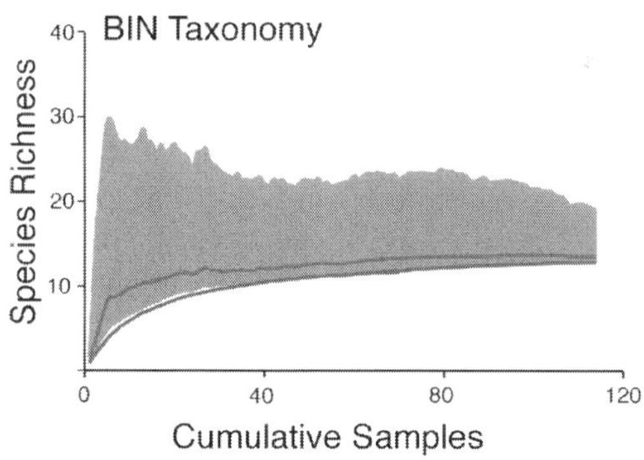

Figure 2. Species accumulation curve (SAC) for immature ticks recovered from Panamanian wild birds based on BIN numerical taxonomy (for SAC based on traditional tick taxonomy). Red line = *S*, mean observed species richness; dark blue line = $\hat{S}$, mean Chao1 *S* estimate; filled blue area defines 95% upper and lower confidence limits (CI) for $\hat{S}$. As Chao1 is downward biased, the 95%

lower CI is probably not useful. The convergence of the S and $\hat{S}$ curves, as well as the asymptotic nature of these curves and the CI curves, suggests that at most only a few more tick species would be recovered from wild birds in Panama, and that their occurrence would be rare.

https://doi.org/10.1371/journal.pone.0155989.g002

Immature ticks show no Species-Level Host Specificity or Ecological Filtering of Avian Hosts

Immature tick species showed no measurable avian species host specificity. Tick species collected from at least two different bird individuals always occurred on at least two different avian host species, and most frequently occurring tick species recovered from wild birds in our samples had the greatest number of avian host species (Pearson's *rho* = 0.984, $P < 0.0000001$). In 5 of 11 birds from which we sequenced multiple individual ticks, we found more than one species or more than one haplotype of tick, suggesting multiple independent colonizations of the host.

In addition to the lack of species-level host specificity, we observed little evidence that individual tick species show ecological preferences in hosts. Instead, networks of specific tick and avian host species demonstrates broad and varied interactions by the ecological traits that were previously shown to influence rates of bird parasitism (Fig 3). We found no difference in the frequency of forest versus non-forest hosts among our 13 tick BIN species ($G = 12.37$, d.f. = 12, $P = 0.42$, Fig 3A), nor in the frequency of arboreal-foraging hosts ($G = 22.18$, d.f. = 12, $P = 0.036$, Bonferonni-corrected $\alpha = 0.013$, Fig 3B). Likewise, neither the proportion of bark insectivorous hosts ($G = 10.287$, d.f. = 12, $P = 0.59$, Fig 3C) nor the proportion of montane hosts ($G = 17.996$, d.f. = 12, $P = 0.12$, Fig 3D) varied among tick species. We found equivalent results using the traditional tick taxonomy.

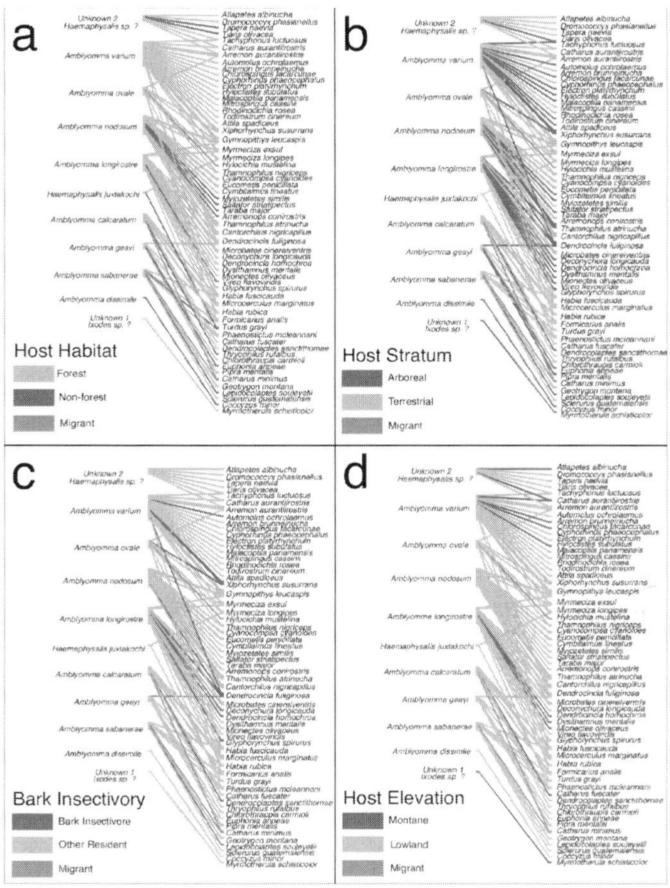

Figure 3. Bird–immature tick quantitative interaction networks. (A) Blue: interactions involving non-forest bird species, yellow: interactions involving forest inhabiting bird species; (B) yellow: terrestrial-foraging bird species; blue: interactions involving arboreal bird species; (C) blue: interactions involving bird species that are bark insectivores; yellow: other species; (D) blue: interactions involving montane bird species, yellow: lowland species. a-d) pink: interactions involving non-breeding migrant bird species. The frequency of hosts among these four ecological traits was not significant among the 11 sampled tick species (See Results for details).

https://doi.org/10.1371/journal.pone.0155989.g003

DISCUSSION

Patterns of tick Parasitism on Panamanian wild birds

Although only a minority of Panamanian land birds are infested by ticks, the broad participation in bird–tick interactions by a diversity of bird and tick taxa indicates that wild birds may play an important role in the life history of many Neotropical tick species and have the potential to play a role in the transmission of tick-borne diseases. While only 6.5% of all birds examined carried ticks, parasitism by ticks includes a diverse taxonomic array of avian hosts. We observed ticks on 93 of the 384 species we examined, which represent 24 avian families. When examining parasitism rates for passerines (11.2%), the Panamanian parasitism rate is in line with rates (9–17%) from other studies that focused almost exclusively on passerine birds [22,32,34].

We found no evidence that tick parasitism by site varied by either absolute or seasonal differences in temperature or rainfall. This indicates that, at the macro-scale, climatological patterns in the Neotropics likely have little influence on the parasitism frequency. Nonetheless, differences in ecological traits among Panamanian birds appear to modulate tick parasitism frequency. We found that lowland, forest inhabiting, ground and bark foraging birds were significantly more likely to be infested with ticks than birds with other ecologies.

Lack of Host Specificity and Ecological Filtering in Bird-Tick Interactions

Historically, it has been argued that adult life stages of Neotropical ticks are strictly or nearly-strictly specific to one or a few mammalian species [20,35,36]. The degree of host specificity in immature Neotropical ticks remains largely unexplored, because of the difficulty of their morphological identification to species. Using our molecular identifications, we found no evidence for strong host specificity in Panamanian tick–bird interactions. Instead, the number of host species increased with the number of sampling events for that tick species, demonstrating that immature Neotropical ticks parasitize a wide variety of birds from diverse families, with no clear preference for particular host species.

Our finding of limited or no host specificity tracks a more general shift in our understanding of the specificity of tick–host relationships. Researchers applying modern statistical approaches generally have found lower levels of host specificity than was assumed in the historical literature [37–40]. Instead of absolute host specificity, research suggests that ecological correlates between potential hosts and ticks may determine the host diversity of tick species [37]. This is a restatement of Combe's "ecological filter" concept of parasitism [41,42]. Our finding of four key ecological traits that are positively correlated with tick parasitism rates in birds allows us to test the ecological filter concept in bird–immature tick interactions.

Here, we found no evidence that certain tick species have ecological preferences to bird species based on any of these four traits. The result was robust using both traditional tick taxonomy and the molecular taxonomy. This latter result is notable because one of the key ways that cryptic tick species might co-occur is by having cryptic host ecological specificities among immature stages [43]. Whether such ecological filtering occurs between adult ticks and mammals in Panama awaits further study. Our finding of neither strict-host specificity, nor the more relaxed ecological filtering type of specificity in bird–tick interactions, coupled with the diversity of bird species involved and the potentially greater dispersal ability of birds, indicates that bird–tick interactions will have meaningful implications for the demographics of Neotropical ticks and also the ecology of tick-borne pathogens [37]. Thus, tick-borne disease transmission models based on patterns of host specificity of adult ticks may require re-examination in order to incorporate the more labile ecology of immature ticks.

DNA Barcoding of Adult and Immature Ticks

Our findings demonstrate that DNA barcodes are a reliable method to identify Panamanian hard ticks (Ixodidae) to species, and can overcome the frequently-cited difficulties in identifying immature forms of Neotropical ticks to species using only morphological characters [22,23,44]. Alternatives to molecular identification of immature ticks include rearing immatures to life stages that can be reliably identified to species [27], but this is time consuming, requires special laboratory conditions that vary among species, and most immature ticks die before reaching an identifiable life stage [23,27]. DNA barcoding appears to yield a much greater percentage of successful species identifications. We were able to identify ~70% of immature ticks, whereas a rearing study from Brazil was able to identify only 12% of the immatures [23].

The role of Wild Birds in the Transmission Ecology of Tick-Borne Pathogens

Our ability to identify immature ticks from Panamanian birds permits insight to the potential role of wild birds in the transmission ecology of Neotropical ticks and tick-borne diseases. We found no evidence that wild birds are involved in the transmission ecology of Rocky Mountain Spotted Fever (RMSF). RMSF is the most virulent tick-borne disease known in the Western Hemisphere and is caused by infection from *Rickettsia rickettsii*. In Panama, RMSF was first reported over 60 years ago [45], although it remained unreported again until 2004 [15], when it resulted in a fatal case in western Panama. Since 2004, RMSF has been regularly reported in central and western Panama [46]. Several species of ixodid ticks are confirmed vectors of *R. rickettsii*, including four species of Panamanian ticks: *Dermacentor nitens*, *Rhipicephalus sanguineus*, *Haemaphysalis leporispalustris*, and *Amblyomma (cajennense) mixtum*. However, members of the *A. cajennense* species complex are considered the primary vector of *R. rickettsii* in tropical America [12], and in Panama *R. rickettsii* has been detected principally in *A. mixtum* [47,48] with a single record in *D. nitens* [49]. We found no examples of *A. mixtum*, *D. nitens*, or *R. sanguineus* parasitizing birds in our study, although we did find five birds infested with *Haemaphysalis sp.* ticks. Our species accumulation curves (Fig 2, S2 Fig) for tick species found on wild birds suggest that at best only a few, rare, species are yet to be recovered from resident wild land birds in Panama. While a couple of studies from other Neotropical regions have reported that immature forms of species in the *A. cajennense* complex parasitize various wild bird species, Labruna et al. [27] challenged the morphological identifications in these cases. Our findings along with others [22,23,25,34] demonstrate that Rocky Mountain Spotted Fever vectors are at best rare parasites of wild birds, and collectively suggest a relatively negligible role for wild birds in Rocky Mountain Spotted Fever transmission in Panama and elsewhere in the Neotropics.

On the other hand, it is likely that immature ticks on wild bird in Panama harbor a diversity of other rickettsial pathogens. Although we did not attempt to isolate *Rickettsia* from sampled ticks, other studies have demonstrated that 10 of the 11 tick species that we recovered from Panamanian wild bird harbor *Rickettsia*. *Rickettsia amblyommii* has been recovered as a parasite of *A. ovale* in Panama [49,50] as well as *A. geayi* and *A. longirostre* in Brazil [22,29] and several species of ticks recovered from Neotropical–Nearctic migrant birds in south Texas, USA [51]. *Rickettsia amblyommii* has been suggested, but not confirmed, to be a cause of spotted fever-like disease in

North America [52]. *Rickettsia parkeri*, which was recently identified as the cause of human spotted fever rickettsiosis in southeastern USA as well as in Brazil, Uruguay and Argentina was recovered from Brazilian samples of *A. nodosum* and *A. ovale* [29]. Other *Rickettsia* species, not yet known to cause human disease, have been recovered from *A. calcaratum* [21], *A. dissimile* [53], *A. varium* [54] and *H. juxtakochi* [29]. Work in the Neotropical regions concerning the pathogenicity of *R. amblyommii* and *R. parkeri* and other rickettsiales is in its infancy. Likewise, spotted fever group rickettsioses are often under-detected, especially in Middle and South America [54]. While the rate of infestation of ticks on wild birds in Panama is relatively modest (~6.5%), as a recent study from south Texas demonstrates, even low relative frequencies of parasitism may have regionally–important consequences for tick and emerging disease ecology given the absolute numbers of wild birds involved [51]. Assuming typical avian densities recorded for Panama [55], as many as 96,000,000 Panamanian wild birds might be infested with ticks. Thus, our finding of a pervasive and diverse relationship between birds and immature ticks in Panama suggests that further consideration of the role of wild birds in the ecology of Neotropical ticks and the pathogens they vector is warranted.

MATERIALS AND METHODS

Bird and tick Specimen Collection

All bird records come from the vouchered collecting program of the STRI Bird Collection, conducted between 2008 and 2012. Collecting occurred year round; however as our collection strategy during this period was the developed as part of a larger program on the ecology of avian-mediated zoonoses, collecting was balanced at most sites between the rainy (May–December) and dry (January–April) season, as well as the across the migratory seasons of Nearctic–Neotropical migrants. Full specimen metadata are available and in the STRI Collections portal (http://stricollections.org).

Our field procedures begin with capturing birds in mistnets (or occasionally collecting with shotgun). Per STRIBC field protocols, wild birds were euthanized in the field and flash frozen on solid CO_2 in individual freezer bags to eliminate the risk of cross-contamination of ectoparasites prior to transportation to the lab. There, the entire ectoparasite assembly is recovered as the first step in the ornithological specimen preparation process via whole body ruffling following [56] who demonstrated that post-mortem ruffling is superior for estimating ectoparasite abundances compared to visual inspection and other live- bird sampling strategies. The number of sampling locations was more than 100, however we combined locations that were within 5 kilometers of each other, resulting in a total of 43 sampling locations throughout Panama (Fig 4) for the purpose of this study (S2 Table). EA separated ticks from other ectoparasites and identified all to age class and sex when possible. Ticks are stored in the STRI Cryological collections in single tubes per avian specimen in 95% ethanol and maintained at -20°C. All specimens collected by the STRI Bird Collection was done with the prior approval of ANAM, Panama's environmental authority (permit numbers: SE/A-60- 10, SE/A-137-10, SE/A-96-09, SE/A-44-10, SE/A-66-11, SE/A-2-12), and collecting methods have been approved by the Smithsonian Tropical Research Institute's Institutional Animal Care and Use Committee (IACUC permits: 2007–03-03-15-07, 2011-0927-2014-03). Complete ornithological specimen metadata is available at: http://stricollections.org.

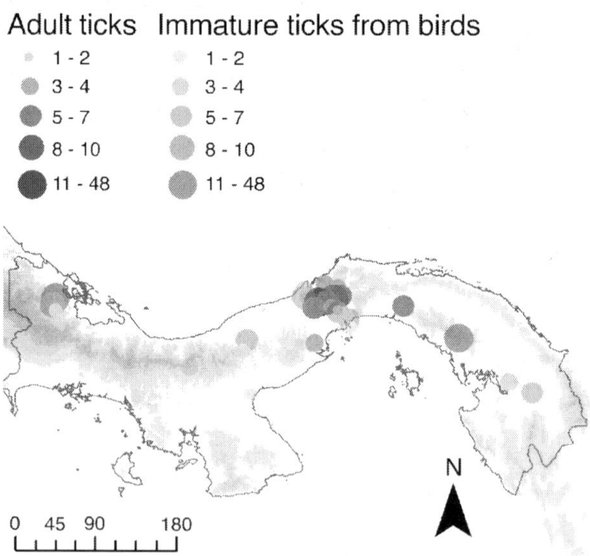

Figure 4. Map of collecting areas for adult and immature ticks. Blue circles represent areas where adult ticks were sampled for the reference library; orange circles represent locations where immature ticks were sampled from birds.

https://doi.org/10.1371/journal.pone.0155989.g004

Ecological Patterns of tick Parasitism on Birds

We generated a tick parasitism data matrix (presence or absence of hard ticks: Ixodidae). Non-land birds from marsh, aquatic, and riverine habitats (e.g. Anseriformes, Charadriiformes, Alcedinidae) were excluded from the analysis, as well as aerial foragers (e.g. Apodidae, Hirundinidae), resulting in 3498 specimen records (S1 Table), from 384 avian species. Using this matrix, we analyzed whether key traits of the host species were correlated with tick parasitism using contingency table analysis. We considered the relationship of eight traits on tick parasitism, some of which have previously been correlated with tick parasitism in birds: residents vs. non-breeding migrants, females vs. males, terrestrial foraging vs. arboreal, ground cavity nesting, tree hole nesting, lowland vs. montane habitats, bark insectivory, forest vs. non-forest habitats. We evaluated the influence of sex and migratory status on the entire dataset, however because ecological traits for many migratory birds are more labile away from their breeding grounds and classifications based on temperate zone ecology may not reflect behavior in Panama, we evaluated the relationship between the remaining six traits and tick parasitism only on resident birds (i.e. those species that breed in Panama). Montane species were defined as those found almost exclusively 600 meters above sea level. Terrestrial foraging species include those species that forage primarily, but not exclusively, on the ground or within 15 centimeters of the ground. Forest inhabitants live in or require mature tropical forests; most edge species were classified as non-forest inhabitants. We evaluated all possible combinations of ecological characters using logistic regression models.

Adult DNA Barcode Reference Library

We generated a DNA barcode reference library for the ticks of central Panama using morphologically pre-identified adult ticks collected as part of on-going research programs (HJE and JRL) on the tick-host interactions in the area surrounding the Panama Canal, and also in a few cases from ethanol preserved museum specimens of adult ticks (S1 Fig). Field collections were accomplished either by removal of ticks from hosts (live-captured animals, road kill, livestock, and pets) or by collecting questing ticks from the free environment via flagging, e.g. sweeping a white cotton cloth along vegetation and leaf litter and harvesting accumulated ticks. Adults were identified using morphological characters and existing taxonomic keys [20,57], and were stored in 95% ethanol and frozen at -20°C prior to molecular analysis (see next section). Our adult reference library included 96 individuals (S3 Table) that were morphologically assigned to 19 of the approximately 37 species of hard ticks recognized for the Republic of Panama, including 14 of 18 species of *Amblyomma* [20]. Unless already part of a museum collection, after DNA extraction adult reference ticks were stored in 95% ethanol and are maintained as voucher specimens in the ectoparasite collection of the STRIBC. A public dataset for these 96 specimens, including geographic details of collection, specimen photographs, and museum voucher information can be found on the BOLD data portal v3 [58] under the name: DS-TICKA (dx.doi.org/10.5883/DS-TICKA).

Immature Ticks from Panamanian Wild Birds

We selected 186 immature ticks from the pool of immature ticks collected from birds for molecular species-level identification using the adult reference DNA barcode library as the basis for identification (S1 Table). We obtained usable DNA barcode sequences for 130 (see Results). Sample details for immature ticks can be found on the BOLD data portal under the name: DS-TICKI (dx.doi.org/10.5883/DS-TICKI).

Molecular Methods

To allow for the preservation of museum vouchers, DNA was extracted from adult specimens from either two legs removed from the specimen, or from a rear quarter section of the abdomen cut from the body, done under an entomological dissecting microscope. We used the entire body of immature ticks for DNA extraction. In all cases, the material being extracted was frozen in a 2 ml tube suspended in liquid nitrogen and pulverized using a sterile micro-pestle to improve DNA yield. We initially obtained poor DNA yield after attempted DNA extractions using DNAeasy spin columns (Qiagen, Valencia, CA), following the manufacturer's instructions (except that we reduced the final elution volume to 50 μl). Subsequently, we switched to the QIAamp DNA Micro kit (Qiagen), which uses similar spin column technology but is optimized for smaller samples and resulted in superior DNA yields.

Amplification of the DNA barcoding region (5' region of the COI mitochondrial gene, 43) was accomplished using the standard invertebrate primers (LCO1490 and HCO2198; 44) following [59], except that we halved the reaction volume (i.e. 25 μL) and raised DNA to 4 μL; we used

Qiagen taq and buffers. Positive and negative controls were run in every reaction. Amplifications were visualized on a low-melting agarose gel from which a single PCR product was extracted using a sterilized scalpel blade, and sequenced at the Naos Molecular Laboratory, Smithsonian Tropical Research Institute. DNA sequences and tracefiles can be examined in BOLD under the DS-TICKA database and DS-TICKI database. Sequences have also been deposited in GenBank under accession numbers KF200076 – KF200171.

Tree Building and Barcode Distance Analysis

In order to understand species limits and confirm morphological identification among our adult reference ticks ($N = 96$), we generated a neighbor-joining tree in MEGA v.5.1 [60] using Kimura-2 parameter (K2P) distances. We assessed branch support by bootstrapping the topology with 500 replicates. We examined the K2P distance matrix and resulting topology for evidence of genetic divergence among our adult reference library that might provide evidence for the presence of cryptic species [43] using both a standard genetic distance approach (3% K2P) as well as looking for the assignation of multiple Barcode Index Numbers (BINs) to a given species. The Barcode Index Number is an alternative, numerical taxonomy that clusters taxa into interim operational taxonomic units using a stage process to employ single linkage clustering [61]. BINs are assigned automatically in the BOLD database portal based on the global dataset of DNA barcode sequences (i.e. including samples not generated in this study). We repeated all tree-building, genetic distance, and clustering analyses for a second, expanded dataset that combined the 96 adult reference sequences with 130 sequences from immature ticks collected from birds.

Species-level Associations between Ticks and wild Birds in Panama

We used the identifications of immature ticks from STRIBC bird specimens to further examine species-specific bird-tick associations in Panama. First, we assessed host-specificity by examining the correlation between the frequency of occurrence of a given tick species (number of birds infested by that species) and the diversity of avian host species (number of host species) for all tick species recovered in our dataset; if immature ticks are non-host specific, this correlation should be strong. While this approach provides one measurement of host–tick specificity, it potentially overlooks the role of ecological filtering by hosts for particular parasite species [41], which we tested for by examining differences in the frequency of parasitism by host ecological traits for each tick species identified using COI barcodes using *G*-tests of independence. We visualized the interaction between immature tick species and specific avian host species via quantitative interaction networks created using the bipartite package following the authors' instructions [62] in the R statistical application [63]. For those birds that had multiple ticks identified by DNA barcoding we scored each tick species separately but only once in the data matrix.

Finally, to estimate the proportion and distribution of the total Panamanian tick species pool that might depend on birds as host vertebrates for immature life stages, we estimated the total species richness of ticks that parasitize wild birds in Panama through species accumulation curves generated in the EstimateS software package [64]. When sampling is exhaustive, the species accumulation curve should reach an asymptote. However, even non-exhaustive sampling can still yield sufficient data to provide a reasonable estimate of the true species richness, which can be assessed by observing an asymptote in the statistical estimate of species richness [65]. We generated species-accumulation curves (SACs) for the 130 ticks where we were successfully able to generate DNA barcodes using both the adult reference library cluster-indicated taxonomy, and using the BIN numerical taxonomy generated in the BOLD database. In both cases, we used the Chao1 species richness estimator [66], which attempts to non-parametrically correct the observed species richness as a function of the proportion of species observed exactly once or twice in the dataset. Mean Chao1 values were obtained from 100 reshuffles of our data set with replacement in EstimateS; sampling with replacement being critical in order to account for sampling error.

ACKNOWLEDGMENTS

We thank Panama's Environmental Agency (MiAmbiente/ANAM) for granting us bird and tick collecting permits, without which this study would not have been possible. We thank the members of the Smithsonian Tropical Research Bird Collection field team, as well as B. Voirin and B. Hirsch for providing adult ticks used in this study, and P. Jansen, K. López, and an anonymous reviewer for comments that improved the manuscript. This study was partially funded by an inter-agency award from the US Centers for Disease Control ("Effect of Anthropogenic Climate Change on the

Ecology of Zoonotic and Vector-borne Diseases"; http://www.cdc.gov/), and a Smithsonian Grand Challenges Award ("Tropical Vertebrate Diversity Loss and the Emergence of Tick-borne Diseases"; https://www.si.edu/consortia). HJE was supported by the graduate school of Production Ecology and Resource Conservation of Wageningen University (https://www.pe-rc.nl/). The STRI Bird Collection was supported by an NIH/NSF Ecology and Evolution of Infectious Diseases Award from the Fogarty International Center (3R01-TW005869-05S1; http://www.fic.nih.gov/). DNA barcode sequencing was supported from a grant from the Smithsonian Institution's DNA Barcoding Network (http://barcoding.si.edu/). The funders had no role in study design, data collection and analysis, decision to publish, or preparation of the manuscript.

AUTHOR CONTRIBUTIONS

Conceived and designed the experiments: MJM HJE JRL EAH EB. Performed the experiments: MJM HJE JRL CA EA. Analyzed the data: MJM HJE JRL. Contributed reagents/materials/analysis tools: MJM HJE JRL DQ. Wrote the paper: MJM HJE JRL EAH CA DQ EA EB.

REFERENCES

1. Rappole JH, Hubálek Z. Migratory birds and West Nile virus. J Appl Microbiol. 2003;94: 47–58. 10.1046/j.1365-2672.94.s1.6.x
2. Kilpatrick AM, Chmura AA, Gibbons DW, Fleischer RC, Marra PP, Daszak P. Predicting the global spread of H5N1 avian influenza. Proc Natl Acad Sci. 2006;103: 19368–19373. 10.1073/pnas.0609227103
3. Ogden NH, Lindsay LR, Hanincová K, Barker IK, Bigras-Poulin M, Charron DF, et al. Role of migratory birds in introduction and range expansion of *Ixodes scapularis* ticks and of *Borrelia burgdorferi* and *Anaplasma phagocytophilum* in Canada. Appl Environ Microbiol. 2008;74: 1780–1790. 10.1128/AEM.01982-07
4. Ginsberg HS, Buckley PA, Balmforth MG, Zhioua E, Mitra S, Buckley FG. Reservoir competence of native North American birds for the Lyme disease spirochete, *Borrelia burgdorferi*. J Med Entomol. 2005;42: 445–449. 10.1093/jmedent/42.3.445
5. Hamer SA, Goldberg TL, Kitron UD, Brawn JD, Anderson TK, Loss SR, et al. Wild birds and urban ecology of ticks and tick-borne pathogens, Chicago, Illinois, USA, 2005–2010. Emerg Infect Dis. 2012;18: 1589–1595. 10.3201/eid1810.120511
6. Lindeborg M, Barboutis C, Ehrenborg C, Fransson T, Jaenson TGT, Lindgren P-E, et al. Migratory birds, ticks, and Crimean-Congo hemorrhagic fever virus. Emerg Infect Dis. 2012;18: 2095–2097. 10.3201/eid1812.120718
7. Jongejan F, Uilenberg G. The global importance of ticks. Parasitology. 2004;129: S3–S14. 10.1017/S0031182004005967
8. Léger E, Vourc'h G, Vial L, Chevillon C, McCoy KD. Changing distributions of ticks: causes and consequences. Exp Appl Acarol. 2013;59: 219–244. 10.1007/s10493-012-9615-0
9. Parola P, Socolovschi C, Jeanjean L, Bitam I, Fournier P-E, Sotto A, et al. Warmer weather linked to tick attack and emergence of severe rickettsioses. PLoS Negl Trop Dis. 2008;2: e338 10.1371/journal.pntd.0000338
10. Randolph SE. To what extent has climate change contributed to the recent epidemiology of tick-borne diseases? Vet Parasitol. 2010;167: 92–94. 10.1016/j.vetpar.2009.09.011
11. Parola P, Paddock CD, Socolovschi C, Labruna MB, Mediannikov O, Kernif T, Abdad MY, Stenos J, Bitam I, Fournier PE, Raoult D. Update

on tick-borne rickettsioses around the world: a geographic approach. Clin Microbiol Rev. 2013; 26(4):657–702. 10.1128/CMR.00032-13

12. Dantas-Torres F. Rocky Mountain spotted fever. Lancet Infect Dis. 2007;7: 724–732. 10.1016/S1473-3099(07)70261-X
13. Nava S, Beati L, Labruna MB, Cáceres AG, Mangold AJ, Guglielmone AA. Reassessment of the taxonomic status of *Amblyomma cajennense* (Fabricius, 1787) with the description of three new species, *Amblyomma tonelliae n. sp.*, *Amblyomma interandinum n. sp.* and *Amblyomma patinoi n. sp.*, and reinstatement of *Amblyomma mixtum* Koch, 1844, and *Amblyomma sculptum* Berlese, 1888 (Ixodida: Ixodidae). Ticks Tick-Borne Dis. 2014;5: 252–276. 10.1016/j.ttbdis.2013.11.004
14. Pinter A, Labruna MB. Isolation of *Rickettsia rickettsii* and *Rickettsia bellii* in cell culture from the tick *Amblyomma aureolatum* in Brazil. Ann N Y Acad Sci. 2006;1078: 523–529. 10.1196/annals.1374.103
15. Estripeaut D, Aramburú MG, Sáez-Llorens X, Thompson HA, Dasch GA, Paddock CD, et al. Rocky Mountain spotted fever, Panama. Emerg Infect Dis. 2007;13: 1763–1765. 10.3201/eid1311.070931
16. Tribaldos M, Zaldivar Y, Bermudez S, Samudio F, Mendoza Y, Martinez AA, et al. Rocky Mountain spotted fever in Panama: a cluster description. J Infect Dev Ctries. 2011;5: 737–741.
17. Argüello AP, Hun L, Rivera P, Taylor L. A fatal urban case of Rocky Mountain spotted fever presenting an eschar in San Jose, Costa Rica. Am J Trop Med Hyg. 2012;87: 345–348. 10.4269/ajtmh.2012.12-0153
18. Hidalgo M, Miranda J, Heredia D, Zambrano P, Vesga JF, Lizarazo D, et al. Outbreak of Rocky Mountain spotted fever in Córdoba, Colombia. Mem Inst Oswaldo Cruz. 2011;106: 117–118.
19. Eremeeva ME, Dasch GA. Challenges posed by tick-borne rickettsiae: eco-epidemiology and public health implications. Front Public Health. 2015;3: 55 10.3389/fpubh.2015.00055
20. Fairchild GB, Kohla GM, Tipton VJ. The ticks of Panama (Acarina: Ixodoidea) Ectoparasites of Panama. Chicago: Field Museum of Natural History; 1966. pp. 167–219.
21. Ogrzewalska M, Martins T, Capek M, Literak I, Labruna MB. A *Rickettsia parkeri*-like agent infecting *Amblyomma calcaratum* nymphs from wild birds in Mato Grosso do Sul, Brazil. Ticks Tick-Borne Dis. 2013;4: 145–147. 10.1016/j.ttbdis.2012.07.001
22. Ogrzewalska M, Uezu A, Labruna MB. Ticks (Acari: Ixodidae)

infesting wild birds in the eastern Amazon, northern Brazil, with notes on rickettsial infection in ticks. Parasitol Res. 2010;106: 809–816. 10.1007/s00436-010-1733-1

23. Ogrzewalska M, Pacheco RC, Uezu A, Richtzenhain LJ, Ferreira F, Labruna MB. Ticks (Acari: Ixodidae) infesting birds in an Atlantic rain forest region of Brazil. J Med Entomol. 2009;46: 1225–1229.
24. Ogrzewalska M, Uezu A, Jenkins CN, Labruna MB. Effect of forest fragmentation on tick infestations of birds and tick infection rates by rickettsia in the Atlantic forest of Brazil. EcoHealth. 2011;8: 320–331. 10.1007/s10393-011-0726-6
25. Pacheco RC, Arzua M, Nieri-Bastos FA, Moraes-Filho J, Marcili A, Richtzenhain LJ, et al. Rickettsial infection in ticks (Acari: Ixodidae) collected on birds in southern Brazil. J Med Entomol. 2012;49: 710–716.
26. Ogrzewalska M, Literak I, Cardenas-Callirgos JM, Capek M, Labruna MB. *Rickettsia bellii* in ticks *Amblyomma varium* Koch, 1844, from birds in Peru. Ticks Tick-Borne Dis. 2012;3: 254–256. 10.1016/j.ttbdis.2012.05.003
27. Labruna MB, Sanfilippo LF, Demetrio C, Menezes AC, Pinter A, Guglielmone AA, et al. Ticks collected on birds in the state of São Paulo, Brazil. Exp Appl Acarol. 2007;43: 147–160. 10.1007/s10493-007-9106-x
28. Labruna MB, McBride JW, Bouyer DH, Camargo LMA, Camargo EP, Walker DH. Molecular evidence for a spotted fever group *Rickettsia* species in the tick *Amblyomma longirostre* in Brazil. J Med Entomol. 2004;41: 533–537.
29. Labruna MB, Mattar V S, Nava S, Bermudez S, Venzal JM, Dolz G, et al. Rickettsioses in Latin America, Caribbean, Spain and Portugal. Rev MVZ Córdoba. 2011;16: 2435–2457.
30. Martins TF, Onofrio VC, Barros-Battesti DM, Labruna MB. Nymphs of the genus *Amblyomma* (Acari: Ixodidae) of Brazil: descriptions, redescriptions, and identification key. Ticks Tick-Borne Dis. 2010;1: 75–99. 10.1016/j.ttbdis.2010.03.002
31. Bermúdez C. SE, Castro A, Esser H, Liefting Y, García G, Miranda RJ. Ticks (Ixodida) on humans from central Panama, Panama (2010–2011). Exp Appl Acarol. 2012;58: 81–88. 10.1007/s10493-012-9564-7
32. Marini MA, Reinert BL, Bornschein MR, Pinto JC. Ecological

correlates of ectoparasitism on Atlantic Forest birds, Brazil. Ararajuba. 1996;4: 93–102.

33. Rocha LA, Aleixo A, Allen G, Almeda F, Baldwin CC, Barclay MVL, et al. Specimen collection: an essential tool. Science. 2014;344: 814–815. 10.1126/science.344.6186.814
34. Ogrzewalska M, Literák I, Capek M, Sychra O, Calderón VÁ, Rodríguez BC, et al. Bacteria of the genus *Rickettsia* in ticks (Acari: Ixodidae) collected from birds in Costa Rica. Ticks Tick-Borne Dis. 2015;6: 478–482. 10.1016/j.ttbdis.2015.03.016
35. Guglielmone AA, Estrada-Peña A, Mangold AJ, Barros-Battesti DM, Labruna MB, Martins JR, et al. *Amblyomma aureolatum* (Pallas, 1772) and *Amblyomma ovale* Koch, 1844 (Acari: Ixodidae): hosts, distribution and 16S rDNA sequences. Vet Parasitol. 2003;113: 273–288.
36. Nava S, Szabó MPJ, Mangold AJ, Guglielmone AA. Distribution, hosts, 16S rDNA sequences and phylogenetic position of the Neotropical tick *Amblyomma parvum* (Acari: Ixodidae). Ann Trop Med Parasitol. 2008;102: 409–425. 10.1179/136485908X278883
37. McCoy KD, Léger E, Dietrich M. Host specialization in ticks and transmission of tick-borne diseases: a review. Front Cell Infect Microbiol. 2013;3 10.3389/fcimb.2013.00057
38. Wells K, O'Hara RB, Pfeiffer M, Lakim MB, Petney TN, Durden LA. Inferring host specificity and network formation through agent-based models: tick–mammal interactions in Borneo. Oecologia. 2012;172: 307–316. 10.1007/s00442-012-2511-9
39. Madinah A, Abang F, Mariana A, Abdullah M t., Mohd-Azlan J. Interaction of ectoparasites-small mammals in tropical rainforest of Malaysia. Community Ecol. 2014;15: 113–120. 10.1556/ComEc.15.2014.1.12
40. Nava S, Guglielmone A. A meta-analysis of host specificity in Neotropical hard ticks (Acari: Ixodidae). Bull Entomol Res. 2013;103: 216–224. 10.1017/S0007485312000557
41. Combes C. Parasitism: the ecology and evolution of intimate interactions. Chicago: University of Chicago Press; 2001.
42. Poulin R. Evolutionary ecology of parasites, 2nd ed. Princeton: Princeton University Press; 2011.
43. Hebert PDN, Penton EH, Burns JM, Janzen DH, Hallwachs W. Ten species in one: DNA barcoding reveals cryptic species in the

neotropical skipper butterfly Astraptes fulgerator. Proc Natl Acad Sci U S A. 2004;101: 14812–14817. 10.1073/pnas.0406166101

44. Bermúdez SE, Miranda RJ, Smith D. Ticks species (Ixodida) in the Summit Municipal Park and adjacent areas, Panama City, Panama. Exp Appl Acarol. 2010;52: 439–448. 10.1007/s10493-010-9374-8
45. de Rodaniche E. C., Rodaniche A. Spotted fever in Panama; isolation of the etiologic agent from a fatal case. Am J Trop Med Hyg. 1950;30: 511–517.
46. Bermúdez CSE, Zaldívar AY, Spolidorio MG, Moraes-Filho J, Miranda RJ, Caballero CM, et al. Rickettsial infection in domestic mammals and their ectoparasites in El Valle de Antón, Coclé, Panamá. Vet Parasitol. 2011;177: 134–138. 10.1016/j.vetpar.2010.11.020
47. De Rodaniche EC. Natural infection of the tick, *Amblyomma cajennense*, with *Rickettsia rickettsii* in Panama. Am J Trop Med Hyg. 1953;2: 696–699.
48. Estrada-Peña A, Guglielmone AA, Mangold AJ. The distribution and ecological "preferences" of the tick *Amblyomma cajennense* (Acari: Ixodidae), an ectoparasite of humans and other mammals in the Americas. Ann Trop Med Parasitol. 2004;98: 283–292. 10.1179/000349804225003316
49. Bermúdez SE, Eremeeva ME, Karpathy SE, Samudio F, Zambrano ML, Zaldivar Y, et al. Detection and identification of rickettsial agents in ticks from domestic mammals in eastern Panama. J Med Entomol. 2009;46: 856–861.
50. Eremeeva ME, Karpathy SE, Levin ML, Caballero CM, Bermudez S, Dasch GA, et al. Spotted fever rickettsiae, *Ehrlichia* and *Anaplasma*, in ticks from peridomestic environments in Panama. Clin Microbiol Infect Off Publ Eur Soc Clin Microbiol Infect Dis. 2009;15 Suppl 2: 12–14. 10.1111/j.1469-0691.2008.02638.x
51. Cohen EB, Auckland LD, Marra PP, Hamer SA. Avian migrants facilitate invasions of Neotropical ticks and tick-borne pathogens into the United States. Appl Environ Microbiol. 2015;81: 8366–8378. 10.1128/AEM.02656-15
52. Apperson CS, Engber B, Nicholson WL, Mead DG, Engel J, Yabsley MJ, et al. Tick-borne diseases in North Carolina: is "*Rickettsia amblyommii*" a possible cause of rickettsiosis reported as Rocky Mountain spotted fever? Vector Borne Zoonotic Dis Larchmt N.

2008;8: 597–606. 10.1089/vbz.2007.0271

53. Miranda J, Portillo A, Oteo JA, Mattar S. *Rickettsia sp.* strain colombianensi (Rickettsiales: Rickettsiaceae): a new proposed *Rickettsia* detected in *Amblyomma dissimile* (Acari: Ixodidae) from iguanas and free-living larvae ticks from vegetation. J Med Entomol. 2012;49: 960–965.
54. Romer Y, Seijo AC, Crudo F, Nicholson WL, Varela-Stokes A, Lash RR, et al. *Rickettsia parkeri* Rickettsiosis, Argentina. Emerg Infect Dis. 2011;17: 1169–1173. 10.3201/eid1707.101857
55. Robinson WD, Brawn JD, Robinson SK. Forest bird community structure in central Panama: influence of spatial scale and biogeography. Ecol Monogr. 2000;70: 209–235. 10.1890/0012-9615(2000)070[0209:FBCSIC]2.0.CO;2
56. Clayton DH, Drown DM. Critical evaluation of five methods for quantifying chewing lice (Insecta: Phthiraptera). J Parasitol. 2001;87: 1291–1300. 10.1645/0022-3395(2001)087[1291:CEOFMF]2.0.CO;2
57. Onofrio V, Labruna M, Pinter A, Giacomin F, Barros-Battesti D. Comentários e chaves as espécies do gênero *Amblyomma* In: Barros-Battesti D, Arzua M, Bechara G, editors. Carrapatos de Importância Médico-Veterinária da Região Neotropical: Un guia ilustrado para identificação de espécie. São Paulo: Instituto Butantan; pp. 53–113.
58. Ratnasingham S, Hebert PDN. BOLD: The Barcode of Life Data System (http://www.barcodinglife.org). Mol Ecol Notes. 2007;7: 355–364. 10.1111/j.1471-8286.2007.01678.x
59. Kumar NP, Rajavel AR, Natarajan R, Jambulingam P. DNA barcodes can distinguish species of Indian mosquitoes (Diptera: Culicidae). J Med Entomol. 2007;44: 1–7.
60. Tamura K, Peterson D, Peterson N, Stecher G, Nei M, Kumar S. MEGA5: molecular evolutionary genetics analysis using maximum likelihood, evolutionary distance, and maximum parsimony methods. Mol Biol Evol. 2011;28: 2731–2739. 10.1093/molbev/msr121
61. Ratnasingham S, Hebert PDN. A DNA-based registry for all animal species: the barcode index number (BIN) system. PloS One. 2013;8: e66213 10.1371/journal.pone.0066213
62. Dormann CF, Fründ J, Blüthgen N, Gruber B. Indices, Graphs and Null Models: Analyzing Bipartite Ecological Networks [Internet]. 2009. Available: http://goedoc.uni-goettingen.de/goescholar/handle/1/5837

63. R Development Core Team. R Development Core Team (2013). R: A language and environment for statistical computing R Foundation for Statistical Computing, Vienna, Austria: ISBN 3-900051-07-0, URL http://www.R-project.org. R Foundation for Statistical Computing, Vienna, Austria; 2013.
64. Colwell R. EstimateS: Statistical estimation of species richness and shared species from samples. Version 9. User's Guide and application published at: http://purl.oclc.org/estimates. 2013.
65. Colwell RK, Coddington JA. Estimating terrestrial biodiversity through extrapolation. Philos Trans R Soc Lond B Biol Sci. 1994;345: 101–118. 10.1098/rstb.1994.0091
66. Chao A. Estimating the population size for capture-recapture data with unequal catchability. Biometrics. 1987;43: 783–791.

SECTION 5:
FLIES

23

A Bayesian Geostatistical Moran Curve Model for Estimating Net Changes of Tsetse Populations in Zambia

Luigi Sedda[1], Cornelius Mweempwa[2], Els Ducheyne[3], Claudia De Pus[4], Guy Hendrickx[3], David J. Rogers[1]

[1] Department of Zoology, University of Oxford, Oxford, United Kingdom,
[2] Department of Veterinary and Livestock Development, Ministry of Agriculture and Cooperatives, Lusaka, Zambia,
[3] Avia-GIS, Zoersel, Belgium,
[4] Animal Health Department, Institute of Tropical Medicine, Antwerp, Belgium

ABSTRACT

For the first time a Bayesian geostatistical version of the Moran Curve, a logarithmic form of the Ricker stock recruitment curve, is proposed that is able to give an estimate of net change in population demographic rates considering components such as fertility and density dependent and density independent mortalities. The method is applied to spatio-temporally

Citation: Sedda L, Mweempwa C, Ducheyne E, De Pus C, Hendrickx G, Rogers DJ (2014) A Bayesian Geostatistical Moran Curve Model for Estimating Net Changes of Tsetse Populations in Zambia. PLoS ONE 9(4): e96002. https://doi.org/10.1371/journal.pone.0096002

referenced count data of tsetse flies obtained from fly-rounds. The model is a linear regression with three components: population rate of change estimated from the Moran curve, an explicit spatio-temporal covariance, and the observation error optimised within a Bayesian framework. The model was applied to the three main climate seasons of Zambia (rainy – January to April, cold-dry – May to August, and hot-dry – September to December) taking into account land surface temperature and (seasonally changing) cattle distribution. The model shows a maximum positive net change during the hot-dry season and a minimum between the rainy and cold-dry seasons. Density independent losses are correlated positively with day-time land surface temperature and negatively with night-time land surface temperature and cattle distribution. The inclusion of density dependent mortality increases considerably the goodness of fit of the model. Cross validation with an independent dataset taken from the same area resulted in a very accurate estimate of tsetse catches. In general, the overall framework provides an important tool for vector control and eradication by identifying vector population concentrations and local vector demographic rates. It can also be applied to the case of sustainable harvesting of natural populations.

INTRODUCTION

There are several ways of analysing animal and plant population data. The simplest involves an actuarial approach as applied to human populations where the likelihood of death and the expectation of future life are assumed to depend only upon age. For many animal and plant populations, however, mortality also comes about from both density independent and density dependent effects and the simple actuarial approach is no longer valid. [1] developed an alternative life-table approach in which the contributions of both density dependent and independent mortalities to changes in total generation mortality were examined over several generations; they defined the 'key factor' as the mortality mostly responsible for changes in the total generation mortality over time. A third approach to analysing population data is due to the fisheries research of [2] who produced stock recruitment curves by plotting successive (usually annual) fish counts against each other on a graph (the present year/generation on the x-axis and the next year/generation on the y-axis). The simplest sort of Ricker curve rises rapidly from the origin (when stock numbers increase rapidly from low densities) but eventually turns over as density dependent population limitation becomes important; in an ideal world fishing removes those individuals that would otherwise die through density dependent losses, without affecting the standing crop. Whilst Varley and Gradwell's life-table approach [1] was developed for species with non-overlapping generations, the Ricker curve [2] can also be applied to species with overlapping generations. Rogers, studying tsetse flies in Africa [3], was the first to connect these two approaches by applying the density dependent and density independent losses concepts of the Varley and Gradwell approach to the logarithmic form of Ricker curve, called a Moran curve (logarithms were necessary because Varley and Gradwell's mortalities were expressed logarithmically, as 'k-values'). Tsetse have continuous overlapping generations and the Moran curve approach was shown, through its application to the output of a simple tsetse population model, to be able to quantify (as k-values) the mean monthly density dependent and density independent losses of tsetse in Nigeria and Zambia.

For any population to persist or increase locally total annual fertility must equal or exceed total annual density independent mortality; at equilibrium, the balance of production (fertility minus density independent mortality) is then accounted for by density dependent mortalities (due to competitors, predators or parasites). If inescapable density independent losses exceed fertility, the population will be in continuous decline until it becomes extinct. Such populations can persist in an area only if they are periodically 'topped up' from elsewhere, by immigration.

Application of the Moran curve approach to the analysis of field data involves making some assumptions about population fertility which, for simplicity, is usually assumed constant. In reality, tsetse fertility depends mostly upon temperature [4], so the assumption of constancy in Moran curve analysis means that variations in fertility are treated as variations in mortality; this is unlikely to be a serious problem because natural variations in fertility are very much less than natural variations in mortality. The net change in the population's demographic rates (hereafter simply "net change") is the difference between fertility and population losses (sum of the density dependent and density independent mortality).

Moran curve analysis (for a complete description of the method and its application to tsetse fly sample data) provoked criticism on several fronts, mainly targeted at the subjectivity in choosing the intercept on the y-axis of the Moran plot (determined by the fertility rate) and the density at which density dependent effects begin to become important – the turnover point in the standard Ricker curve [5]. In this paper, we aim: (i) to develop a less subjective Moran curve analysis; (ii) to quantify the tsetse population net change; (iii) to estimate the strength of density dependent effects, the importance of which for tsetse has previously been questioned [6]; and (iv) to demonstrate the advantages for control techniques in identifying both the times (seasons) and places in which to concentrate tsetse suppression activities.

There are other options to analyse population models, for example the Bayesian state-space "normal dynamic linear models" proposed by [7] and applied to the Ricker model. This advanced approach allows the inclusion in the model of observational error (related to the sample data) and process error (related to the functions applied in the model), and the extension of time series analysis to nonstationary and non linear models [8]–[11]. Bayesian state-space models are at present one of the best options for population models. Our aim here, however, is to consider a model with explicit spatio-temporal covariance and population parameters estimated from the Moran curve.

METHODS

Ethics Statement

The sampling at each location was authorized by the Department of Veterinary and Livestock Development, Zambia. The field studies did not involve endangered or protected species but only tsetse flies. This is not a vertebrate study.

Field Methods

Analysis was applied to data for tsetse flies (*Glossina morsitans morsitans* Westwood) from South-Eastern Zambia (field data provided in Appendix S6). The sampling scheme was both spatial and longitudinal, and involved man-baited black screens carried on fixed fly-rounds [12], which are commonly used for sampling this species of tsetse (see Appendix S2 for a full description of the sampling method and the geographic area analysed). Sampling was carried out in four sites (listed from North to South): Lusandwa, Kasamanda, Zinaka and Chisulo (Table 1 and Appendix S2). In each site there were two sampling transects arranged in opposite directions and with stopping points at 100 m intervals. *G. morsitans* are attracted to moving black screens and follow the catching party without necessarily landing on the humans or the screen. Periodic stops are therefore used to sample with small fly nets both the following flies and flies landing on either the screen or the surrounding vegetation.

Table 1. Details of the sample sites and data collection periods (from – to).

Site information			Fly-round sampling		Tsetse catches		
Name	Geographic centre (degrees)	Length (m)	Date (month/year)	Stops	Males	Females	Total
Lusandwa	31.81(long);-13.70(lat)	6,400	12/06–11/07	64	1,310	523	1,833
Kasamanda	31.90(long);-13.78(lat)	4,200	12/06–11/07	94	9	0	9
Zinaka	31.78(long);-13.86(lat)	8,000	07/06–06/07	116	855	249	1,104
Chisulo	31.86(long);-13.96(lat)	4,000	12/06–11/07	74	218	34	252

'Stops' are the number of fly catching points on each fly-round.

https://doi.org/10.1371/journal.pone.0096002.t001

Sampling was repeated on average seven times each month. For the present analysis, the monthly catches for each stop were used without distinguishing the sexes, because some catches were very small.

Although the data from Kasamanda contain only few non-zero tsetse catches, this site had a similar number of space and time sampling points (Table 1, column "Stops") as the other three sites and hence it contributes equally in determining the spatio-temporal correlation.

It is common in analysis of tsetse populations to model only the average catches at each site. Here, in a full spatio-temporal framework, the dataset is not averaged, either in space or time. This enables us to retain all of the population variability within sampling sites. Finally, the same number of visits was used at each stop per month, so the data did not need to be standardised.

Environmental Correlates

The variables used to interpret the level and variation of the density independent mortality are the day-time land surface temperature (DLST), night-time land surface temperature (NLST) and monthly cattle density.

Statistical Method

The Bayesian Geostatistical model applied for estimating tsetse population net changes is a modification of the Moran curve method. Starting from the Moran curve model equation, which is assumed to be applied to each of the stopping points on each fly round:

$$\log_{10}(y_{t+1}) = (b-(di_t+dd_t))+\log_{10}(y_t) \quad \textbf{(1)}$$

where b is the monthly logarithmic fertility rate; y_t and y_{t+1} are the numbers of flies caught at n locations in successive months t and $t+1$; di is the density independent mortality and d is the density dependent mortality acting at the S locations at time t, and usually different from one fly round stop to another. The first term on the right hand side $(b-(di_t+dd_t))$ is the net change (difference between fertility rate and population losses) which is zero when the population is in equilibrium.

The model has been extended to a linear regression form with the addition of a spatio-temporal correlation effect, Z, and an error component ε (observation error):

$$\log_{10}(y_{t+1})=(b-(\bar{di}_t+dd_t)+\log_{10}(y_t))+Z+\varepsilon \quad \textbf{(2)}$$

with the following conditionings:

$$\bar{di}_t=\beta_1+\beta_2 X$$

$$Z\sim N\left(0,\sigma_z^2\Omega(\varphi,\rho,\delta)\right)$$

$$\varepsilon\sim N\left(0,\sigma_e^2 I\right)$$

where $\bar{di}_t$ is the density independent mean response from a linear regression of the (matrix of) environmental covariates, X, with intercept β_1 and regression coefficients β_2 (a vector of length equal to the number of environmental covariates). In other words, estimates of di are used to fit an environmental linear regression from which β_1 and β_2 are obtained. The coefficients are then used to back calculate the mean density independent mortality $\bar{di}$. We are interested in the environmental signal of di, while its stochasticity will contribute to the other model components (i.e. the spatio-temporal effect and/or the error term).

Z is a one column vector with spatio-temporal normally distributed random effects with mean zero and a covariance matrix given by the product of the spatial variance, σ_z^2, and the correlation matrix, Ω; the latter is expressed as a function of the spatial correlation parameter, φ, the temporal correlation parameter, ρ, and the interaction term, δ, as appearing in the double exponential non-separable spatio-temporal function [13] chosen for this analysis:

$$\Omega(\varphi,\rho,\delta) = \exp\left(\frac{\Delta_s}{\varphi} - \delta\frac{\Delta_s}{\varphi}\frac{\Delta_t}{\rho} - \frac{\Delta_t}{\rho}\right) \tag{3}$$

where Δ is the Euclidean distance in space, s, or time, t, as indicated by the subscript. Finally, ε is the independent and identically normally distributed error.

$\log_{10}(y_{t+1})$ is Gaussian distributed. Fixing $NC = (b - (d\bar{i}_t + dd_t) + \log_{10}(y_t))$ and following the re-parameterisation of the variance parameter of Yan et al. (2007), the model can be re-written as:

$$\log_{10}(y_{t+1}) \sim N(NC, \sigma^2[(1-\kappa)\Omega(\varphi,\rho,\delta) + \kappa I])$$

$$\sigma^2 = \sigma_e^2 + \sigma_z^2 \tag{4}$$

$$\kappa = \sigma_e^2 / \sigma^2$$

Hence, the likelihood is:

$$p(\log_{10}(y_{t+1})|\sigma^2,\varphi,\rho,\delta,\kappa) = (\sigma^2)^{-n/2}|\Omega(\varphi,\rho,\delta,\kappa)|^{-1/2} \times \exp \left[-\frac{1}{2\sigma^2}[\log_{10}(y_{t+1}) - NC]^T \Omega(\varphi,\rho,\delta,\kappa)^{-1}[\log_{10}(y_{t+1}) - NC]\right] \tag{5}$$

The posterior density of the model's unknown parameters, given the observed data y, is defined as:

$$p(\varphi,\rho,\delta,\kappa,\sigma^2|\log_{10}(y_{t+1})) \sim p(\log_{10}(y_{t+1})|\sigma^2,\varphi,\rho,\delta,\kappa)\pi(\sigma^2|\kappa)\pi(\kappa)\pi(\rho)\pi(\varphi)\pi(\delta) \tag{6}$$

where $\pi(\cdot)$ denotes the prior probability of its argument.

In the model the parameters associated with net change are treated differently from those associated with the variance. *di* and *dd* are computed from the fertility rate, *b*, the population size at which the density dependent mortality starts, *a*, and the intensity of density dependent losses expressed as the slope, α. Hence, given a set of b, a and α, di and dd are estimated from the Moran curve model:

$$dd_t = \begin{cases} 0, & if\ \log_{10}(y_t) \le a \\ [\log_{10}(y_t) - a][1 + \tan(90 - \alpha')] & otherwise \end{cases}$$

$$di_t = b + \log_{10}(y_t) - \log_{10}(y_{t+1}) - dd_t$$

$$\alpha' = 135 - \alpha \quad \textbf{(7)}$$

The losses due to density dependence, *dd*, operate in addition to any density independent losses only above a threshold population size of *a* (the derivation of equation (7)). Thus the analysis assumes density dependent and independent effects are mutually exclusive and separable (i.e. additive in their effects). This is likely to be an over-simplification; in reality both density dependent and independent mortalities are operating simultaneously. Therefore, as with fertility and mortality, it is likely that there is in reality some mixing of the density dependent and independent mortalities estimated in the way outlined above; our approach will identify only the equivalent density dependent and independent mortalities required additively to bring about the population change actually recorded from one month to the next.

We have produced 27,000 net change candidates from the combinations of values (30) of the parameters:

b=sequence of values from 0.2 to 0.4 at intervals of 0.0069.

a=sequence of values from 0 to 1.5 at intervals of 0.051.

α=sequence of values from 0 to 90 at intervals of 3.1.

For the second set of parameters φ, ρ, δ, κ, σ_ε^2 we assumed independent priors. The two variances (σ_z^2, σ_e^2) are both Inverse Gamma distributed (this distribution is the conjugate prior for the variance of a Gaussian distribution) of the form IG(θ,γ) with θ=5 and γ=1 for and θ=0.9 and γ=0.25 for σ_z^2. Uniform priors over valid positive limits are defined for the spatio-temporal parameters: spatial range (km), φ Uniform (0,3); temporal range (months), ρ Uniform(0,6); interactive term δ Uniform(0,1). These priors are not the best choice for mixed models without spatial effects [14], [15] because they can lead to an improper posterior, but they are a common choice for their properties (and computation results) in Bayesian mapping [16]. In fact, choosing uniform priors for the spatio-temporal function parameters allows the restriction of their estimate to a small number of them, which leads to a faster and better control of the spatio-temporal covariance by avoiding values associated with the absence of autocorrelation. Alternatively, informative priors can be applied (and need to be investigated), such as Jeffreys priors or Maximal Data Information priors. The posterior samples of φ, ρ, δ, σ^2 and κ given y are obtained from the RAMPS algorithm [17] and the convergence of the MCMC chains was tested with the Gelman-Rubin convergence statistic [18]. The MCMC was run until convergence and until the Monte Carlo error for all the parameters was equal or less than 5% [19]. At the end of MCMC sampling, 200 samples for each parameter were extracted with regular thinning from the respective chains.

In practice, for each net change candidate the RAMPS algorithm finds the Bayesian estimates of the spatio-temporal covariance parameters and observational error. Once the MCMC chains have converged, a sample of the converged chains is taken and used to calculate the model DIC. Therefore the lowest DIC within the 27,000 models (27,000 is the number of net change candidates) was selected and the associated parameters were then employed in the mapping predictions.

The algorithm was built using a series of functions described in Appendix S4.

Mapping

The fertility rate, the regression coefficients for the density independent effect and the average density dependent effect acting at each month in the model with the lowest DIC were employed to predict the net population change on a grid of points within the study area. Simple kriging was then applied to add a spatio-temporal covariance to this net change. The latter is an interpolator, the estimates of which are obtained from weighted linear combination of the data. Weights are obtained from the spatio-temporal covariance function. In simple kriging the mean is assumed to be known.

Out-Of-Sample Validation

The model was run on two sites (Lusandwa and Zinaka, the sites with a larger number of tsetse catches) in order to make predictions for a third site (Chisulo). The statistics applied were: mean difference between original and predicted logarithmic tsetse counts (ME) and the mean squared difference between original and predicted logarithmic tsetse counts (MSE). For $s=1,2,\ldots,S$ and $t=1,2,\ldots,T$ points. The formulae are:

$$ME = \frac{1}{S * T} \sum_{t=1}^{T} \sum_{s=1}^{S} y_{s,t} - \hat{y}_{s,t} \quad (8)$$

$$MSE = \frac{1}{S * T} \sum_{t=1}^{T} \sum_{s=1}^{S} (y_{s,t} - \hat{y}_{s,t})^2 \quad (9)$$

RESULTS

The principal aim of this analysis was to estimate and understand the net changes in tsetse populations over space and time. The impact and therefore importance of density dependent mortality can be tested by re-doing the analysis without it, thus considering only density independent mortality. Removing the density dependent component from the model yielded a DIC value of 938, a drastic reduction in model fit when compared with a value of 560 for the full model. This suggests that density dependent losses explain a great deal of the variation of the population data.

The net change in abundance for the study year was slightly greater than zero (0.05), with the largest seasonal value in the hot-dry season (Table 2). Therefore population losses balanced gains and the modelled total population in the area was apparently in equilibrium (increasing by only one individual).

Table 2. Mean values of the net change during the three seasons in the four sites in Zambia.

Locations	***Rainy***	***Cold-dry***	***Hot-dry***	***Year mean***
Lusandwa	**0.080**	**0.003**	**0.105**	**0.061**
Kasamanda	**−0.022**	**NA**	**NA**	**−0.022**
Zinaka	**0.006**	**0.057**	**NA**	**0.032**
Chisulo	**−0.007**	**0.033**	**0.107**	**0.049**
All sites	**0.018**	**0.028**	**0.106**	**0.054**

Rainy (January to April), cold-dry (May to August) and hot-dry (September to December) seasons; and the mean values throughout the year.

https://doi.org/10.1371/journal.pone.0096002.t002

The spatio-temporal net change predictions for the three seasons in Zambia are shown in Fig. 1. All three maps show a mixture of high and low net change in both space and time. There is a substantial reduction in net change in the North-West part of the study area between the rainy and cold-dry seasons. The positive large values of net change in the other areas, however, balance these losses and produce an overall positive value of net change for the cold-dry season that is larger than that for the rainy season. The maximum net change occurred in the hot-dry season. The model allows us to identify areas where the tsetse population is always doing well, that is, where the net changes are zero or positive in all three seasons (green areas in the black and green picture of Fig. 1). These areas may be regarded as tsetse 'sources' (reservoirs), although a longer sampling period would be needed to determine if this is permanently the case.

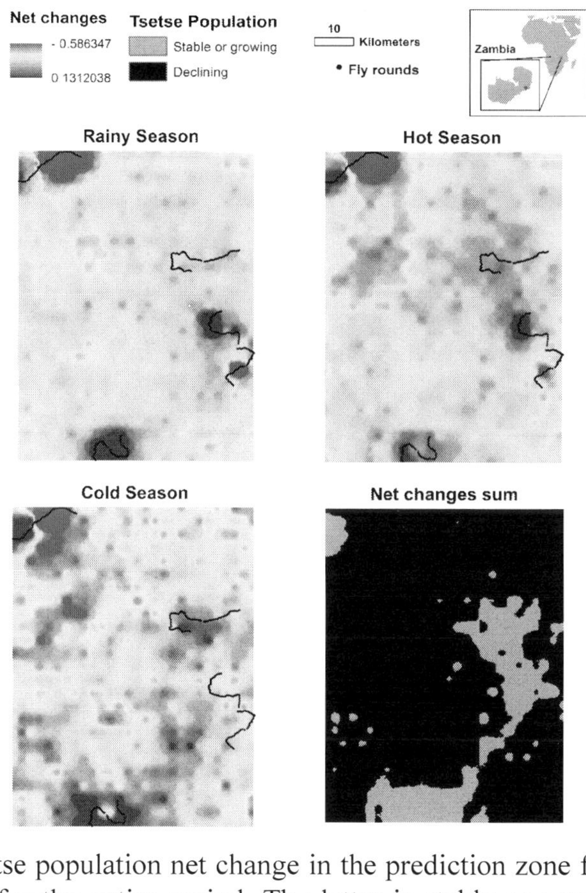

Figure 1. Tsetse population net change in the prediction zone for each season and summed for the entire period. The latter is stable or growing if the net change is equal to or larger than 0; otherwise it is declining. The upper-right corner map shows the position of the predicted area (red square) in Zambia.

https://doi.org/10.1371/journal.pone.0096002.g001

The Moran curve parameters associated with the lowest DIC are: 0.23 for fertility rate, 30° for the intensity of density dependence (slope), and 0.46 for the logarithmic population size at which density dependent mortality starts (Table 3).

Table 3. Estimates of the different parameters in the model.

Moran curve optimal parameters (4 best models)				
	1	*2*	*3*	*4*
DIC	560	683	693	701
b	0.230	0.401	0.297	0.297
a	0.459	0.3380	0.338	0.621
α	30	30	30	30
Coefficients for the density independent linear regression				
	Estimate	***Std. error***	***T value***	***Probability***
β_1	0.642	0.087	7.304	5.19e-13
β_2 ***(LSTD)***	0.011	0.003	3.834	<0.001
β_2 ***(LSTN)***	−0.020	0.005	−4.292	1.92e-05
β_2 ***(CT)***	−0.004	<0.001	−6.531	9.80e-11
Posterior estimates from the Bayesian components				
	Median	***Mean***	***Std. deviation***	***95% HPD***
σ^2_z	0.267	0.283	0.130	(0.184,0.424)
σ^2_e	0.060	0.061	0.009	(0.047,0.075)
φ (km)	1.568	1.744	1.894	(0.965,3.157)
ρ (months)	0.975	0.943	0.082	(0.720,0.999)
Δ	0.864	0.836	0.142	(0.467,0.999)

b, fertility rate; a, population size at which the density dependent mortality effect starts; ***α***, angle of the density dependent effect; β_1, intercept of the linear regression for the density independent mortality effect; β_2 **regression coefficients;** ***φ***, spatial range; ***ρ***, temporal range; ***δ***, interactive term for the spatio-temporal effect; σ^2_z, spatial variance; σ^2_e, error variance.

https://doi.org/10.1371/journal.pone.0096002.t003

The value of 0.23 (the antilog value is 1.69) indicates a maximum rate of population increase in the study area of just under 70% per month, less than the maximum possible value (at slightly higher temperatures) of about 100%, that is, a population doubling per month [3] (female flies produce one full grown larva every 6/7 or more days which pupates more or less immediately and emerges as an adult fly after 20 or more days; all development times are temperature dependent). The slope of the density dependent effect (30 degrees) is characteristic of a population showing under-compensation [20]. This is to be expected of insect species such as tsetse that have very comparatively low fertility rates. Finally, the density dependent mortality starts at a population density of around 2.87 (antilog of 0.46) tsetse flies per stop on the fly rounds. This reflects the usually low values of the catches (90% of the catches were of between zero and three flies per stop; the minimum and maximum catches were, respectively, 0 and 18 tsetse).

The beta coefficients obtained from the regression of density independent losses on climate variables (Table 3) suggest lower mortalities as night-time surface temperatures increased, and when cattle numbers increased, and higher mortalities when day-time land surface temperatures increased. These results are in accordance with the literature on the biology of tsetse, but this analysis quantifies these relationships more clearly than previously.

Finally, the model predicts relatively rapid declines in both spatial and temporal correlations with a spatial range of 1.7 km and a temporal range of 1 month. The average values used to correct the space-time interaction was significantly different from zero, indicating that the two dimensions of space and time cannot be separated, and that the spatial and temporal ranges co-varied (Table 3).

Validation based on applying the net change parameters estimated from Lusandwa and Zinaka to estimate the logarithmic catch data from Chisulo produced accurate statistics: the ME was 0.1 (equivalent to 1.25 tsetse flies) and the MSE 0.35 (equivalent to 2.23 tsetse flies).

DISCUSSION

The study sites included areas with mean total population losses that were both lower and higher than the estimated fertility rate (Figure 2). Between rainy and cold-dry seasons net changes were negative for most of the sites (except from Zinaka) indicating that the total population was sustained only through immigration; for example in Lusandwa, Chisulo and Kasamanda. In particular the Lusandwa population appears to be maintained only by considerable immigration (from the North-West) occurring mainly in the cold-dry and hot-dry seasons. Such immigration could be from an animal game park near to the northern border of the area [21]. Despite the high immigration rate, the Lusandwa tsetse population is very low, due to low survival rates of both immigrant and resident flies, perhaps because of the limited availability of hosts in Lusandwa. In Zinaka, the partial results (the survey does not cover the whole year) indicates a more stable population with population losses always lower than the fertility rate. Zinaka is a site far from game parks and is unlikely to be affected by immigration. Hence environmental conditions make this area suitable for low density tsetse populations.

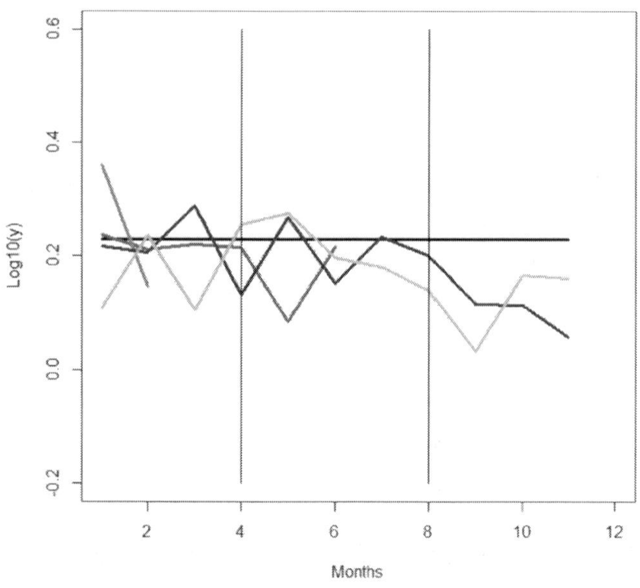

Figure 2. Tsetse population losses (*dd +di*) for the different locations: Lusandwa (sky-blue line), Zinaka (red line), Kasamanda (green line) and Chisulo (dark-blue line). The horizontal bold black line at 0.23 is the estimated fertility rate level (assumed constant). The vertical lines delineate the different seasons: rainy (months 1 to 4), cold-dry (months 5 to 8) and hot-dry (months 9 to 11).

https://doi.org/10.1371/journal.pone.0096002.g002

Net changes are highest in the hot-dry seasons in Zambia (Table 2). These results are in agreement with previous studies on tsetse populations in Zambia and elsewhere [22], [23], where *G. m. morsitans* is most abundant between the end of the hot dry season and the beginning of the rainy season and less abundant during the cold-dry season because of a combination of climate, vegetation and host grazing patterns.

During the rainy season, density dependent mortality provides an upper limit to fly population density. Density dependent losses from one area may be due not to death but to emigration to other areas which results in immigration (which may also be density dependent) into those areas. Thus density dependent population movement [24], [25] may result in the persistence of tsetse throughout much larger areas. Considerable, periodic large scale movement of tsetse flies has been recorded in the Nguruman area of SW Kenya [26], with possibly the same effects (maintaining populations that otherwise would become seasonally extinct). It appears therefore that in this region of Zambia tsetse seem to show all the characteristics of a metapopulation [27].

It is a matter of some debate whether tsetse populations should be targeted by control when they are most or when they are least numerous; the approach described here can identify these periods in the annual cycle and the additional level of control that needs to be applied to reduce the tsetse population's rate of increase to the low level required for eventual population extinction.

CONCLUSIONS

The modelling approach suggested here has three key innovations: (i) estimation of the tsetse population's fertility rate and the levels of density independent and density dependent losses, (ii) implementation of spatio-temporal autocorrelation methods; and (iii) mapping net change, which can be related to environmental or host variables. Even though the optimal search between 27,000 models and the use of Bayesian geostatistics required considerable computer resources and a long processing time, it is an improvement on previous population mapping methods because it quantifies both birth and death rates in a spatio-temporal framework, thus highlighting the levels of additional mortality required, and the times and places to apply them, to reduce the target population to any desired level, and even to extinction.

The present analysis could be improved by adapting the model for simultaneous operation of density independent and density dependent effects. Density dependent losses are likely continuously to be adjusting their impacts as populations rise and fall each month through density independent effects. Obviously, the total losses cannot be less than those actually observed but the allocation of these losses to density dependent or independent causes may change if a continuous approach is adopted.

In conclusion, this research offers a novel approach to estimating key demographic parameters in a geostatistical Bayesian framework. Referring to the origin of the Ricker/Moran curve approach we suggest that this approach could also usefully inform policies for the rational management and harvesting of natural populations such as fish stocks. In the case of tsetse flies, population loss maps are fundamental tools for tsetse control and management. In the case of harvesting, analogous maps can indicate where and when harvesting is most sustainable. Estimating the accuracy and uncertainty of such maps is now possible with widely available, enhanced computational power.

ACKNOWLEDGMENTS

This paper is dedicated to the memory of Peter Van Den Bossche (16/03/1962–11/11/2010) who contributed much to our present knowledge of tsetse and trypanosomiasis in Africa.

The authors are grateful to David Benz (TALA-Oxford) for the supply of MODIS imagery, and to the editor and three anonymous reviewers for their constructive comments, which helped us to improve the manuscript. We thank Sarah Randolph for useful suggestions and Eva De Clerq and Hilde Verniewe for carrying out the land cover and fragmentation classifications.

AUTHOR CONTRIBUTIONS

Conceived and designed the experiments: CM. Performed the experiments: CM. Analyzed the data: LS DJR. Contributed reagents/materials/analysis tools: LS ED CDP GH. Wrote the paper: LS DJR.

REFERENCES

1. Varley GC, Gradwell GR (1962) The interpretation of insect population changes. 142–156.
2. Ricker WE (1954) Stock and recruitment. Journal of the Fisheries Board of Canada 11: 559–623.
3. Rogers D (1979) Tsetse Population-Dynamics and Distribution - New Analytical Approach. Journal of Animal Ecology 48: 825–849.
4. Hargrove JW (1994) Reproductive Rates of Tsetse-Flies in the Field in Zimbabwe. Physiological Entomology 19: 307–318.
5. Hargrove JW (2004) Tsetse population dynamics. The trypanosomiases 2004: 139–180.
6. Hargrove JW (2001) Mark-recapture and Moran curve estimates of the survival probabilities of an island population of tsetse flies Glossina morsitans morsitans (Diptera : Glossinidae). Bulletin of Entomological Research 91: 25–35.
7. Calder C, Lavine M, Muller P, Clark JS (2003) Incorporating multiple sources of stochasticity into dynamic population models. Ecology 84: 1395–1402.
8. Cressie N, Calder CA, Clark JS, Hoef JMV, Wikle CK (2009) Accounting for uncertainty in ecological analysis: the strengths and limitations of hierarchical statistical modeling. Ecological Applications 19: 553–570.
9. West M, Harrison J (1997) Bayesian Forecasting and Dynamic Models: Springer.
10. Clark JS (2003) Uncertainty and variability in demography and population growth: A hierarchical approach. Ecology 84: 1370–1381.
11. .Clark JS, Bjornstad ON (2004) Population time series: Process variability, observation errors, missing values, lags, and hidden states. Ecology 85: 3140–3150.
12. Ford J, Glasgow J, Johns D, Welch J (1959) Transect fly-rounds in field studies of Glossina. Bulletin of Entomological Research 50: 275–285.
13. Cressie N, Huang H-C (1999) Classes of nonseparable, spatio-temporal stationary covariance functions. Journal of the American Statistical Association 94: 1330–1339.

14. Natarajan R, Kass RE (2000) Reference Bayesian methods for generalized linear mixed models. Journal of the American Statistical Association 95: 227–237.
15. Kass RE, Natarajan R (2006) A Default Conjugate Prior for Variance Components in Generalized Linear Mixed Models (Comment on an Article by Browne and Draper). Bayesian Analysis 1: 535–542.
16. Wakefield J (2007) Disease mapping and spatial regression with count data. Biostatistics 8: 158–183.
17. Yan J, Cowles MK, Wang S, Armstrong MP (2007) Parallelizing MCMC for bayesian spatiotemporal geostatistical models. Statistics and Computing 17: 323–335.
18. Smith BJ (2007) boa: An R package for MCMC output convergence assessment and posterior inference. Journal of Statistical Software 21: 1–37.
19. Brooks SP, Gelman A (1998) General methods for monitoring convergence of iterative simulations. Journal of Computational and Graphical Statistics 7: 434–455.
20. Varley GC, Gradwell GR, Hassell MHP (1974) Insect population ecology: an analytical approach: University of California Pr.
21. Mwape L, Mweemba P, Kasonde J (2012) Strengthening the Health System to Enhance Mental Health in Zambia: A Policy Brief. International Journal of Technology Assessment in Health Care 28: 294–300.
22. Hargrove JW (2001) The effect of temperature and saturation deficit on mortality in populations of male Glossina m. morsitans (Diptera : Glossinidae) in Zimbabwe and Tanzania. Bulletin of Entomological Research 91: 79–86.
23. Van den Bossche P, De Deken R (2002) Seasonal variation in the distribution and abundance of the tsetse fly, Glossina morsitans morsitans in eastern Zambia. Medical and Veterinary Entomology 16: 170–176.
24. Randolph SE, Rogers DJ, Kuzoe FAS (1984) Local Variation in the Population-Dynamics of Glossina-Palpalis-Palpalis (Robineau-Desvoidy) (Diptera, Glossinidae).2. The Effect of Insecticidal Spray Programs. Bulletin of Entomological Research 74: 425–438.
25. Rogers DJ, Randolph SE, Kuzoe FAS (1984) Local Variation in the Population-Dynamics of Glossina-Palpalis-Palpalis (Robineau-

Desvoidy) (Diptera, Glossinidae).1. Natural-Population Regulation. Bulletin of Entomological Research 74: 403–423.

26. Dransfield RD, Brightwell R, Kyorku C, Williams B (1990) Control of Tsetse-Fly (Diptera, Glossinidae) Populations Using Traps at Nguruman, South-West Kenya. Bulletin of Entomological Research 80: 265–276.
27. Hanski I (1999) Metapopulation ecology: Oxford University Press.

INDEX

A

Abundance estimation 298
Accumulation 413, 418, 424
Accurate detection 256
Adaptability 382
Administration 265, 277, 278, 279
Adult mosquitoes 318
Aetiological 26, 37
Agricultural facility 172
Akaike Information Criterion (AIC) 374
Algorithm 443
Alphavirus 169, 178, 181
Ambiguous 146
Analysing population 436
Anaplasmosis 154
Animal Experimentation Commission 266
Animal health 406
Antagonistic 4, 5, 7, 12, 18
Anthropogenic 210
Arboviruses 234, 235, 254, 255
Arthropod-borne Bacterial Diseases 50
Arthropod-borne disease 73
Arthropods 108, 109, 110, 111, 114, 120, 122, 123
Arthropod vectors 158, 159
Atmospheric 373
Atmospheric dynamics 210
Avaritia Culicoides 234

B

Babesia 382, 383, 388, 391, 392, 395, 397, 399, 400, 401, 402, 403
Bacterium 334
Barcode of Life Data Systems (BOLD) 138

Bayesian analysis 288, 290, 296, 311
Bayesian geostatistical 435
Behavior 113
Beta diversity 120
Binomial confidence 79
Biodiversity 137, 142, 148
Biogents Sentinel (BGS) 318
Bioinformatics analysis 123
Bioinformatics platform 138
Biology 333, 334, 335
Biosecurity 208, 210, 217, 218, 219
Biosystem 30
Bird Collection 408, 419, 420, 424
Bloodmeals 136, 137, 138, 139, 141, 142, 143, 145, 148, 149
Bluetongue 187, 188, 189, 196, 197, 199, 202, 203, 205
Bluetongue virus (BTV) 188, 234
Borreliosis 158, 159
Brain-heart infusion (BHI) 29
Brucellosis 159

C

Carbapenemase resistance 60, 63
Ceratopogonidae 143, 144
Climatic phenomena 361
Climatic suitability 210, 212, 216
Clinical symptoms 39, 40, 271, 273, 279
Collection 72, 74, 82
Communicable Diseases 339, 346, 349
Complex ecology 339
Complexity 305
Confidence intervals (CIs) 78
Consortium 137
Correlation coefficients 191
Countryside Information System 369
Crop-livestock production system 50
Cross-contamination 61
Crosscorrelation 191
Cross correlation map (CCM) 191, 192
Cross-correlations analyses 94
Culicoides 187, 188, 189, 190, 202, 203, 204, 205, 233, 234, 235, 236, 237, 238, 243, 244, 245, 247, 248, 251, 252, 253, 254, 255, 256, 257, 258, 259, 260, 261, 262
Culicoides genus 264
Culture infective 267
Cytochrome 136, 137, 141
Cytochrome oxidase 372
Cytokine 270, 275, 276, 278

D

Data acquisition 103
Databases 137, 142
Day-time land surface temperature (DLST) 439
Demographic 137
Dengue fever 288
Density 435, 436, 437, 438, 439, 440, 441, 442, 443, 444, 446, 447, 448, 449, 450
Deoxyuracil triphosphates 29
Desensitization 179
Deviance Information Criterion 242
Diagnosis 26, 27, 40
Diagnostic 119, 122, 123, 128
Digestion mass spectrometry 142
Discordance 123
Discrepancy 373
Disease ecologists 3, 13

Disease ecology 135, 142
Diversity 363, 368, 371
Domestic 368, 371, 377, 378
Domestic animals 46, 47, 48, 51, 53, 55, 57, 58, 60, 62, 63, 64
Dynamic bioclimatic 212
Dysregulated cytokine 180

E

Eastern equine encephalitis virus (EEEV) 14
Ecoclimatic Index (EI) 212, 215
Ecological 370, 377, 406, 407, 408, 414, 415, 416, 417, 421, 424, 430
Ecology 208, 212, 221, 222, 223, 325
Economic significance 48
Ectoparasite 341, 419, 420, 422, 430
Ectoparasite infestation 26, 28
Ectoparasites 362, 374, 379
Ectoparasite-vertebrate 4, 5, 6, 7, 8, 10, 12, 13, 15
Ectoparasitic 168, 169, 171
Ehrlichiosis 154, 160
Electropherograms 136, 138, 139, 142, 146
Embryonic 109
Emerging infectious 135
Endosymbiont-related species 123
Endosymbionts 108, 119, 120, 122, 123, 130
Enzootic cycle 316
Epidemic 188, 189, 192, 199, 200, 202, 203
Epidemiology 47
Episterum 267
Ethical approval 53
Ethical Committee 292

F

Fiocruz conducts 292
Food and Agriculture Organization (FAO) 155
Frequency 209, 408, 410, 414, 415, 416, 423, 424

G

Generalised additive model (GAM) 374
Generalized Linear Model (GLM) 294
Genomic Service 373
Geographic 81, 317, 324
Geographical 79, 83, 85, 338, 339, 341, 342, 352, 357
Global local reaction 274
Global warming 316
Growth medium 173

H

Habitat data 369
Habitats 317, 320, 322, 323, 324
Habitat suitability 339
Haemaphysalis 383, 395, 396, 397, 401, 402, 403
Haematophagous 136, 142, 146, 264, 265, 282
Health certification 159
Health challenges 153
Health Committee 154, 161
Hematophagous 169, 171, 179, 185
Heteroplasmy 140
Hippoboscidae 108, 110, 114, 115, 116, 117, 120, 121
Hoplopleura 55
Host-pathogen system 169

Human behaviour 354
Human demography 210
Human environment 396
Human immunodeficiency virus (HIV) 156
Human medicine 48
Human population 27, 40, 42, 344
Human Resources 398
Human social system 11
Humanus 52, 56, 60, 61, 67
Humidity 194, 239, 240, 249, 250, 252, 254
Humoral immune response 178, 179
Humoral immune system 179
Hybridization 325, 328, 329
Hybrids 315, 316, 317, 320, 321, 322, 323, 324, 325, 328, 329, 330
Hymenoptera 114
Hyperaemia 265, 269
Hypersensitivity 278, 334
Hypothesis 370
Hypothetical 9

I

Illumina manual 123
Immature 406, 407, 408, 411, 412, 413, 415, 416, 417, 418, 419, 420, 422, 423, 424
Immigration 437, 447, 448, 449
Immune evasion 334
Immune recognition 180
Immune responses 168, 169, 179, 180
Immunocompromised 47, 159
Immunologically 168, 170
Immunology 334
Indirect immunofluorescence assay (IFA) 29, 30
Infection 108, 122, 129, 130
Infection efficiency 30
Infectious diseases 135, 136, 148
Inflammatory 264, 265, 269, 274, 278, 279
Innate immune responses 265
Institute for Environmental Protection and Research (ISPRA) 110
Institutional Animal Care and Use Committee (IACUC) 387
Intensity 362, 363, 364, 365, 366, 367, 368, 369, 370, 371, 374
Interaction networks 415, 424
Interaction Web Database 8, 15, 21
Intergovernmental Panel on Climate Change (IPCC) 189, 210
Intra-thoracic inoculation 267, 271
Ischnocera 56
Ixodes hexagonus 361, 362, 363, 364, 367, 372, 375, 377, 378
Ixodid 381, 384

L

Landscape 363
Linognathus 45, 46, 51, 54, 56, 63
Local inflammatory 264, 269, 274, 275, 277, 278
Logarithmic fertility 440
Logistic regression 78
Louping ill virus (LIV) 81
Lyme borreliosis (LB) 339
Lyme disease 406, 426

M

Manipulations 335
Mapping Authority provided altitude (MASL) 344
Mapping tick distribution 339

Mark-release-recapture (MRR) 287
Mechanism 323, 397
Mechanistic 180
Medical profession 161
Metabarcoding 108, 109, 110, 112, 113, 122, 123, 124
Metabarcoding analysis 123
Metagenomics 112, 113, 123
Meteorological 239, 240, 241, 259, 291, 306
Methodology 375
Metrics data 90
Microbial composition 109, 117, 120
Microbiota 107, 108, 109, 110, 117, 120, 123, 125
Microclimate 213
Microhabitat 324
Microinjector 267
Microorganism 30
Migratory 107, 108, 109, 113, 120, 122, 124, 125, 130, 406, 407, 409, 419, 421, 426
Misclassification issue 120
Mitochondrial 136, 137, 138, 140, 148, 372, 379
Moisture Index 212
Molecular 47, 51, 52, 58, 60, 61, 66, 67, 68, 69
Molecular analysis 14
Molecular detection 26, 29, 36, 38, 39
Molestus 316, 322
Monocytic ehrlichiosis 159, 160
Monte Carlo Markov Chain (MCMC) 295
Morphological 45, 46, 51, 54, 61, 238, 410, 411, 416, 417, 418, 421, 423
Morphologically 51, 54, 55, 61
Morphology 238
Mosquito abundance 288, 289, 300, 305, 306, 307
Mosquito bloodmeals 14
Mosquito data 290, 297
Mosquitoes 6, 9, 10, 11, 12, 14, 16, 21, 109, 122, 288, 289, 290, 291, 292, 293, 294, 295, 296, 297, 298, 300, 301, 302, 303, 304, 305, 306, 307, 309, 310
Mosquito-host networks 11, 14, 16
Mosquito-Magnet Liberty 316
Mosquito Magnet Liberty Plus (MMLP) 318
Multiscribe 270
Municipality 342, 343, 344, 345, 346, 348, 350, 351, 354
Murine typhus 25, 26, 27, 30, 37, 38, 39, 40, 41, 43, 44
Murine typhus (MT) 25
Museum 421, 422
Mutualistic network 4, 5, 11, 18, 20

N

Natural infection 264, 280
Natural transmission 325
Needle transmission 271, 272, 273, 275, 277, 278
Nestedness 4, 5, 6, 7, 8, 9, 10, 11, 12, 13, 14, 15, 18, 19, 20, 21
Neurological 90
New South Wales (NSW) 90, 92
Night-time land surface temperature (NLST) 439
Normalized Difference Vegetation Index (NDVI) 213
North Atlantic Oscillation (NAO) 361, 362, 365, 368, 373

Norwegian Cattle Health Recording system (NCHRS) 339
Norwegian Institute of Public Health (NIPH) 342
Norwegian Veterinary Institute (NVI) 340
Nosocomial bloodstream 48
Nuclear insertions 140
Nucleotide 140, 141, 144, 146

O

Observation 290, 293, 294, 295
Onderstepoort Veterinary Institute (OVI) 237
Optical density (OD) 174
Organism 5
Organization for Animal Health (OIE) 155, 235
Oropouche virus 264, 282

P

Pathogen 334
Pathogen transmission 264, 265, 266, 280
Peridomestic cycle 25, 27, 37
Pharmacological 334
Phenological 213, 234, 236, 237, 242, 243, 250, 251, 252, 254
Phenology 234, 236, 237, 243, 251, 252, 253, 254, 255, 256, 259, 260
Phenotypic 11
Phosphate buffered saline (PBS) 173
Phosphate-buffered saline (PBS) 29
Photoperiod 241, 248, 250, 251
Phylogenetic 385, 392
Phylogenetic information 11
Phylogeny 61, 397, 400, 402
Physiopathological 280
Piroplasm 382, 384, 388, 391, 395, 396, 397
Piroplasmoses 382, 398
Plant-pollinator network 8, 12, 15
Plant population 436
Poisson regression 188, 191, 195, 199
Polymerase Chain Reaction (PCR) 14, 136
Positive influence 249, 251
Possibility 297, 306
Potential distribution 208, 210, 211, 214, 215, 217, 218, 221
Potential epidemiological 397
Predominance 393
Principal component analysis (PCA) 338, 344
Probability 290, 294, 295, 296, 297, 302, 303, 305, 306, 307
Programmable Thermal Controller 144
Prokaryotic 110, 124
Prophylaxis 103
Prophylaxis application 103
Psychological 162
Public databases 140, 142
Public health 91, 102, 135, 136, 137
Pyrosequencing 142

Q

Queensland 89, 90, 92, 94, 106

R

Random expectation 9
Randomisation procedure 74
Random selection 79, 82
Relative humidity (RH) 291

Reticulatus 383, 395, 397, 399
Reverse transcription polymerase chain reaction (RT-PCR) 173
Ribosomal sequence variants 112, 117
Ricinus 339, 357
Rickettsiales 419
Rickettsioses 26
Risk assessment 187, 188, 189, 199
Rocky Mountain Spotted Fever (RMSF) 407, 418

S

Schmallenberg virus (SBV) 235
Seasonality 201
Seasonal timing metric 241, 242
Seasonal vector free period (SVFP) 235
Sensitivity 189, 193, 196, 203
Seroprevalence 26, 27, 38, 39
Spatial correlation 441
Spatio-temporal correlation 439, 440
Species-accumulation curves (SACs) 424
Species identification 52
Specific amplification 141, 144
Spectrophotometer 110
Spotted fever group (SFG) 30
Subclinical infection 27, 38
Surveillance 89, 90, 91, 92, 101, 102, 103, 104
Susceptibility 168, 169, 180
Symbiont 108, 117, 120, 121, 122, 123, 130
Symbiont organism 123
Symbiont species 108, 120, 122
Symptom 91
Synanthropic 383
Synchronise 237
Synchronous 90, 101

T

Taxonomic 114, 120, 123
Terrestrial habitat 215
Tetramethylbenzidine (TMB) 174
Thermocycler 173
Tick-borne encephalitis (TBE) 81, 339
Tick infestation 78, 79, 81, 82, 84, 86
Transcriptase 270
Transillumination 52
Transmission 4, 12, 13, 17
Transmission cycle 406
Transmission efficiency 279
Transstadial transmission 397
Tuskegee National Forest (TNF) 12

U

Ubiquitous nature 362

V

Vasoconstriction 265
Vector-borne diseases 316, 339
Vector densities 188, 189, 192, 193, 197, 198, 199, 200
Vector population 188, 189, 196, 197
Venatorum 391
Vertical transmission 62
Veterinary 28, 29, 31, 37, 39
Veterinary clinics 26
Veterinary medicine 91, 103
Viral infection 278, 280
Virus transmission 265
Visualization 201

W

Web-based questionnaire 343
West Nile virus (WNV) 14, 108
World Health Organization (WHO) 155
World Small Animal Veterinary Association (WSAVA) 161
World Wide Web 92
Wrapped Normal (WN) 241

Z

Zoonotic infectious disease 154, 161